Basistexte Geographie

Herausgegeben von
Martin Coy, Anton Escher
und Eberhard Rothfuß

Band 3

Tabea Bork-Hüffer / Anke Strüver (Hg.)

Digitale Geographien

Einführungen in sozio-materiell-technologische
Raumproduktionen

Franz Steiner Verlag

Bibliografische Information der Deutschen Nationalbibliothek:
Die Deutsche Nationalbibliothek verzeichnet diese Publikation in der Deutschen
Nationalbibliografie; detaillierte bibliografische Daten sind im Internet über
<http://dnb.d-nb.de> abrufbar.

© Franz Steiner Verlag, Stuttgart 2022
Satz: DTP + TEXT, Eva Burri
Druck: Druckerei Steinmeier GmbH & Co. KG, Deiningen
Gedruckt auf säurefreiem, alterungsbeständigem Papier.
Printed in Germany.
ISBN 978-3-515-13038-7

INHALTSVERZEICHNIS

III WIE DATEN, ALGORITHMEN UND SENSOREN AGIEREN

DIGITALE GEOGRAPHIEN – PERSPEKTIVEN AUF DEN GESELLSCHAFT-TECHNOLOGIE-UMWELT-NEXUS

Tabea Bork-Hüffer / Anke Strüver

1 EINLEITUNG: WIESO DIGITALE GEOGRAPHIEN?

Dieses analoge Buch erscheint in einer Zeit, in der die Welt neben (und aufgrund) einer globalen Pandemie von einem massiven „Digitalisierungsschub" erfasst ist. Wir möchten die Sichtweise des plötzlichen Schubs direkt relativieren, da wir im Anschluss an vielzählige Arbeiten von Digitalen Geograph*innen die Historizität der Ko-Produktionen von Raum und sozio-materiell-technologischen Artefakten und Prozessen hervorheben möchten. Algorithmische, automatisierende und robotische Systeme hatten schon lange vor der Pandemie disruptive und transformative Effekte (Kitchin 2017: 15 → in diesem Band, vgl. auch Del Casino et al. 2020) und ganz unterschiedliche digital-geographische Arbeiten haben bereits bestehende und zugleich hoch dynamische und brüchige Durchdringungen gesellschaftlicher und mehr-als-menschlicher Kontexte und ihrer Ko-Konstitutionen durch Daten, Codes, Algorithmen, Automatisierungen und robotische Technologien aufgezeigt. Das Jahr 2020 und die Pandemie betten sich also keinesfalls in eine lineare Geschichte der Digitalisierung und Computerisierung ein. Digitale Geographien waren und sind von Wellen und Friktionen technologischer Innovationen geprägt und zugleich eng verknüpft mit wissenschaftstheoretischen Paradigmen und Umbrüchen sowie gesellschaftlichen Ereignissen und politischen Agenden.

Unbestritten ist, dass die Digitalisierung unabhängig von der Pandemie eine zuvor unbekannte Beschleunigung und ein Ausmaß von Veränderungen, einen „digital turn" (Ash et al. 2018 → in diesem Band, Ash et al. 2019), hervorgerufen hat, der sämtliche Bereiche von Gesellschaft, Politik, Wirtschaft, Kultur, Wissenschaft und Umwelt umfasst. Diese Veränderungen sind von divergierenden Graden gesellschaftlicher und individueller Reflexion sowie Beteiligung geprägt und sie schreiben sich durch digitale Alltagstechnologien teilweise subtil und gesellschaftlich fast unbemerkt in Körper, Objekte, Praktiken, (Macht-)Aushandlungen, Diskurse, Sozialräume, Naturkonstruktionen und Gesellschaft-Umwelt-Beziehungen ein. Kitchin betont zudem, wie insbesondere die an digitale Interaktionen gebundenen Algorithmen „shape how we understand the world and they do work in and make the world through their execution as software, with profound consequences" (Kitchin 2017: 19 → in diesem Band).

Angesichts dieser Allgegenwärtigkeit des Digitalen ist es von besonderer Bedeutung, die jeweils spezifischen und kontextuellen Wechselwirkungen mit Raum und Räumlichkeit bzw. den Gesellschaft-Technologie-Umwelt-Nexus herauszuarbeiten (Ash et al. 2019). Genau hier setzen Digitale Geographien und daher dieser

Band der Steiner Basistexte an. Diese Textsammlung gibt einen kleinen, flüchtigen Einblick in die inzwischen rasant zunehmende Forschung zur Digitalisierung, die digitale Geograph*innen, geleitet von höchst verschiedenen epistemologischen und ontologischen Perspektiven, und mit einem breiten sowie sich dauernd erweiternden Repertoire an Methoden, verfolgen. Der Band umfasst elf einschlägige Veröffentlichungen zum Thema, die richtungsweisende Impulse, zentrale theoretisch-konzeptionelle Debatten und Kontroversen Digitaler Geographien eröffnet haben. Er richtet sich an Studierende, Lehrende und Forschende der Geographie, die eine kondensierte Zusammenstellung einschlägiger Beiträge für Lehrveranstaltungen oder als Hintergrund über eigene Forschungen hinaus suchen.

Eine Kompilation geographischer Arbeiten zu einer derart weiten Thematik in dem vorgegebenen Rahmen dieser Reihe kann nur unvollständig sein. Wir sehen den Band daher als Versuch einer notwendigerweise lückenhaften, aber dabei breiten Auffächerung der Vielfalt Digitaler Geographien. Diese Sammlung von Texten kann allein Ausgangspunkt für eine vertiefte Beschäftigung mit den hier angesprochenen und vielen weiteren Schwerpunkten digital-geographischer Forschung sein. Durch unsere eigene wissenschaftliche Sozialisation und unsere sprachlichen Barrieren und mit Blick auf die Zielgruppe dieser deutschsprachigen Reihe, umfasst dieser Band v. a. Werke auf Deutsch und Englisch. Damit spiegelt der Band in seiner Zusammenstellung bestehender Beiträge auch die ungleiche globale (digitale) Wissensproduktion und Dominanz des Wissenschaftssystems des Globalen Nordens wider. Diese prägt auch massiv die digital-geographische Forschung, in der Publikationen zum und aus dem Globalen Süden extrem unterrepräsentiert sind (vgl. Warf 2013, 2019).

Die hier zusammengestellten Beiträge haben wir in drei Abschnitte gegliedert, die weder der Logik einer zeitlichen Abfolge noch einer thematischen Systematisierung folgen. Vielmehr haben wir uns drei zentrale Fragen gestellt, die alles andere als abschließend durch die Basistexte behandelt werden, die aber dennoch helfen, die angedeutete Ambivalenz aus Flüchtigkeit und Persistenz digitaler Geographien zu strukturieren. Die Auswahl der Beiträge des ersten Abschnitts ist von der Frage geleitet, *wie das Digitale anhaltend die Geographie erobert*. Die Aufsätze führen die Leser*innen in fachspezifische thematische und methodische Schwerpunkte ein. Die Selektion der Veröffentlichungen des zweiten Abschnitts ist orientiert an der Frage, *wie das Digitale mit Gesellschaftsprozessen verschränkt ist*. Die ausgewählten Beiträge konzentrieren sich auf digital vermittelte verräumlichte Gesellschaftsprozesse und -dynamiken. Die Publikationen des dritten Abschnitts fokussieren auf die Frage, *wie Daten, Codes, Sensoren und Algorithmen in/mit Sozialem wie Räumlichem interagieren und darüber gesellschaftliche Raumproduktionen und verräumlichte Gesellschaftsprozesse produzieren*.

Diese Einleitung haben wir als Ergänzung zu den hier versammelten Beiträgen und den darin angesprochenen Thematiken entwickelt. Wir nehmen den Leser*innen die eigene Auseinandersetzung mit und die Reflexion über die Beiträge nicht durch eine iterative Zusammenfassung ab. Stattdessen erweitern wir mit dieser Einleitung die Sammlung an Texten mit einer Diskussion darüber, was Digitale Geographien überhaupt sind, und wie sie (nicht) gefasst werden können (Kapitel 2). Kapitel 3 spricht einige der dominanten metatheoretischen Perspektiven digitaler Geographien an und ordnet zugleich die Beiträge in diesem Band darin ein. Wir schließen die Einführung mit einer Skizzierung einiger zentraler Herausforderungen digital-geographischer Forschung zum Gesellschaft-Technologie-Umwelt-Nexus (Kapitel 4).

2 KEINE SUBDISZIPLIN: WAS SIND DIGITALE GEOGRAPHIEN (NICHT)?

Digitale Geographien sind keine (Meta-)Theorie oder Denkweise – wie es die Bezeichnung *Digital Turn* (Ash et al. 2018 → in diesem Band) im ordnenden Vergleich mit dem *Cultural Turn* oder dem *Spatial Turn* vermuten ließe. Während der *Cultural Turn* das konstruktivistische und theoretisch fundierte Denken zunächst epistemologisch und mittlerweile auch ontologisch etabliert hat, stellt *das Digitale* einen technologischen wie gesellschaftlichen Rahmen für diverse und kaum eingrenzbare raumbezogene Forschungsthemen, -methoden und -theorien dar (s. Kapitel 3). *Das Digitale* hat die räumliche Organisation, das Strukturieren und Funktionieren von Gesellschaft sowie gesellschaftliche Raumproduktionen verändert: Wie und wo wir Räume wahrnehmen und nutzen, was wir alles als Raum verstehen, wie, wo und für wen räumliche Visualisierungen wirkmächtig werden, ist nicht nur sozial vermittelt, sondern soziodigital bzw. technosozial. Problematisch ist die ubiquitäre Verwendung des Begriffs *des Digitalen* und seine Verschleierung (Ash et al. 2019: 4). Was in geographischen Arbeiten theoretisch-konzeptionell als *das Digitale* gefasst wird, ist höchst vielfältig. Nach Ash et al. (2019) kann eine willkürliche Verwendung nur durch eine klare Benennung und Herausarbeitung des spezifisch Digitalen, das in konkreten Forschungszusammenhängen und -kontexten betrachtet wird, vermieden werden.

Digitale Geographien sind keine Methodologie im Sinne einer einheitlichen wissenschaftlichen Praxis und sie sind nicht über die Verwendung „digitaler Methoden" abzugrenzen. Mit Raunig und Höferl (2018: 13) verstehen wir hier Methodologien „im Sinne von methodisch geleiteten (d.h. reflektierten, theoretisch fundierten oder zumindest thematisierten) wissenschaftlichen Praktiken". Digital-geographische Arbeiten sind durch ein besonders breites Repertoire solcher wissenschaftlicher Praktiken geprägt, da sie sich aus der Geographie als Gesellschaft-Umwelt-Wissenschaft entwickelt haben, die bereits Praktiken der Sozial-, Geistes-, Natur-, Technik- und Lebenswissenschaften aufgenommen, kombiniert, adaptiert und teilweise weiter- und neuentwickelt hat. Zugleich sei darauf verwiesen, dass Digitale Geographien keinesfalls im Sinne einer digitale Technologien und „digitale Methoden" einsetzenden Wissenschaft einer etwaig „traditionellen", nicht-digitalen Geographie entgegen zu stellen sind. Denn eine solche Unterscheidung zwischen analog und digital ist „methodologisch betrachtet [...] ziemlich irreführend" (Raunig und Höfler 2018: 14) bis hin zu aberwitzig. Denn einerseits ist wissenschaftliche Praxis ohne computergestützte Verfahren und digitale Informationsverarbeitung heute kaum noch vorstellbar, andererseits involviert eine solche Praxis heute meist noch menschliche Akteure, ihre Sinngebungen, Selektionen, Interpretationen und Repräsentationen. Wissenschaftliche Praxis ist in hohem Maße durch Hybridität des produzierten

Wissens gekennzeichnet – in (mindestens) dreierlei Hinsicht: zum einen durch die
Hybridität zwischen Situiertem Wissen und Objektivitätsanspruch, zum zweiten die
Kombinationen positivistischer und partizipativer Methoden (z. B. in People's GIS;
siehe Schuurman und Pratt 2002; Pavlovskaya 2018 sowie Boeckler 2014 → in diesem Band) und zum dritten die Wissensproduktionen, die sich aus dem verkörperten
Erleben durch Verschränkungen mit *mobile devices* wie Smartwatch oder Smartphone im Alltag ergeben (Lupton 2020; vgl. hierzu auch Kapitel 3).

Auch jenseits dieser drei Aspekte darf wissenschaftliche Praxis nicht auf den
Einsatz von Technologien reduziert werden, sondern bedarf der wissenschaftstheoretischen wie gesellschaftlichen Kontextualisierung. Computergestützte Verfahren
haben ursprünglich die Geographie als (Teil der) quantitative(n) Revolution und
eingebettet in den kritischen Rationalismus erobert und im Umgang mit Statistiken
und kartographischen Visualisierungen immer stärker das (raumwissenschaftliche)
empirische Arbeiten bestimmt. Raumwissenschaftliche Ansätze wurden gesellschaftlich – in den wachstumsstarken 1950er und 60er Jahren – gesteuert durch den
Wunsch nach verbesserter Prognostizier- und Steuerbarkeit sowohl umwelt- als
auch gesellschaftsbezogener Prozesse. Gleichwohl wurden Fragen nach dem „warum" dieser Gesetzmäßigkeiten jenseits von vermeintlichen Kausalbeziehungen
nicht be(tr)achtet. Der Frage nach dem „warum" wird mittlerweile vor allem über
das „wie" der Raum-Gesellschafts-Wechselwirkungen als Effekte des *Cultural
Turn* wie auch der *Science and Technology Studies* (Knorr-Cetina 1989, 2005)
nachgegangen. Digitale Erhebungs- und Analysewerkzeuge haben sich mittlerweile
auch im Bereich der qualitativen Methoden etabliert. Sie reichen von banaler Unterstützung durch digitale Audio- und Bilddokumentationen sowie so genannte
QDAS (Qualitative Datenanalyse Software wie Atlas.ti, nvivo oder MAXQDA), zu
Critical & Participatory GIS, dem Smartphone als Forschungswerkzeug (Kaufmann 2018, 2020), Elizitationsstrategien mittels mobiler Medien (Kaufmann 2019)
und digitalen Ethnographien, die qualitativ-exploratives empirisches Arbeiten ermöglichen (Elwood 2010, Pavlovskaya 2016, Pink et al. 2016). Computergestützte
Verfahren werden dennoch teilweise als Vorläufer oder Beginn einer Digitalen Geographie betrachtet (vgl. Barnes 2013). Vor dem Hintergrund der angeführten Kritik
eines solchen vereinfachten Verständnisses wissenschaftlicher Praxis als Einsatz
computergestützter Verfahren, kann dies unseres Erachtens kein Abgrenzungskrite-
rium für Digitale Geographien sein.

*Digitale Geographien sind nicht fixiert und damit weder klar definierbar noch
leicht fassbar.* Die vorangegangenen Ausführungen zu einigen Einflüssen auf Digitale Geographien deuten bereits auf die Veränderlichkeit digital-geographischer
Forschungsschwerpunkte und die Vielfalt von Methoden und meta-theoretischer
Perspektiven hin, die im nächsten Kapitel weiter ausgeführt werden. Digitale Geographien sind gleichermaßen flüchtig und persistent; sie sind dynamisch und immer
„im Prozess" (vgl. für eine ausführliche Diskussion von Raum und Zeit im digitalen
Zeitalter Bork-Hüffer et al. 2020b). Digitale Geographien erscheinen oftmals materiell nicht direkt „fassbar" durch die alltägliche, subtile und banale Einschreibung
von Daten, Algorithmen und Automatisierungen in räumliche Repräsentationen
und Manifestationen gesellschaftlicher Strukturen, Diskurse und Subjektivierungen

sowie mehr-als-menschliche Assemblagen. Digitale Technologien sind im Alltag in Form von Hardware jedoch durchaus auch sichtbar und erfahrbar. Sie liegen in Form von Smartphones mittlerweile in fast jeder Hand- oder Hosentasche („persistent"). Die Formation des Technosozialen ist zugleich durch die Software, ihre Nutzung und die daraus resultierende Mutation (durch Updates aber auch durch lernende Algorithmen) wiederum veränderlich und flüchtig. Die Ambivalenz von Flüchtigkeit und Persistenz macht einerseits diesen Band wichtig, um die Persistenzen und Historizitäten aufzuspüren und zu dokumentieren. Andererseits ist die Entwicklung der Themen und Methoden so rasant, dass dieser Band nur eine flüchtige Momentaufnahme (im Spätsommer 2020) darstellt.

Digitale Geographien sind keine Teildisziplin; sie durchziehen alle „Schubladengeographien". Die Geographie definiert sich durch ihre starke Interdisziplinarität, Offenheit zu und Entlehnung von anderen Disziplinen und ist zunehmend gekennzeichnet durch eine Bearbeitung von intersektoralen bzw. Querschnittsthemen (vgl. Gebhardt und Reuber 2020[3]). Die Geographie ist zudem höchst divers über ihr Verständnis als Wissenschaft von Gesellschaft-Umwelt-Beziehungen und der (zumindest versuchten) Integration von Natur- und Sozialwissenschaften (von physischer und Humangeographie, vgl. für Diskussionen dazu z. B. Müller-Mahn 2005). Beides steht im starken Kontrast zur weiter dominierenden Einteilung des Faches von innen und außen in Subdisziplinen wie Bevölkerungs-, Stadt-, Klima-, Vegetations- und Verkehrsgeographie etc. Eine solche Schubladenkategorisierung steht auch im starken Kontrast zu der Aufweitung von Konzepten zu Raum und Räumlichkeit (d. h. sozial-räumlicher Relationen, Leszczynski 2019: 13) – den Kerngegenständen der Geographie. Bis weit in die 2000er Jahre wurde die Geographie von vier Raumkonzepten dominiert: Einem *raumwissenschaftlichen*, das den Fokus v. a. auf die Verteilung von X im Raum Y legt – und das auch für Arbeiten mit GIS lange Zeit vorherrschend war (Goodchild 2009); einem *sozialwissenschaftlichen*, das Raum als sozial konstruiert und als symbolische wie materielle Ressource des Gesellschaftlichen adressiert (Werlen 1995, 2000), einem *polit-ökonomischen*, das im Anschluss an u. a. Lefebvre und Harvey Raum als Produkt gesellschaftlicher Praxis thematisiert und vor allem materielle und institutionelle Raumproduktionen fokussiert (Belina 2013) und ein diskurstheoretisches, das Raum als Effekt von gesellschaftlichen Normen und Diskursen und als performativ materialisiert begreift (Bauriedl 2009). Den drei letzten Konzepten ist gemein, dass sie Raum als relational konzeptionalisieren, d. h. Raum als durch Beziehungen konstituiert betrachten. Im Gegensatz zu raumwissenschaftlichen Ansätzen fassen relationale Konzeptionen Raum als gleichermaßen spezifisch und offen, als unabgeschlossen, veränderbar und nicht fixiert (Massey 1993, 2012[7]; Dirksmeier et al. 2014). Raum ist eben nicht auf einen gesellschaftlichen oder physischen Teilbereich, wie Bevölkerung, Stadt, Klima, Wasser, Vegetation, Boden, etc., ein- und begrenzbar, sondern durch die Relationalität von Mensch und Umwelt hochkomplex und konstruiert. Die aktive Rolle von Technologien bei der Produktion von Raum darf dabei jedoch nicht vernachlässigt werden, wie wir im Folgenden darlegen.

Digitale Geographien verbinden ihr Interesse an der Untersuchung sozio-materiell-technologischer Raumproduktionen. Das Digitale durchdringt, verbindet und produziert Raum, Räumlichkeit und Gesellschaft-Umwelt-Beziehungen zunehmend. Raum als Kerngegenstand der Geographie muss vor diesem Hintergrund grundlegend neu gedacht und konzeptionalisiert werden. Wir gehen so weit, zu postulieren, dass angesichts der Ubiquität *des Digitalen* Konzepte, die digitale Raumproduktionen gar nicht mitdenken, grundsätzlich überholt sind. Es gibt inzwischen keinen Fleck der Erde, der nicht mit Lagen von Daten erfasst und vermessen wird – wenn zugleich auch die Dichte dieser Daten massiv variiert (Graham 2013, Graham et al. 2013) und ihre Bedeutung für die Formation sozialer Realität je nach Kontext geringer oder stärker sein kann. Umfassend kritisiert wurden ebenfalls techno-deterministische, frühe Ansätze der 1980er und 1990er, die digitalen Raum als Parallelwelt oder Szenarien einer zukünftig den physisch-materiellen Raum ersetzenden digitalen Sphäre skizziert haben (vgl. hierzu ausführlicher Graham 2004). Neuere, *das Digitale* integrierende Raumkonzepte, die ab Mitte der 2000er vorgeschlagen wurden, betonen, wenn auch im Detail sehr unterschiedlich, dass die digitale und physische Sphäre inzwischen untrennbar miteinander verwoben sind. Diese Ansätze unterscheiden sich in ihrer Schwerpunktsetzung. Sie haben beispielsweise einen Fokus eher auf Konzeptionalisierungen des Mikro- oder Makroraums, einen Fokus mehr auf die Rolle von Gesellschaft und Menschen oder stärker auf mehr-als-menschliche Aktanten und Assemblagen, darunter auf die Bedeutung von Technologien als aktive Aktanten in der Ko-Produktion von Raum (eine Diskussion verschiedener Ansätze finden Leser*innen zum Beispiel in Leszczynski 2019, Bork-Hüffer et al. 2020b).

Mit ihrem Konzept von *code/space* rücken Kitchin und Doge beispielsweise den Einfluss von Codes und Software und deren kontingente, ontogenetische und performative Funktion in der Ko-Produktion von (eher Makro-)Raum ins Zentrum. Graham (2013) fokussiert mit seinem Konzept der *digital shadows* auf digitale Daten und Inhalte, die sich wie Teppiche über insbesondere den städtischen Raum legen. Ganz ähnlich sprechen Rabari und Storper (2015) von „the digital skin of cities" und umschreiben damit die zunehmende Ausstattung von Städten und Haushalten mit Sensoren und überall verfügbaren mobilen Kommunikationstechnologien, die sowohl gezielte Kommunikation als auch automatisiert Nutzerdaten übertragen (siehe auch Gabrys 2015 → in diesem Band). In einer Weiterentwicklung zu *Augmented Realities* legen Graham et al. (2013) sowie Graham (2017) zudem dar, wie insbesondere Codes (städtischen) Raum *augmentieren*, also erweitern, Räumlichkeit verändern und Ungleichheiten ko-produzieren.

Aus der Soziologie und den Medien- und Kulturwissenschaften stammt der Vorschlag des *datafied space* Konzepts (Sumartojo et al. 2016), das, Doreen Masseys (2012[7]) relationale Konzeption von Raum erweiternd, die Einschreibung von Daten über (tragbare) Vermessungstechnologien (wie Smart Watches oder Tracker) in Körper, materielle und immaterielle Umwelten und damit in die Ko-Produktion von (eher Mikro-)Raum beschreibt. Mit einer noch stärkeren Konzeptionalisierung untrennbarer Verschränkungen von Mensch, Umwelt und Technologien über Karen Barads (2007) Konzept der ‚entanglements' (s. Kapitel 3.2) haben Bork-Hüffer et al. (2020a, b) das *cON/FFlating spaces* Konzept weiterentwickelt. Das Konzept verschneidet algorithmische mit mehr-als-menschlichen und oben genannten relationalen Perspektiven. Mit einem Fokus auf Alltagsraum und Mikroraum greift es dabei auch die vielfach geäußerte Kritik an einer Vernachlässigung der Rolle von menschlichen Subjekten, ihren Praktiken und sozialer Einbettungen in neueren Konzeptionen von Raum (vgl. Rose 2017, Elwood 2020) auf. Angesichts der Komplexität, Vielfältigkeit, Veränderbarkeit, und Multiskalarität von Raum sowie kontinuierlicher technologischer Innovationen, muss die Debatte um theoretisch-konzeptionelle Rahmungen von Raum kontinuierlich weitergeführt werden.

Digitale Geographien verbinden ihr Interesse an den Wechselwirkungen von Digitalisierung und Gesellschaft-Umwelt-Beziehungen im weitesten Sinne. Digitale Geographien sind also keine Teildisziplin und sie betrachten Raumproduktionen unter dem Einfluss eines untrennbaren Wechselspiels von Mensch, Technologie und Umwelt. Digitale Geographien können unseres Erachtens daher am ehesten beschrieben werden als Gesellschaft-Technologie-Umwelt-Wissenschaft. Wir sind uns der Weite dieses Postulats bewusst, denn ein solcher Entwurf zielt auch auf eine neue (bzw. kontinuierliche) Debatte zur wiederum – unseres Erachtens – offen zu denkenden und weiter zu öffnenden, interdisziplinär ausgerichteten und auszurichtenden, das Digitale intrinsisch mitdenkenden Geographie. Wenn Raum nicht ohne den Einfluss von Daten, Codes, Algorithmen, Automatisierungen und Technologien gedacht werden kann, dann auch nicht die Geographie. Bisher hat die (Mit-)Erforschung des Digitalen in der deutsch-sprachigen aber auch anglophonen Geographie einen untergeordneten Stellenwert. Viel zu spät ist gerade in der deutsch-sprachigen Geographie die Relevanz des Digitalen erkannt worden und zu wenig wird derzeit dazu geforscht. Ganz besonders betrifft dies den Bereich von Arbeiten, die es schaffen den tatsächlich integrativen Mehrwert der Geographie mitzudenken und den Nexus von Gesellschaft-Technologie-Umwelt explizit zu adressieren. Unseres Erachtens ist dies aber auch eine zentrale Chance für die Positionierung und Profilierung einer Geographie des 21. Jahrhunderts.

Entsprechend der hier nur angedeuteten thematischen Weite Digitaler Geographien sind die wissenschaftlichen Praktiken zur Durchführung dieser Untersuchung und die sie leitenden (meta-)theoretischen Perspektiven vielfältig. Im nächsten Kapitel stellen wir eine Auswahl aktueller einflussreicher Strömungen vor.

3 PERSPEKTIVEN: WELCHE AKTUELLEN EINFLUSSREICHEN METATHEORETISCHEN STRÖMUNGEN LASSEN SICH IDENTIFIZIEREN?

Gerade angesichts der stark anwachsenden Zahl und inhaltlich-thematischen Bandbreite digital-geographischer Arbeiten ist eine Möglichkeit ihrer systematischen Differenzierung die nach den dominierenden metatheoretischen Perspektiven. Diese Einteilung nehmen wir gleichermaßen als Ergänzung und als Alternative vor zu jener von Ash et al. (2018: 27) in die „geographies produced *through*, produced *by*, and *of* the digital". Angesichts der großen Bedeutung neuer methodischer Zugänge und der Methodenentwicklung durch digitale Geograph*innen, ist die folgende Unterteilung de facto wohl eine vorwiegend nach onto-epistemologischen

Positionen, teilweise nach methodologischen Herangehensweisen (insbesondere im Bereich der algorithmischen Geographien). Mit „Ontoepistemologien" (der Begriff geht zurück auf Barad 2007) sind Perspektiven gemeint, die zum einen die Untrennbarkeit von verschiedenen Seinsarten, z. B. Natur und Kultur oder Mensch und Technologie, betonen und zum anderen Wissensformen praktizieren, die statt auf Abgrenzungen vor allem auf die Verschränkungen von diesen Seinsarten fokussieren. Es handelt sich bei der folgenden Differenzierung zum einen um Momentaufnahmen und zum anderen aber auch um ineinanderfließende und veränderliche Kontinuen sowie Verbindungen von Sichtweisen.

Eine Auseinandersetzung mit diesen Perspektiven mag gerade für diejenigen schwierig und komplex erscheinen, die erst beginnen, sich mit Wissenschaft zu beschäftigen. Für umso wichtiger erachten wir es, einen Überblick auch gerade für Studierende und Nachwuchswissenschaftler*innen zu schaffen, die sich ein Grundverständnis für diese forschungsleitenden Perspektiven erarbeiten wollen. Sie sind Grundlage für eine kritische Hinterfragung von und Sensibilisierung für vielfältige, niemals objektive und neutrale, sondern normativ-geleitete Wissensproduktionen.

Wie bereits erwähnt, fand eine kritische Debatte um bis hin zur Abwendung von Objektivitätsansprüchen in der Geographie vor allem im Zuge des *Cultural Turn* statt (siehe Glasze 2017 → in diesem Band für konstruierte Weltbilder durch analoge wie digitale Kartographien). Felgenhauer & Gäbler (2018) haben dies im Sinne von Geographien digitaler Kulturen gleich in mehrfacher Hinsicht aufgegriffen, z. B. mit Blick auf subjektive Raumwahrnehmungen und Alltagsroutinen, die durch digitale Echtzeitkommunikation und mobile Suchanfragen kulturell gerahmt werden:

> […] it is helpful to focus on subjective individual perceptions and appropriations of space and place. This requires a constructivist, hermeneutic view assuming that spatial formations are basically cultural constructions. Space and place can be considered to be both the matter and the product of individual perceptions and experience, cultural and linguistic convention, public discourse, and economic/political action. In this sense, space and place are not pre-given entities in which the cultural and the digital are contained. Instead space and place depend in their existence on culturally evoked meanings and everyday action. (Felgenhauer & Gäbler 2018: 9 f.)

Von dieser Feststellung „als kleinstem gemeinsamen Nenner" ausgehend stellen wir nun drei derzeit dominante Perspektiven vor.

3.1 Poststrukturalistische, postkoloniale und feministische Perspektiven

Im Fokus der sehr weiten Perspektiven poststrukturalistischer, postkolonialer und feministischer digitaler Geographien steht eine Kritik an und Dekonstruktion von hegemonialen und heteronormativen techno-sozialen Wissensproduktionen, die Herausbildung von Wissens(chafts)systemen und Wahrheitskonstruktionen (Elwood 2010, 2020). Digital-geographische Arbeiten, die insbesondere an den *Cultural Turn* in den Geistes- und Sozialwissenschaften, der Etablierung der Neuen Kulturgeographie und der Diskursforschung in der Geographie anschließen (vgl.

Glasze & Mattissek 2009), konzentrieren sich auf das Aufzeigen von gesellschaftlich produzierten Ungleichheiten und die Dekonstruktion techno-sozialer und diskursiver Machtverhältnisse.

Postkoloniale Studien, wie u. a. diejenige von Ayona Datta (2018 → in diesem Band), fokussieren dabei auf die subtile Ermöglichung und Verstärkung globaler Ungleichheiten durch techno-soziale Assemblagen. Sie weisen auf die Dominanz bestimmter Staaten in der infrastrukturellen Bereitstellung des Internets, der damit divergierenden Möglichkeiten der Wissensproduktion und -verbreitung und der resultierenden digitalen Klüfte und Ungleichheiten. Sie beleuchten eine damit einhergehende Reproduktion des Machtgefälles zwischen dem Globalen Norden und Süden und der Persistenz kolonialer oder auch der Schaffung neokolonialer Strukturen auf verschiedenen Maßstabsebenen (vgl. ausführlicher Young et al. 2020).

Poststrukturalistische Arbeiten hinterfragen stärker die Rolle von Codes, Software und Alltagstechnologien in der Produktion von gesellschaftlichen Diskursen und diskursiven Territorialisierungen und Grenzziehungen. In ihrer diskurstheoretischen Analyse von Twitter-Tweets während der Bundestagswahl 2017 diskutieren beispielsweise Wiertz und Schopper (2019), wie sich diskursive Logiken mit technischen Prozessen in der sprachlich-symbolischen Konstitution von Bedeutungen verbinden. Sie identifizieren, wie aufgrund dieser Logiken und ihrer Dynamiken mehr oder weniger stark abgegrenzte Tweet-Cluster entstehen, die sich insbesondere in Bezug auf den Einfluss rechtspopulistischer und extrem rechter Äußerungen unterscheiden. Zugleich unterscheiden sich die Cluster durch ihre diskursiven Konstruktionen von Zugehörigkeit, insbesondere im Kontext von Debatten über Geflüchtete.

In ihrem Entwurf von *Digital Feminist Geographies* heben Elwood und Leszczynski (2018: 631) vier aktuelle Kernthemen hervor: die Entwicklung feministischer digitaler Epistemologien und Methodologien, Analysen von digital-mediatisierten vergeschlechtlichten Formen der sozialen Reproduktion (vgl. hierzu auch Marquardt 2018, Bauriedl und Strüver 2020), der digitalen Repräsentation und Produktion von Körpern und Verkörperung (z. B. Cockayne et al. 2017) sowie feministische Theoretisierungen von Raum und Räumlichkeit (z. B. Cockayne und Richardson 2017). Ein Beispiel für die Rolle digitaler Repräsentation und der Produktion von Körpern und Verkörperung ist Nasts (2017) psycho-analytische und geopolitische Arbeit zur Zunahme der Nutzung von weiblichen Spiel-, Komfort- und Sex-Puppen bzw. zu ihren fembot-Weiterentwicklungen durch robotische Technologien. Aus einer polit-ökonomischen Perspektive diskutiert sie die Zunahme der Nutzung der Puppen durch japanische Männer im Kontext von Japans schrumpfender Bevölkerung, sinkender Fertilität, Widerstand gegen traditionelle und Neuaushandlung von Rollen von Frauen und Männern auch im Kontext des „demise of Japan's salaryman" (Nast 2017: 761) im Zuge der Finanzkrise der 1990er. Sie beleuchtet wie fembots neue bzw. veränderte Formen der maternalen und ödipalen Sozialität, Intimität, Sexualität produzieren.

Elwood (2020) diskutiert neuere digital-geographische Arbeiten aus den Perspektiven der *Queertheory*, der *Critical Race Studies* und der *Black* und *Queer Code Studies*, die thematisieren, wie bestehende Ungleichheiten durch digital-mediatisierte Prozesse der Exklusion, Überwachung und Enteignung oftmals reproduziert

und teilweise verschärft werden. Diese Studien sind durch Arbeiten im weiten Feld
der feministischen Intersektionalitätsforschung beeinflusst worden, die multiple
Identitätskonstruktionen, beispielsweise von Gender, Sexualität, Ethnie, Nationali-
tät und Alter, im Kontext von Aus- und Abgrenzungsprozessen adressieren (vgl.
dazu z. B. Valentine 2007, Strüver 2013, Marquardt und Schreiber 2015, Hopkins
2018). Die neuen digital-geographischen Beiträge zielen dabei explizit auf eine
Überwindung normativer und exkludierender Kategorisierungen und Grenzziehungen – „[they] refuse/elude hegemonic digital-social-spatial orders and mediate for
ways of thriving otherwise" (Elwood 2020: 2).

3.2 Mehr-als-menschliche und mehr-als-repräsentationale Perspektiven

Ein sich zunehmend ausdifferenzierender Zweig sind digital-geographische Arbeiten aus mehr-als-repräsentationalen Perspektiven und affektiven Geographien sowie
mehr-als-menschlichen Perspektiven. Eine Gemeinsamkeit dieser im Detail voneinander abweichenden Ansätze ist ihr Fokus auf Untersuchungen des Mikroraums, der
Berücksichtigung der Rolle von Körpern und Materialitäten sowie der Versuch der
Rekonstruktion der Komplexität sozio-materiell-technologischer Raumproduktionen. Sie teilen zudem das Anliegen, die Grenzziehungen zwischen Mensch/Gesell-
schaft, Natur/Umwelt und Technik bzw. Technologien zu überwinden.

Ansätze, die die Rolle von Verkörperung, unterbewussten, prä-kognitiven und
unspezifischen Affekten in Raumproduktionen hervorheben, wurden von Nigel
Thrift zunächst als „nicht-repräsentationales" Paradigma eingeführt (Thrift 2008),
später als „mehr-als-repräsentationale" Geographien (Lorimer 2005, Schurr 2014,
Schurr und Strüver 2016) relativiert. Diese Strömungen verbinden sich explizit mit
der Forderung nach mehr-als-repräsentationalen Method(ologi)en. Repräsentatio-
nalen Method(ologi)en, beispielsweise der qualitativen und quantitativen Sozial-
forschung sowie der Diskursforschung, werfen sie vor, die sozio-(technologisch-)
materielle Realität auf Repräsentationen in Form von Text und Daten reduzieren,
und ihr damit Aktivität und Komplexität zu entziehen, d. h. „deadening effects on
an otherwise active world" (Cadman 2009: 456) zu produzieren. Mehr-als-repräsentationale Methoden im Bereich digitaler Geographien nutzen einerseits mobile
Technologien wie Smartphones, Smartwatches, Videokameras oder mobile Datenbrillen um Aktivität, Körperlichkeit, Lebendigkeit, Materialität und Komplexität
soziomaterieller und technosozialer Realitäten zu erfassen und „more-than human,
more-than-textual, multisensual worlds" zu adressieren (Lorimer 2005: 83). Zugleich reflektieren sie die subtile Einschreibung von Daten und Vermessungstechnologien in Körper, Affekte und Alltagspraktiken kritisch – beispielsweise jene
Arbeiten dieser Perspektive, die aufbauend auf Ansätzen des *Quantified Self* untersuchen. Inspiration für die Untersuchungen ergeben sich insbesondere aus den Arbeiten der Soziologin und Medien- und Kulturwissenschaftlerin Deborah Lupton
und ihrem Team (Lupton 2016, 2018, 2020, Sumartojo et al. 2016). Es gehören
dazu auch die kritische Reflexion von Versuchen die affektive Wirkung der Nutzung digitaler Technologien in Alltagssituation über neue biosensorische Methoden

messbar zu machen (wie Elektroenzephalographie [EEG], Galvanic Skin Response [GSR], Eye-Tracking über mobile Datenbrillen, Spinney 2015).

Mehr-als-menschliche Perspektiven haben sich insbesondere in Anlehnung und Weiterentwicklung von *Science and Technology Studies* (STS) und Akteur-Netzwerk-Theorien (ANT) herausgebildet; ihre ontoepistemologischen Grundlagen liegen vor allem in den feministischen Technowissenschaften (Haraway 1995, 1997, 2008, 2016, Barad 2003, 2007), die – anders als Thrift mit der NRT – dualistische Grenzziehungen wie bspw. zwischen Körper und Geist nicht manifestieren, sondern (versuchen) auf(zu)lösen. Zudem stehen sie in der Tradition verschiedener (post-)strukturalistischer Theorien und berücksichtigen daher makropolitische Diskurse, Gesellschafts- und Machtstrukturen im Zusammenwirken mit Mikropraktiken. Dieses Zusammenwirken versteht die Wissenschaftshistorikerin Karen Barad (2007) als radikal-relational, als Verschränkungen von Bedeutungen und Materialität, von Mikro und Makro, von Technologien und Körpern etc. Barad (2007: 89f., 185, 409) – und in gewisser Weise auch Haraway (2016) – betreiben daher die zu Beginn des Kapitels erläuterte Ontoepistemologie, die auf eine Überwindung des Mensch-Umwelt-Dualismus zielt. Derartige Überlegungen zu Verschränkungen („entanglements") basieren zumeist auf Donna Haraways Figur des Cyborgs bzw. ihrem „Cyborg-Manifest" (1995 [1985]), das seit mehr als dreißig Jahren die Überlegungen zu Mensch-Technik-Beziehungen sowie technosoziale Ansätze entscheidend inspiriert (vgl. Militz et al. 2021). Haraways Cyborg als *cybernetic organism*, als Verschränkung von Technologien und menschlichem Organismus, spricht die Auflösung des Natur-Technik-Dualismus an. Es geht ihr damit um die aktive Aneignung und wechselseitige Gestaltung der Grenz(be)ziehungen von cyber und organism, denn „es ist unklar, wer macht und wer gemacht wird" (Haraway 1995: 181). Dies wurde u.a. von Wilson (2009 → in diesem Band) für seine „Cyborg Geographies" aufgegriffen. Er schlägt anhand der Figur des Cyborgs vor, die ontologische Hybridität erneut um eine epistemologische zu erweitern, die mehr als das (ebenfalls von Haraway, 1995) entwickelte Situierte Wissen umfasst. In Bezug auf die Mensch-Technologie-Interaktion spricht er von ko-produzierten Wissensformen und folgert: „If ontological hybridity is concerned with what it means to be hybrid, I suggest that epistemological hybridity considers what it means to know hybridly (Wilson 2009: 504; Hervorh. i. O. → in diesem Band).

Gemein ist all diesen Ansätzen eine Betonung mehr-als-menschlicher Raumproduktionen, die nicht nur durch menschliche Aktanten und ihre Interaktionen sowie gesellschaftliche Diskurse und Strukturen, sondern von vernetzten Objekten, Tieren, Daten, Stoffflüssen etc., geprägt werden. Dazu gehören auch die Arbeiten aus dem Bereich der Tiergeographien, die die zunehmende Datafizierung von (Wild-)Tieren zum Zwecke des Tracking, der Sammlung von Daten nicht nur über Tiere, sondern auch ihre Habitate avisieren und dabei sowohl auf den Schutz der Tiere als auch den Schutz der Menschen abzielen. Mithilfe von GPS-Daten aus Senderhalsbändern von Wildtieren können darüber hinaus geschützte mehr-als-menschliche Begegnungen im Sinne von Mensch-Tier-Technologie-Assemblagen entstehen (Poerting et al. 2020).

3.3 (Kritische) Algorithmische und Big Data Perspektiven

Perspektiven der kritischen algorithmischen und Big Data Geographien sind sehr stark epistemologisch-methodologisch geleitet, da sie sich Verfahren der Computer Sciences und Informatik zunutze machen, um computationale und Programmier-perspektiven um eine dringend benötigte kritische Reflexion der Arbeitsweisen von und Raumproduktionen durch Algorithmen sowie Big Data zu ergänzen (Kitchin und Dodge 2011, Kitchin 2017 → in diesem Band, Straube und Belina 2018). Dieser Zweig hat teilweise starke Überschneidungen mit mehr-als-menschlichen Geographien in den metatheoretischen Fundamenten; es besteht aber ein signifikanter Unterschied in Bezug auf die Skalenebenen, auf die die jeweiligen Arbeiten fokussieren, sowie in den methodologischen Herangehensweisen. Im Vordergrund algorithmischer Geographien steht der Makroraum und das Erfassen von Raumproduktionen durch Big Data, algorithmische Systeme und Assemblagen. Obwohl es sich hierbei um Analysen mit dem Zweck der Erfassung von Geographien *durch* das Digitale handelt, stehen aufgrund der aufwendigen Analysen oftmals der virtuelle Raum und seine technologische Produktion im Vordergrund.

In dem auch in diesem Band enthaltenen Beitrag, differenziert Rob Kitchin (2017: 16) Fokusse algorithmischer Geographien: erstens in Fallstudien einzelner oder einer Klasse von Algorithmen, zweitens über detaillierte Analysen von Algorithmen einer Domain sowie, drittens, durch abstraktere, kritische Herausarbeitungen von Algorithmen, ihrer Natur und Performanz. Die damit verbundenen Methoden sind im Falle der ersten beiden Schwerpunkte Methoden der Computerwissenschaften zur Auslesung und Rekonstruktion von Algorithmen sowie Big Data Analysen, während der letzte Zweig der Arbeiten solche Analysen oft zusätzlich mit (digital-) ethnographischen Methoden, qualitativen (Expert*innen-)Interviews und teilweise diskursanalytischen Methoden ergänzt. Ein Beispiel für den dritten Zweig dieser Arbeiten zu Algorithmen ist Lizzie Richardsons (2020) Analyse der algorithmischen Berechnungen von Gütern und Diensten am Beispiel der Deliveroo Plattform. Mit ihrer Beleuchtung der komplexen sozialräumlichen Assemblagen der beteiligten menschlichen und mehr-als-menschlichen Aktanten, geht sie zugleich bedeutend über eine reine algorithmische Rekonstruktion hinaus und zeigt vor diesem Hintergrund den bedeutenden Mehrwert einer kritischen human- und digital-geographischen Untersuchung und Kontextualisierung der Arbeitsweise und Einbettung von Algorithmen in sozialräumliche Assemblagen (siehe auch Richardson 2018 → in diesem Band).

Mei-Po Kwan (2016: 274) erweitert den Einblick in algorithmische Geographien durch eine explizite Beleuchtung der Produktionen geographischen Wissens und Forschung durch Algorithmen und Daten und schlussfolgert: „knowledge about the world generated with big data might be more an artifact of the algorithms used than the data itself". Dies würde sich auch für den gesellschaftlichen Alltag bzw. das „geographische Alltagswissen" als relevant erweisen, sofern wir die Kritik von Shaw und Graham (2018: 181) ernst nehmen, dass die Algorithmen von Google beeinflussen „wohin Menschen gehen, wie und wann sie dort ankommen und was sie dort tun".

Wie Kitchin (2017 → in diesem Band) hervorhebt, sind bedeutende Grenzen geographischer algorithmischer Untersuchungen die Komplexität, Ontogenese und dauerhafte flexible Veränderung und Anpassung von Algorithmen, die ihre Rekonstruktion sehr aufwendig und schwierig machen. Er bemängelt aber auch zu Recht, dass eine weitere sehr gravierende Hürde das Fehlen digitaler Geograph*innen ist, die einen Hintergrund in Computer- und Programmierwissenschaften haben und die damit in der Lage sind algorithmische Analysen überhaupt durchzuführen. Auch Till Straube (2018: 156) merkt an, dass angesichts einer Dominanz qualitativer Forschung „[t]he devices and technologies themselves remain largely ‚black-boxed‘ and are considered only in regards to their effects on society (or as cultural expressions)".

4 ZUKÜNFTE:
WELCHE CHANCEN BIETET EINE GESELLSCHAFT-TECHNOLOGIE-UMWELT-FORSCHUNG UND WELCHE HERAUSFORDERUNGEN BESTEHEN?

Nicht nur kritische Arbeiten sehen in der Digitalisierung eine potentielle Gefahr für eine friedliche, demokratische und gerechte Zukunft der Menschheit und die ökologische Nachhaltigkeit des Planeten. Algorithmisierung, Automatisierung, Sensor- und robotische Technologien sowie künstliche Intelligenz werfen neue Fragen der politischen und technologischen (Del Casino et al. 2020), aber auch der individuellen Souveränität (Zuboff 2018) auf. Fragen der Gestaltung, Steuerung und der Governance der Digitalisierung und damit zusammenhängender Machtaushandlungen, -verschiebungen und Beteiligungsprozesse werden in der digital-geographischen Literatur allerdings höchst kontrovers diskutiert und bewertet. Einerseits werden neue Formen von Crowdsourcing, Citizen Science, Volunteered Geographic Information und der Laiengeographien (vgl. für eine Differenzierung dieser Konzepte Haklay 2013, See et al. 2016) als nicht nur neue Formen der Partizipation, sondern der Kooperation bis hin zur Ko-Kreation zelebriert. Noch weiter gehen Perspektiven, die in algorithmischen, robotischen Technologien und künstlicher Intelligenz neue Möglichkeiten einer technologischen Steuerung sehen, die Formen menschlichen Regierens ersetzen. Unter den jüngeren Arbeiten in diesem Kontext sind beispielsweise die von Casey Lynch (2020) zu nennen, die die Bewegung zur „technologischen Souveränität" in Barcelona untersucht und diese als alternative postkapitalistische, anti-hegemoniale Möglichkeit der Entwicklung städtischer Organisation porträtiert.

Zugleich sehen andere Perspektiven in der Digitalisierung eine „zutiefst antidemokratische soziale Kraft" (Zuboff 2018: 110). Datenökonomien ermöglichen einen Überwachungskapitalismus (Zuboff 2018), der zu einer Monopolisierung bis hin zur Entmenschlichung von Macht führt. Die inzwischen in der (Digitalen) Geographie weit rezipierte Wirtschaftswissenschaftlerin Shoshana Zuboff urteilt:

So wie die Industriezivilisation auf Kosten der Natur florierte und uns heute die Erde zu kosten droht, wird eine vom Überwachungskapitalismus und seiner instrumentären Macht geprägte

Informationszivilisation auf Kosten der menschlichen Natur florieren, womit sie uns unser Menschsein zu kosten droht. […] Es handelt sich hier um eine Form der Tyrannei, die sich vom Menschen nährt, aber nicht vom Menschen ist. […] Wie der instrumentäre Schwarm ist die Tyrannei die Auslöschung von Politik. Der Überwachungskapitalismus herrscht mittels instrumentärer Macht, die wie der Tyrann außerhalb der Menschheit existiert, während sie paradoxerweise menschliche Gestalt annimmt. (Zuboff 2018: 106, 110)

Für zentral in diesen Machtverschiebungen wird in den Sozialwissenschaften die zunehmende Plattformisierung des Alltagslebens erachtet, d. h. die Praktiken des Austauschs von Informationen, Gütern und Dienstleistungen über digitale Plattformen, einschließlich der Effekte von Datenflüssen und Algorithmen auf private und öffentliche Interaktionen (Srnicek 2018, van Dijck et al. 2018, Zuboff 2018, Richardson 2020, Sadowski 2020). Die kritische Stadtforschung bezeichnet dieses Phänomen mittlerweile als „Plattformurbanismus" (Barns 2019, 2020) – denn die Plattformökonomie ist im Kern eine urbane Ökonomie, die sich in Folge der Finanzkrise seit 2008 als Urbanisierungsprozess manifestiert (Graham 2020, Lee et al. 2020). Jonathan Cinnamon (2017) hat mithilfe von Nancy Frasers (2009) Gerechtigkeitskonzeption – und damit der *Verknüpfung* von ungerechter ökonomischer Verteilung und fehlender kultureller Anerkennung – untersucht, inwiefern die Plattformökonomie als Grundlage des Plattformurbanismus soziale Teilhabe im Globalen Norden unter bestimmten Bevölkerungsgruppen erschwert: Erstens werden Verhaltensdaten und Persönlichkeitsprofile, wie auch die Rohdaten, aus denen sie gewonnen werden, als Waren gehandelt und besitzen somit ökonomischen Wert. Die Benutzer*innen, auf deren Daten diese Profile beruhen, haben hingegen keinen Zugriff auf ihre Daten. D. h. die asymmetrische Akkumulation persönlicher Daten ist eine Fehlverteilung ökonomisch wertvoller Ressourcen – eine Dimension von Verteilungsungerechtigkeit: „The separation of people from their data […] is a clear obstacle to parity of participation and therefore an injustice" (Cinnamon 2017: 614). Zweitens bildet die Praxis der ungerechten Daten-Akkumulation die Grundlage für eine weitere Praxis des Überwachungskapitalismus, die Kategorisierung und Klassifikation von Menschen auf Basis der akkumulierten Daten. Dies ist im Hinblick auf die Anerkennung ungerecht, da die Praktiken der algorithmischen Klassifikation in Gruppen soziokulturelle Diskriminierung oder gar Marginalisierung reproduzieren. Wie Young et al. (2020) darlegen, reproduzieren Datenökonomien nicht nur soziale, sondern auch bestehende globale Ungleichheiten, indem Datenlandschaften die englisch-sprachigen, städtischen, wohlhabenden Gesellschaften des Globalen Norden ungleich stärker abbilden. Auf der Mikroebene induzieren digitale Technologien zudem neue Formen der (panoptischen) Selbst-Kontrolle und sozialen Governance bis hin zu einem damit verbundenen Eindringen des Staates in den Privatraum und die „intime Häuslichkeit" (Datta 2020). Es gibt jedoch aktuell auch postkapitalistische Visionen von Plattformzukünften, die Hannes Gerhardt (2020) in seinen Überlegungen zu Collaborative Commons auf Basis von Peer-to-Peer Interaktionen diskutiert.

Während sich die Untersuchung der Rolle von Daten(-ökonomien) auf Machtverhältnisse als ganz zentraler Fokus jüngerer digital-geographischer Arbeiten herauskristallisiert hat, wird eine kritische (human-)geographische Erforschung der

Auswirkungen von robotischen Technologien und künstlicher Intelligenz in diesem Prozess gefordert (Del Casino 2016, McDui-Ra und Gulson 2019, Del Casino et al. 2020). Portraits von futuristisch anmutenden, jedoch sich immer häufiger manifestierenden Mensch-Maschine-Hybriden in Formen von Cyborgs, Transhumanen und Robotern stehen vielfältige Forderungen einer Neuaushandlung und ethischen sowie gerechten Gestaltung der Gesellschaft-Umwelt-Technologie-Verhältnisse entgegen. Der WBGU (2019: 8) sieht in der Digitalisierung die Gefahr der Entgrenzung und des Missbrauchs im Verhältnis von Mensch und Maschine. Robotische Technologien ersetzen zunehmend menschliche Arbeit und führen zu einer Neubewertung menschlicher Wissenssysteme (vgl. WBGU 2019). Diese neuen Assemblagen und Aktanten intervenieren dabei in bestehende Achsen von Ungleichheit, die entlang von sozial konstruierten Kategorien und Hierarchisierungen – beispielsweise von Geschlecht, Klasse, Ethnie und/oder Alter – bestehen (Nast 2017).

Des Weiteren benötigen die bereits jetzt greifenden Veränderungen von Mensch-Umwelt-Interaktionen durch robotische Technologien, beispielsweise im Bereich der Landwirtschaft, nach Del Casino et al. (2020: 609) kritische integrative Forschung: „there is a need for geographers to examine and theorise how the rise of robots is reorganising ways of knowing, seeing, producing, and intervening in nature, and the political economic and environmental justice implications of this reorganisation". Die Autor*innen betonen dabei, dass auch Umweltmonitoring zunehmend durch komplexe, autonome und vernetzte Interaktionen von Sensoren, mobilen robotischen Plattformen und Drohnen ersetzt wird, die wiederum künstliche Intelligenz, machine learning, Automatisierung kombinieren (Del Casino et al. 2020: 608). Damit gehen neue Formen der Wissensproduktion, der Überwachung und Kontrolle, aber auch der Wahrnehmung von „Natur" und Umwelt sowie der Intervention und Steuerung einher. Bakker und Ritts (2018 → in diesem Band) unterstreichen gleichwohl explizit, dass „better data does not necessarily lead to better [environmental] governance".

Die Digitalisierung fördert und ist Teil der Zunahme von „kontrollierten Umwelten", beispielsweise von Forschungslaboren, einer zunehmend kontrollierten Landwirtschaft, von künstlich geschaffenen Inseln und Schneelandschaften bis hin zu städtischen Konsum- und Produktionslandschaften. Die Bestrebungen der Automatisierung, Simulierung und Kontrolle von Prozessen in diesen Umwelten sind dabei immer weitreichender, wie Lockhart und Marvin (2020) darlegen. Sie zeigen mit ihrer Untersuchung zu Automatisierungsprozessen in drei verschieden extensiv und intensiv kontrollierten Umwelten (ein botanisches Glashaus, ein Luxushotel und ein universitäres Forschungslabor) Logiken, Imperative und Grenzen der Kontrollierbarkeit von Automatisierungen in neuen Ökosystemen und Mikroklimaten auf (Lockhart und Marvin 2020). Dabei belegen sie eindrucksvoll, wie diese nur vermeintlich abgeschlossenen und kontrollierten Umwelten in einer engen und intensiven Austauschbeziehung mit den sie umgebenden Umwelten stehen und dass politische Imperative essentiell zum Verständnis ihrer Entwicklung sowie der Automatisierung von Filtrierung und materiellen Flüssen und deren Grenzen sind.

Neben Forschungen zu neuen, flüchtigen und sich rasant ändernden Themen im Gesellschaft-Technologie-Umwelt-Nexus ist schließlich ein Transfer und Ko-Kreation geographischen Wissens in transdisziplinären Initiativen zentrale Aufgabe digitaler Geograph*innen. Bereits jetzt sind Geograph*innen intensiv beteiligt an der Entwicklung ethischer Prinzipien, beispielsweise in den Bereichen der Digitalisierung von Arbeit und Gig Economy. Das Team um Mark Graham vom Oxford Internet Institute hat die Entwicklung von 5 „Fairwork Principles" (Fair Pay, Fair Conditions, Fair Contracts, Fair Mangement, Fair Representation; siehe https://fair.work/principles/) zur Förderung fairer Arbeit im Bereich der Gig Economy mitgestaltet (siehe außerdem für den Bereich plattformvermittelter Sorgearbeit: Schwiter und Steiner 2020 sowie Bauriedl und Strüver 2020). Dorothea Kleine und ihr Team an der Sheffield University haben die Entwicklung der „Ethical standards for the ICTD/ICT4D community" vorangetrieben (Dearden and Kleine 2019). Im Bereich der Weiterentwicklung von Lehr- und Lernmethoden sind Digitale Geograph*innen aus dem deutsch-sprachigen Raum zentral engagiert. Sie schlagen richtungsweisende Anstöße für einen *Spatial Citizenship* (Gryl und Jekel 2012) vor – eine Perspektive, die Lernende in einer aktiven Rolle in der kollektiven Ko-Kreation von Technosozialraum unterstützen möchte, und diskutiert konstruktive Ansätze einer neuen *Mündigkeit* in der Kultur der Digitaliät (vgl. Dorsch und Kanwischer 2019, Dorsch und Kanwischer 2020 → in diesem Band).

Wie wir bereits erwähnt haben, hat die COVID-19-Pandemie die bereits zuvor bestehenden zentralen Herausforderungen der Digitalisierung nicht erzeugt, aber sie hat sie jüngst mehr ins Licht gerückt (Kligler-Vilenchik und Literat 2020, Rose-Redwood et al. 2020). Dies muss unseres Erachtens dringend, aktiv und umfassend als Chance der sozialökologisch-technologisch nachhaltigeren Gestaltung von Gesellschaft-Technologie-Umwelt-Interaktionen genutzt werden. Denn die Beschleunigung, das Ausmaß, und die zunehmende Loslösung des Digitalen von menschlicher Kontrolle und Kontrollierbarkeit rücken Fragen der Macht(-aushandlung) im Nexus Gesellschaft-Technologie-Umwelt ins Zentrum und machen sie zur entscheidenden Frage der Zukunft der Menschheit im 21. Jahrhundert.

BIBLIOGRAPHIE

ASH, J., KITCHIN, R. & LESZCZYNSKI, A. (2018): Digital Turn, Digital Geographies? *Progress in Human Geography*, 42, 25–43.

ASH, J., KITCHIN, R. & LESZCZYNSKI, A. (2019): Introducing Digital Geographies. In: ASH, J., KITCHIN, R. & LESZCZYNSKI, A. (eds.): *Digital Geographies*. London: SAGE, 1–10 (pre-print version).

BAKKER, K. & RITTS, M. (2018): Smart Earth: A Meta-review and Implications for Environmental Governance. *Global Environmental Change*, 52, 201–211.

BARAD, K. (2003): Posthumanist Performativity: Toward an Understanding of How Matter Comes to Matter. *Signs: Journal of Women in Culture and Society*, 28, 801–831.

BARAD, K. (2007): Meeting the Universe Halfway. Quantum Physics and the Entanglement of Matter and Meaning, Durham: Duke University Press.

BARNS, S. (2019): Negotiating the Platform Pivot: From Participatory Digital Ecosystems to Infrastructures of Everyday Life. *Geography Compass*, 13, 1–13.

BARNS, S. (2020): *Platform Urbanism*. Singapore: Palgrave.

BAURIEDL, S. (2009): Impulse der geographischen Raumtheorie für eine raum- und maßstabskritische Diskursforschung. In: GLASZE, G. & MATTISSEK, A. (Hg.): *Handbuch Diskurs und Raum: Theorien und Methoden für die Humangeographie sowie die sozial-und kulturwissenschaftliche Raumforschung.* Bielefeld: transcript Verlag, 219–233.

BAURIEDL, S. & STRÜVER, A. (2020): Platform Urbanism: Technocapitalist Production of Private and Public Spaces. *Urban Planning*, 5, 267–276.

BELINA, B. (2013): Raum. Zu den Grundlagen eines historisch-geographischen Materialismus, Münster: Westfälisches Dampfboot.

BORK-HÜFFER, T., MAHLKNECHT, B. & KAUFMANN, K. (2020a): (Cyber)Bullying in Schools – When Bullying Stretches Across cON/FFlating Spaces. *Children's Geographies.*

BORK-HÜFFER, T., MAHLKNECHT, B. & MARKL, A. (2020b): Kollektivität in und durch cON/FFlating spaces: 8 Thesen zu Verschränkungen, multiplen Historizitäten und Intra-Aktionen in sozio-materiell-technologischen (Alltags-)Räumen. *Zeitschrift für Kultur- und Kollektivwissenschaften*, 6, 131–170.

CADMAN, L. (2009): Non-representational Theory/Non-representational Geographies. In: KITCHIN, R. & THRIFT, N. (eds.): *International Encyclopedia of Human Geography.* Oxford: Elsevier, 456–463.

CINNAMON, J. (2017): Social Injustice in Surveillance Capitalism. *Surveillance & Society*, 15, 609–625.

COCKAYNE, D., LESZCZYNSKI, A. & ZOOK, M. (2017): #HotForBots: Sex, the Non-human and Digitally Mediated Spaces of Intimate Encounter. *Environment and Planning D: Society and Space*, 35, 1115–1133.

COCKAYNE, D.G. & RICHARDSON, L. (2017): Queering Code/Space: The Co-production of Sociosexual Codes and Digital Technologies. *Gender, Place & Culture*, 24, 1642–1658.

DATTA, A. (2018): The Digital Turn in Postcolonial Urbanism: Smart Citizenship in the Making of India's 100 Smart Cities. *Transactions of the Institute of British Geographers*, 43, 405–419.

DATTA, A. (2020): Self(ie)-governance: Technologies of Intimate Surveillance in India under COVID-19. *Dialogues in human geography*, 10, 234–237.

DEARDEN, A. & KLEINE, D. (2019): Ethical Standards for the ICTD/ICT4D Community: A Participatory Process and a Co-created Document. Proceedings of the Tenth International Conference on Information and Communication Technologies and Development, January 2019, Article No.: 38, 1–15 [Online]. Available: https://dl.acm.org/doi/10.1145/3287098.3287134.

DEL CASINO, V.J. (2016): Social Geographies II: Robots. *Progress in Human Geography*, 40, 846–855.

DEL CASINO, V.J., HOUSE-PETERS, L., CRAMPTON, J.W. & GERHARDT, H. (2020): The Social Life of Robots: The Politics of Algorithms, Governance, and Sovereignty. *Antipode*, 52, 605–618.

DIRKSMEIER, P., HELBRECHT, I. & MACKRODT, U. (2014): Situational Places: Rethinking Geographies of Intercultural Interaction in Super-diverse Urban Space. *Geografiska Annaler: Series B, Human Geography*, 96, 299–312.

DORSCH, C. & KANWISCHER, D. (2019): Mündigkeitsorientierte Bildung in der geographischen Lehrkräftebildung – Zum Potential von E-Portfolios. *Zeitschrift für Geographiedidaktik*, 47, 98–116.

DORSCH, C. & KANWISCHER, D. (2020): Mündigkeit in einer Kultur der Digitalität – Geographische Bildung und „Spatial Citizenship". *Zeitschrift für Didaktik der Gesellschaftswissenschaften*, 11, 23–40.

ECKER, Y., ROWEK, M. & A. STRÜVER (zur Veröffentlichung angenommen): *Care on Demand*: Geschlechternormierte Arbeits- und Raumstrukturen in der plattformbasierten Sorgearbeit. In: ALTENRIED, M., DÜCK, J. & WALLIS, M. (Hg.): *Plattformkapitalismus und die Krise der sozialen Reproduktion.* Münster: Westfälisches Dampfboot.

ELWOOD, S. (2010): Thinking Outside the Box: Engaging Critical Geographic Information Systems Theory, Practice and Politics in Human Geography. *Geography Compass*, 4, 45–60.

ELWOOD, S. (2020): Digital Geographies, Feminist Relationality, Black and Queer Code Studies: Thriving Otherwise. *Progress in Human Geography*, 30913251989973.

ELWOOD, S. & LESZCZYNSKI, A. (2018): Feminist Digital Geographies. *Gender, Place and Culture: a Journal of Feminist Geography*, 25, 629–644.

FELGENHAUER, T. & GÄBLER, K. (2018): *Geographies of Digital Culture: An Introduction.* Abingdon, New York: Routledge.

FRASER, N. (2009): *Scales of Justice.* Cambridge, Polity Press.

GEBHARDT, H. & REUBER, P. (2020³): Gesellschaftliche Raumfragen und die Rolle der Humangeographie. In: GEBHARDT, H., GLASER, R., RADTKE, U., REUBER, P. & VÖTT, A. (Hg.) *Geographie: Physische Geographie und Humangeographie.* Berlin, Heidelberg: Springer, 660–668.

GERHARDT, H. (2020): Engaging the Non-Flat World: Anarchism and the Promise of a Post-Capitalist Collaborative Commons. *Antipode*, 52, 681–701.

GLASZE, G. & MATTISSEK, A. (2009): Diskursforschung in der Humangeographie: Konzeptionelle Grundlagen und empirische Operationalisierungen. In: GLASZE, G. & MATTISSEK, A. (Hg.) *Handbuch Diskurs und Raum: Theorien und Methoden für die Humangeographie sowie die sozial-und kulturwissenschaftliche Raumforschung.* Bielefeld: transcript Verlag, 11–59.

GOODCHILD, M. (2009): What Problem? Spatial Autocorrelation and Geographic Information Science. *Geographical Analysis* 41, 411–417.

GRAHAM, M. (2013): The Virtual Dimension. In: ACUTO, M. & STELLE, W. (eds.): *Global City Challenges: Debating a Concept, Improving the Practice.* London: Palgrave, 117–139.

GRAHAM, M. (2020): Regulate, Replicate, and Resist the Conjunctural Geographies of Platform Urbanism. *Urban Geography*, 41, 453–457.

GRAHAM, M., ZOOK, M. & BOULTON, A. (2013): Augmented Reality in Urban Places: Contested Content and the Duplicity of Code. *Transactions of the Institute of British Geographers*, 38, 464–479.

GRAHAM, S. (2004): Beyond the ‚dazzling light': From Dreams of Transcendence to the ‚remediation' of Urban Life: A Research Manifesto. *New Media & Society*, 6, 16–25.

GRYL, I. & JEKEL, T. (2012): Re-centrring Geoinformation in Secondary Education: Toward a Spatial Citizenship Approach. *Cartographica*, 47, 18–22.

HAKLAY, M. (2013): Citizens Science and Volunteered Geographic Information. Overview and Typology of Participation. In: SUI, D., ELWOOD, S. & GOODCHILD, M. (eds.) *Crowdsourcing Geographic Knowledge: Volunteered Geographic Information (VGI) in Theory and Practice.* Dordrecht: Springer, 105–122.

HARAWAY, D.J. (1995): *Die Neuerfindung der Natur,* Frankfurt am Main: Campus.

HARAWAY, D.J. (1997): Modest Witness – Second Millenium. FemaleMan Meets OncoMouse; Feminism and Technoscience, New York: Routledge.

HARAWAY, D.J. (2008): *When Species Meet,* Minneapolis: University of Minnesota Press.

HARAWAY, D.J. (2016): Staying with the Trouble. Making Kin in the Chthulucene, Durham: Duke University Press.

HOPKINS, P. (2018): Feminist Geographies and Intersectionality. *Gender, Place & Culture*, 25, 585–590.

KAUFMANN, K. (2018): Mobil, vernetzt, geräteübergreifend: Die Komplexität alltäglicher Smartphone-Nutzung als methodische Herausforderung. In: KATZENBACH, C., PENTZOLD, C., KANNENGIESSER, S., ADOLF, M. & TADDICKEN, M. (Hg.): *Neue Komplexitäten für Kommunikationsforschung und Medienanalyse: Analytische Zugänge und empirische Studien.* Berlin, 139–158.

KAUFMANN, K. (2019): The Smartphone as a Snapshot of its Use: Mobile Media Elicitation in Qualitative Interviews. *Mobile Media & Communication*, 6, 233–246.

KAUFMANN, K. (2020): Mobile Methods: Doing Migration Research with the Help of Smartphones. In: SMETS, K., LEURS, K., GEORGIOU, M., WITTEBORN, S. & GAJJALA, R. (eds.) *The SAGE Handbook of Media and Migration.* London: SAGE, 167–179.

KITCHIN, R. (2017): Thinking Critically About and Researching Algorithms. *Information, Communication & Society: The Social Power of Algorithms*, 20, 14–29.

KITCHIN, R. & DODGE, M. (2011): *Code/Space: Software and Everyday Life*. London, Cambridge: MIT Press.

KLIGLER-VILENCHIK, N. & LITERAT, I. (2020): Youth Digital Participation: Now More than Ever. *Media and Communication*, 8, 171–174.

KNORR-CETINA, K. (1989): Spielarten des Konstruktivismus. *Soziale Welt*, 40, 86–96.

KNORR-CETINA, K. (2005): Science and Technology. In: CALHOUN, C., ROJEK, C. & TURNER, B.S. (eds.) *The SAGE Handbook of Sociology*. London: SAGE, 546–560.

KWAN, M.-P. (2016): Algorithmic Geographies: Big Data, Algorithmic Uncertainty, and the Production of Geographic Knowledge. *Annals of the American Association of Geographers: Geographies of Mobility*, 106, 274–282.

LEE, A., MACKENZIE, A., SMITH, G.J.D. & BOX, P. (2020): Mapping Platform Urbanism: Charting the Nuance of the Platform Pivot. *Urban Planning*, 5, 116–128.

LESER, H. & SCHNEIDER-SLIWA, R. (1999): Geographie – eine Einführung: Aufbau, Aufgaben und Ziele eines integrativ-empirischen Faches, Braunschweig: Westermann.

LESZCZYNSKI, A. (2019): Spatialities. In: ASH, J., KITCHIN, R. & LESZCZYNSKI, A. (eds.) *Digital Geographies*. London: SAGE, 13–23.

LOCKHART, A. & MARVIN, S. (2020): Microclimates of Urban Reproduction: The Limits of Automating Environmental Control. *Antipode*, 52, 637659.

LORIMER, H. (2005): Cultural Geography: The Busyness of Being More-than-representational. *Progress in Human Geography*, 29, 83–94.

LUPTON, D. (2016): Personal Data Practices in the Age of Lively Data. In: DANIELS, J., GREGORY, K. & McMILLAN COTTOM, T. (eds.): *Digital Sociologies*. London: Policy Press, 335–350.

LUPTON, D. (2018): How Do Data Come to Matter? Living and Becoming with Personal Data. *Big Data & Society*, 5, 2053951718786314.

LUPTON, D. (2020): *Data Selves. More-than-human Perspectives,* Cambridge: Polity Press.

LYNCH, C.R. (2020): Contesting Digital Futures: Urban Politics, Alternative Economies, and the Movement for Technological Sovereignty in Barcelona. *Antipode*, 52, 660–680.

MARQUARDT, N. (2018): Digital assistierter Wohnalltag im smart home. Zwischen Care, Kontrolle und vernetzter Selbstermächtigung. In: BAURIEDL, S. & STRÜVER, A. (Hg.) *Smart City. Kritische Perspektiven auf die Digitalisierung in Städten*. Bielefeld: transcript Verlag, 285–297.

MARQUARDT, N. & SCHREIBER, V. (2015): Mothering Urban Space, Governing Migrant Women: The Construction of Intersectional Positions in Area-based Interventions in Berlin. *Urban Geography*, 36, 44–63.

MASSEY, D. (1993): Power-geometry and a Progressive Sense of Place. In: BIRD, J., CURTIS, B., PUTNAM, T., ROBERTSON, G. & TICKNER, L. (eds.): *Mapping the Futures: Local Cultures, Global Changes*. London, New York: Routledge, 60–70.

MASSEY, D. (2012⁷): *For Space*. Los Angeles, London, New Delhi, Singapore, Washington DC: SAGE [first edition: 2005].

McDUIE-RA, D. & GULSON, K. (2019): The Backroads of AI: The Uneven Geographies of Artificial Intelligence and Development. *Area*.

MILITZ, E., DZUDZEK, I. & SCHURR, C. (2021): Feministische Geographien der Technowissenschaften. In: AUTOR*INNENKOLLEKTIV GEOGRAPHIE UND GESCHLECHT (Hg.): *Handbuch Feministische Geographien*. Opladen: Budrich, 190–214.

MÜLLER-MAHN, D. (Hg.) (2005): Möglichkeiten und Grenzen integrativer Forschungsansätze in physischer Geographie und Humangeographie. Leipzig: Selbstverl. Leibniz-Inst. für Länderkunde.

NAST, H.J. (2017): Into the Arms of Dolls: Japan's Declining Fertility Rates, the 1990s Financial Crisis and the (Maternal) Comforts of the Posthuman. *Social & Cultural Geography*, 18, 758–785.

PAVLOVSKAYA, M. (2016): Digital Place-making: Insights from Critical Cartography and GIS. In: TRAVIS, C. & VON LÜNEN, A. (eds.) *The Digital Arts and Humanities: Neogeography, Social Media and Big Data Integrations and Applications.* Cham: Springer, 153–167.

PAVLOVSKAYA, M. (2018): Critical GIS as a Tool for Social Transformation. *The Canadian Geographer*, 62, 40–54.

PINK, S., HORST, H., POSTILL, J., HJORTH, L., LEWIS, T. & TACHHI, J. (2016): *Digital Ethnography. Principles and Practice,* London, SAGE.

POERTING, J., VERNE, J. & KRIEG, L. (2020): Gefährliche Begegnungen: Posthumanistische Ansätze in der technologischen Neuaushandlung von Mensch und Wildtier. *Geographische Zeitschrift*, 108, 3, 153–175.

RABARI, C. & STORPER, M. (2015): The Digital Skin of Cities. Urban Theory and Research in the Age of the Sensored and Metered City, Ubiquitous Computing and Big Data. *Cambridge Journal of Regions, Economy and Society*, 8, 27–42.

RAUNIG, M. & HÖFLER, E. (2018): Digitale Methoden? Über begriffliche Wirrungen und vermeintliche Innovationen. *Digital Classics Online*, 4, 12–22.

RICHARDSON, L. (2020): Platforms, Markets, and Contingent Calculation: The Flexible Arrangement of the Delivered Meal. *Antipode*, 52, 619–636.

Rose-Redwood, R., Kitchin, R., Apostolopoulou, E., Rickards, L., Blackman, T., Crampton, J., Rossi, U. & Buckley, M. (2020): Geographies of the COVID-19 Pandemic. *Dialogues in Human Geography*, 10, 97–106.

ROSE, G. (2017): Posthuman Agency in the Digitally Mediated City: Exteriorization, Individuation, Reinvention. *Annals of the American Association of Geographers*, 107, 779793.

SADOWSKI, J. (2020): Cyberspace and Cityscapes: On the Emergence of Platform Urbanism. *Urban Geography*, 41, 448–452.

SCHURR, C. (2014): Emotionen, Affekte und mehr-als-repräsentationale Geographien. *Geographische Zeitschrift*, 102, 148–161.

SCHURR, C. & STRÜVER, A. (2016): „The Rest": Geographien des Alltäglichen zwischen Affekt, Emotion und Repräsentation. *Georgaphica Helvetica*, 71, 87–97.

SCHUURMAN, N. & G. PRATT (2002): Care of the Subject: Feminism and Critiques of GIS. *Gender, Place & Culture*, 9, 291–299.

SCHWITER, K. & STEINER, J. (2020): Geographies of Care Work: The Commodification of Care, Digital Care Futures and Alternative Caring Visions. *Geography Compass*, 14/e12546, 1–16.

See, L., MOONEY, P., FOODY, G., BASTIN, L., COMBER, A., ESTIMA, J., FRITZ, S., KERLE, N., JIANG, B., LAAKSO, M., LIU, H.-Y., MILCINSKI, G., NIKSIC, M., PAINHO, M., PODÖR, A., OLTEANU-RAIMOND, A.-M. & RUTZINGER, M. (2016): Crowdsourcing, Citizen Science or Volunteered Geographic Information? The Current State of Crowdsourced Geographic Information. *International Journal of Geo-Information*, 5, 1–23.

SHAW, J. & GRAHAM, M. (2018): Ein informationelles Recht auf Stadt? In: BAURIEDL, S. & STRÜVER, A. (Hg.): *Smart City. Kritische Perspektiven auf die Digitalisierung in Städten.* Bielefeld: transcript Verlag, 177–204.

SPINNEY, J. (2015): Close Encounters? Mobile Methods, (Post)phenomenology and Affect. *Cultural Geographies*, 22, 231–246.

SRNICEK, N. (2018): *Plattform-Kapitalismus*, Hamburg: Hamburger Edition.

STRAUBE, T. (2018): Situating Data Infrastructures. In: KITCHIN, R., LAURIAULT, T. P. & MCARDLE, G. (eds.): *Data and the City.* London: Routledge, 156–170.

STRAUBE, T. & BELINA, B. (2018): Policing the Smart City. In: BAURIEDL, S. & STRÜVER, A. (Hg.): *Smart City. Kritische Perspektiven auf die Digitalisierung in Städten.* Bielefeld: transcript Verlag, 223–235.

STRÜVER, A. (2013): „Ich war lange illegal hier, aber jetzt hat mich die Grenze übertreten" – Subjektivierungsprozesse transnational mobiler Haushaltshilfen. *Geographica Helvetica*, 68, 191–200.

SUMARTOJO, S., PINK, S., LUPTON, D. & LABOND, C. H. (2016): The Affective Intensities of Datafied Space. *Emotion, Space and Society*, 21, 33–40.

THRIFT, N.J. (2008): Non-representational Theory: Space, Politics, Affect, New York, London: Routledge.

VALENTINE, G. (2007): Theorizing and Researching Intersectionality: A Challenge for Feminist Geography. *The Professional Geographer*, 59, 10–21.

VAN DIJCK, J., POELL, T. & DE WAAL, M. (2018): *The Platform Society. Public Values in a Connective World.* Oxford: Oxford University Press.

WARF, B. (2013): *Global Geographies of the Internet.* Dordrecht, New York: Springer.

WARF, B. (2019): Teaching Digital Divides. *Journal of Geography*, 118, 77–87.

WBGU (2019): Unsere gemeinsame digitale Zukunft: Empfehlungen [Online]. Available: https://www.wbgu.de/de/publikationen/publikation/unsere-gemeinsame-digitale-zukunft [Accessed 12 May 2019].

WERLEN, B. (1995): Sozialgeographie alltäglicher Regionalisierungen. Band 1: Zur Ontologie von Gesellschaft und Raum, Stuttgart: Franz Steiner Verlag.

WERLEN, B. (2000): *Sozialgeographie. Eine Einführung*, Bern, Stuttgart, Wien: Verlag Paul Haupt.

WIERTZ, T. & SCHOPPER, T. (2019): Theoretische und methodische Perspektiven für eine Diskursforschung im digitalen Raum: Die Bundestagswahl 2017 auf Twitter. *Geographische Zeitschrift*, 107, 4, 254–281.

WILSON, M.W. (2009): Cyborg Geographies: Towards Hybrid Epistemologies. *Gender, Place & Culture*, 16, 499–516.

YOUNG, J.C., LYNCH, R., BOAKYE-ACHAMPONG, S., JOWAISAS, C., SAM, J. & NORLANDER, B. (2020): Volunteer Geographic Information in the Global South: Barriers to Local Implementation of Mapping Projects Across Africa. *GeoJournal.*

ZUBOFF, S. (2018): Der dressierte Mensch. Die Tyrannei des Überwachungskapitalismus. *Blätter für deutsche und internationale Politik*, 2018, 101–111.

I WIE DAS DIGITALE DIE GEOGRAPHIE EROBERT

DIGITALE GEOGRAPHIEN:
NEOGEOGRAPHIE, ORTSMEDIEN UND DER ORT
DER GEOGRAPHIE IM DIGITALEN ZEITALTER

Marc Boeckler

Neue Begriffe wie Geoweb, Neogeographie, Geomedien, geosoziale Netzwerke, Volunteered Geographic Information, Geobrowsing und augmentierte Geographie deuten es an: Mit den Technologien des Web 2.0 hat sich der Umgang mit geographischen Informationen und die Produktion geographischen Wissens grundlegend verändert. Dieses Themenheft diskutiert einige der Herausforderungen des digitalen Medienumbruchs aus geographischer Perspektive.

In anthropologischen Monographien stellte die „arrival story" den Habitus des Ethnographen sicher. Gleichzeitig wurde mit der Schilderung des Ankommens und Eindringens in eine fremde Kultur der Nachweis der Glaubwürdigkeit und Authentizität erbracht. In der Geographie übernahm diese Funktion für lange Zeit – und das eher implizit und unausgesprochen – die Karte. Sie hatte die Wirkung eines Echtheitszertifikats, das den Lesern die Botschaft des Autors übermittelte: „Ich bin ein echter Geograph und habe es mit eigenen Augen gesehen" (vgl. *Wardenga* 2010). Spätestens mit der Popularisierung und Kommerzialisierung des Kartenhandelns durch digitale Medien ist dieser heimliche Baustein geographischer Identität bedroht. „Google was here" ist der Leitspruch der digitalen Welterkundung, und Google war nicht nur auf den Straßen dieser Welt, sondern hat längst auch Landschaften wie den Grand Canyon vermessen (vgl. *Abb. 1*). Damit nicht genug. Über eine offene Schnittstelle hat Google alle Internetnutzer dazu eingeladen, die kartographierte Welt weiterzuentwickeln, zu kommentieren, zu personalisieren und letztlich eigene Karten und Karten des eigenen Lebens zu entwerfen.

„Neogeographie", wie es *Andrew Turner* (2006, S. 3) als Erster formuliert hatte, „besteht aus verschiedenen Techniken und Instrumenten außerhalb traditioneller Geographischer Informationssysteme (…), die es jedem ermöglichen, selbständig eigene Karten zu erstellen, raumbezogene Informationen mit Freunden und Bekannten zu teilen und zu einer Verbreitung von Ortskenntnissen und geographischem Wissen beizutragen. Lastly, Neogeography is fun."

Eine „neue Geographie", die Spaß machen darf und zur Mehrung und Verbreitung geographischen Wissens beiträgt, ohne wissenschaftlich abgesichert zu sein? Es überrascht wenig, dass die akademische Geographie diesen neuen Geographen zunächst mit Distanz und Abwertung begegnet ist. Neogeographen wurden zu Laiengeographen und „Citizen Sensors" degradiert, die zwar Koordinaten registrieren können, ihnen aber keine Bedeutung beizumessen haben. Den Status vollwertiger Geographen wollte man diesen neuen Geschöpfen jedenfalls nicht zusprechen, im besten Fall wurden sie als (ungewollte) Lieferanten freiwilliger geographischer

Abb. 1: Das Google Street View Team auf dem Colorado
(Google Maps, Streetview, in der Nähe des Lake Mead). Quelle: Google Maps
(https://www.google.com/maps/@36.129489,113.970507,3a,75y,127.81h,90t/data=!3m5!
1e1!3m3!1snBCaPz01WhmHOSq4EUw95w!2e0!3e5 (abgerufen am 10.03.2014).

Informationen (volunteered geographic information) geduldet (*Goodchild* 2007).
Erst in jüngerer Zeit wird versucht, „Neogeographie" als neue Teildisziplin der
Geographie gewissermaßen „nach Hause" zu holen (*Wilson & Graham* 2013). Dieser Beitrag geht jedoch einen anderen Weg und schlägt vor, „Neogeographie" als
jenen Ausschnitt der umfassenden Digitalisierung der Gesellschaft zu verstehen,
der sich aufgrund seiner Raumbezüge als spezifischer geographischer Gegenstand
für eine noch zu entwickelnde „Digitale Geographie" anbietet. Nach der Vorstellung einiger empirischer Bausteine dieser „neuen Geographie" wird in spekulativer
Absicht gefragt, wie sich Ort und Raum sowie die wissenschaftliche Praxis der
Geographie unter dem Einfluss digitaler Medien verändern könnten.

„NEOGEOGRAPHY IS FUN":
BAUSTEINE EINER NEUEN GEOGRAPHIE

Die vermutlich kurz währende Epoche der „Neogeographie" beginnt im Jahr 2004.
Im Oktober vor genau zehn Jahren hatte *Tim O'Reilly* zur ersten Web 2.0 Konferenz
geladen und dort auf visionäre Weise einige soziotechnische Elemente der Digitalisierung von Gesellschaft skizziert. Im selben Jahr starteten Google Maps, Open
Street Map, Facebook, Flickr und wenig später folgte Youtube. In Deutschland
markiert das Jahr 2004 den Übergang zur flächendeckenden Versorgung mit stationären und mobilen Breitbandzugängen ins Internet. Apple initiierte 2004 das ge-
heime „Project Purple", aus dem zweieinhalb Jahre später das iPhone, die „App"

und neue Informations- und Kommunikationspraktiken hervorgehen sollten. 2012 wurden weltweit bereits 1,7 Mrd. Mobiltelefone verkauft, knapp die Hälfte davon waren Smartphones. In kürzester Zeit hatten sich einige der Aufforderungen *O'Reillys* an die Technologieindustrie des Silicon Valley realisiert: „design for hackability", „architecture of participation", „software above the level of a single device" (Apple und iOS). Diese hier nur angedeuteten soziotechnischen Entwicklungen haben zur Entstehung der „neuen Geographie" beigetragen, deren konstitutiven Elemente sich fünf Bausteinen zuordnen lassen.

Erstens zeichnet sich Neogeographie durch eine „Beteiligungsarchitektur" aus. Die partizipative Grundstruktur des Web 2.0 war für kurze Zeit mit Demokratisierungshoffnungen verbunden, die sich nicht erst mit dem Bekanntwerden der staatlichen Überwachungsapparaturen oder der Rekonstitution autoritärer Strukturen im Anschluss an Ägyptens „Facebook-Revolution" zerschlagen haben. Auch die Kartenproduktion lief in den ersten Jahren unter Bezeichnungen wie „counter-mapping" oder „the people's geography" der Plattform „Platial". Inzwischen hat Platial den Dienst eingestellt und das Demokratisierungsversprechen der Neogeographie wird als problematische Irreführung bezeichnet (*Haklay* 2013). Nichtsdestotrotz hat die Bereitstellung der Google Maps API (eine offene Programmierschnittstelle, mit der andere Programme an ein Softwaresystem angedockt werden können – „design for hackability") die Zahl aktiver Geographen und Geographinnen exponentiell vervielfältigt. Nicht nur die private Herstellung und öffentliche Distribution von Karten und sogenannten Map-Mashups, auch die generelle Zuweisung von ortsbezogenen Informationen zu Datensätzen (Geotagging) ist zu einer digitalen Alltagspraxis geworden: Wikipedia-Einträge, geokodierte Tweets und Fotos, Hotel- und Restaurantbewertungen, Open Street Map, Running-Apps, die Laufstrecken aufzeichnen und nicht zuletzt die automatische Verortung aller Suchanfragen. Die Beteiligungsarchitektur hat nicht zu einem Abbau sozialer Ungleichheit beigetragen. Georeferenzierungen sind weltweit ebenso wenig gleich verteilt wie der Zugang zum Internet (vgl. *Abb. 2*). An der soziotechnischen Grundstruktur dieser „neuen Geographie" ändert das aber nichts: Neogeographie ist ein crowdsourced, user-generated, kollaborativer und bisweilen performativer Prozess.

Zweitens beruhen Neogeographien auf der zunehmenden Verbreitung und Miniaturisierung „ortsbewusster" technischer Apparaturen, die sich allgemein unter dem Begriff „Geomedien" zusammenfassen lassen. Einerseits ist damit das „Geoweb" angesprochen als Gesamtmenge aller online verfügbaren Inhalte, die durch Koordinatenpaare Orten auf der Erdoberfläche zugeordnet sind. Im engeren Sinn sind unter Geomedien jedoch „Ortsmedien" („locative media") zu verstehen, die einen mobilen Internetzugang mit „Verortungstechnologien" verbinden, wie beispielsweise Smartphones und Tablet-PCs. Durch die automatische Lokalisierung über GPS-Sensoren, WiFi-, Mobilfunktriangulation, über RFID (Radio-frequency identification) oder NFC (Near Field Communication) hat sich der Vorgang des Geotaggings verändert. Das stationäre Geoweb war darauf angewiesen, dass bestimmten Informationen manuell eine Position zugewiesen wurde. Ortsmedien automatisieren diesen Prozess, wie beispielsweise beim Versenden eines geocodierten Tweets.

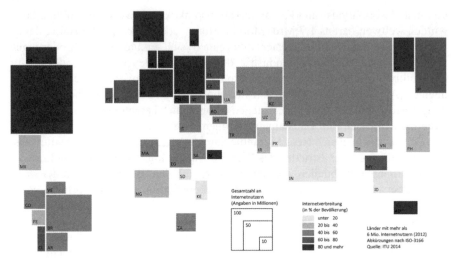

*Abb. 2: Treemap der weltweiten Internetnutzung nach absoluter Anzahl (Größe der Felder)
und Internetverbreitung in % der Bevölkerung.
Alle Länder mit mehr als 6 Mio. Internetnutzern (2012).
Graphik: M. Boeckler*

Weil Ortsmedien technisch ortsunabhängig sind, ist die Produktion ortsbezogener Inhalte mobil geworden. Zukünftig werden daher verschiedene Formen „vernetzter wearables" an der Ausweitung des Geowebs beteiligt sein: Brillen, Uhren, Armbänder, Kleidung etc. Neogeographen „gehen" dann nicht mehr online, sie sind „gehend" online, immer und überall. „Alte Geographie" hatte nicht nur die Tendenz sesshaft und statisch auf Orte zu blicken, sie war auch für die Gewinnung geographischen Wissens immer auf stationäre Prozesse der Datenverarbeitung und -darstellung angewiesen. Neogeographie hingegen ist mobil, dynamisch und vernetzt. Ortsbezogene Informationen werden nicht nur mobil geschaffen, sie werden auch in Bewegung abgefragt.

Drittens befindet sich die „neue Geographie" auch deswegen in ständiger Veränderung, weil mit dem Web 2.0 ein harter Wettbewerb um den monopolistischen Zugriff auf bestimmte Datensätze entstanden ist, die zuvor wenig bis keinen ökonomischen Wert hatten. Das Rennen um allgemeine Information (Was?) wurde von Google und seinem überlegenen Suchalgorithmus gewonnen. Digitale Identitäten (Wer?) sind (vorläufig) im Besitz von Facebook (und der NSA). Die Kommodifizierung von Geodaten (Wo?) ist noch im Gange. Ab 2009 wurde beispielsweise mit großen Anstrengungen und wenig Erfolg die Integration von Ortsmedien und sozialen Netzwerken vorangetrieben, um über geokodierte Bewegungsmuster neue Geodaten sammeln und vermarkten zu können. Neue Strategien setzen daher auf räumliche Orientierung und Gamification – wie Googles jüngste Produkte „Waze" (soziale Navigation) und „Ingress" (Mobile Urban Gaming) –, weil die Nutzung entsprechender Plattformen die Veröffentlichung von Geokoordinaten zwingend erfordert. Warum ist die diversifizierte Sammlung ortsbezogener Informationen für

Google so wichtig? Wenn die kontinuierliche Verortung von allem und jedem zu einem zentralen Merkmal des „Internets der Dinge" werden wird, dann könnte sich die Karte zu dieser „location aware future" so verhalten wie sich Windows zum PC verhalten hat. Google Maps würde zum Betriebssystem unserer vernetzten digitalen Zukunft (*Fisher* 2013).

Viertens ist Neogeographie mit einem neuen Verständnis von Handlung verbunden, da Algorithmisierung – als die jüngste Form gesellschaftlicher Technisierung nach Habitualisierung und Mechanisierung (*Rammert* 2007) – die Auslagerung menschlicher Praktiken von Körpern zu digitalen Geräten ermöglicht. Handlungen finden zunehmend als „distribuierte Praktiken" verteilt zwischen Menschen und Apparaturen statt. Erinnerung, Kalkulation, räumliche Orientierung, Freundschaft, Präferenzen, Geschmack sind längst Teil von Software-Applikationen geworden. In den „Science and Technology Studies" wird diese Position schon seit dreißig Jahren diskutiert (*Latour* 2005). Inzwischen stellt sich aber selbst die Europäische Union der Rekonfiguration von Akteuren in einer digitalen Welt und fragt ergebnisoffen „What does it mean to be human in the computational era?" und „how can we endorse and attribute responsibilities in a world where artefacts become agents?" (EU 2014).

In dieser Welt, in der Artefakte zu Akteuren geworden sind, synchronisiert der „Smart Fridge" (z. B. Samsung RF4289) nicht nur den Inhalt des Kühlschranks per WLAN mit dem digitalen Notizblock „Evernote", er weist auch auf den Ablauf von Haltbarkeitsdaten hin, unterbreitet Menü-Vorschläge auf der Grundlage des Kühlschrankinhalts und versendet Spam (als Teil eines gehackten Bot-Netzes). „Smart" sind aber nicht mehr nur „Phones" und „Kühlschränke". Klug geworden sind auch „smart cities", „smart homes", „smart factories", „smart grids", „smart pipes" usw. Mit dem Übergang zum Internetprotokoll IPv6 im September 2013 wurde der Adressraum für die Identifikation kommunizierender Geräte im Netz auf 48 Quadrillarden Adressen für jeden der sieben Milliarden Menschen auf der Erde erweitert. Dieses „Internet der Dinge" ermöglicht die eindeutige Identifikation eines jeden Elements und die selbstständige Kommunikation dieser Elemente untereinander. Kurz: Neogeographie wird auch von einer großen Zahl nicht-menschlicher Geographen betrieben, die als Sensoren automatisiert Geodaten registrieren, prozessieren und repräsentieren.

Mit der Zunahme distribuierter Praktiken eröffnen sich *fünftens* auch menschlichen Akteuren über Selbsttechnologien neue Differenzierungsoptionen. Diese Geographien des Selbst haben mindestens zwei Seiten: Die strategische Verteilung in sozialen Netzwerken, um einen präferentiellen Zugang zu Informationen, Beratung und Unterstützung zu erhalten und die spezifische Form der sozio-technischen Distribution, bei der bestimmte individuelle Aufgaben an Algorithmen ausgelagert werden. Prominente Beispiele sind das „Algo-Trading", bei dem Algorithmen im Hochfrequenzhandel den Kauf und Verkauf von Wertpapieren automatisiert ausführen sowie Selbstgeographien im Rahmen der „Quantified Self" Bewegung. Mit „self-tracking", „body data and life-hacking" wird „self knowledge through numbers" angestrebt (http://quantifiedself.com/), wobei nicht nur Körperfunktionen ununterbrochen überwacht werden, um durch die Berechnung des optimalen

Biorhythmus eine besonders effiziente Strukturierung des Alltags zu ermöglichen, sondern auch Gewohnheiten aufgezeichnet werden können und gezielte Anreize zur Überwindung subeffizienter Routinen gesetzt werden. Die Vermarktung eines klugen Armbands („Smartband") durch Sony in Verbindung mit der „Lifelog"-App zeigt, dass das quantifizierte Selbst längst zu einem Massenphänomen geworden ist: „Es zählt deine Schritte und deinen Kalorienverbrauch. Damit du morgens fit in den Tag startest, weckt dich das SmartBand zur optimalen Aufstehzeit" (Sony 2014). Neogeographie konstituiert sich aus immer mehr und immer weiter individualisierten Mensch-Technik-Assoziationen – es bedarf nur kleinerer Umstellungen und aus Geographien werden Egographien.

DIGITALE GEOGRAPHIE UND VERNETZTE LOKATIVITÄT

Geographie ist schon immer ein erfreulich diffuser Zusammenhang gewesen. Trotz aller Widersprüchlichkeit konkurrierender Definitionen scheint der genuine Forschungsgegenstand der Geographie unverändert die Erde (Geo) mitsamt ihren Landschaften und Orten geblieben zu sein. Dieses Verhältnis von Erde und Ort wird meist als Raum adressiert, Raum entweder als materielle Anordnung natürlicher und anthropogener Elemente oder Raum in seiner symbolischen Bedeutung (vgl. *Gebhardt* et al. 2011: 11). Mit neogeographischen Praktiken der „Geo-Referenzierung" fallen diese beiden geographischen Räume wieder zusammen (vgl. *Abb. 3*). Dafür bietet sich der Begriff „Digitale Geographie" als paradoxer Gegenentwurf zu gängigen Vorstellungen einer binären Räumlichkeit an, paradox, weil er seinerseits auf eine grundlegende Binarität abstellt – jedoch nicht in räumlichem Sinn.

Medientheoretisch wird gerne zwischen dem Medium als Informationsträger und der Vermittlungsform unterschieden und damit eine Doppelräumlichkeit begründet: Der Raum des Mediums wird dem Raum der Medialität gegenübergestellt (*Günzel* 2013). Das Kino und die Kinoleinwand, dunkel und hell, das Smartphone und der Inhalt des Smartphone-Bildschirms. Diese analytische Differenzierung fällt jedoch in die Anfangszeit der Computerära zurück, in der man von einer Differenz zwischen einem vermeintlichen realen Raum und einem von der materiellen Wirklichkeit unabhängigen Cyberspace ausgegangen war.

Eine Dualität, die längst der Einsicht gewichen ist, dass die digitale Dimension ein untrennbarer Bestandteil der einen räumlichen Wirklichkeit geworden ist. Den Ausgangspunkt für „Digitale Geographien" bilden daher nicht Raumvorstellungen, sondern Konzeptualisierungen des Digitalen selbst (*Horst & Miller* 2012). Drei Dimensionen des Digitalen sind hier wichtig:
- *Materialität*. Das Digitale ist materiell und unterscheidet sich ontologisch nicht von anderen greifbaren Dingen. Bits (binary digits), Binärziffern zusammengesetzt aus 0 und 1, stellen den fundamentalen Baustein digitaler Technologien dar. Sie existieren und wirken nur in materieller Form – seien es Löcher in Stempelkarten oder als optische Signale, die durch Kabel transportiert werden. Besser wäre es von digitalen Materialien zu sprechen, die in unterschiedlichen Formen als Technologien greifbar werden.

Abb. 3: Space to Space. „Spacefon" war ab 1996 einer der ersten Mobilfunkanbieter in Ghana.
Foto: M. Boeckler

– *Technologie.* Das Digitale lässt sich daher als Technologie verstehen, die binäre
  Zustände erzeugt und nutzt, um im Zusammenspiel mit Apparaturen auf Menschen und Dinge einzuwirken, Aufgaben auszuführen und sichtbare Praktiken
  zu erzeugen.
– *Abstraktes Zeichensystem.* Wenn heterogene Elemente und Prozesse durch die
  Einführung eines binären Codes auf die gleiche Grundstruktur zurückgeführt
  werden können, ermöglicht die Verwendung des abstrakten Zeichensystems
  eine raum-zeitliche Distanzierung und eine beschleunigte Ausdifferenzierung
  von Gesellschaft. Die Logik ist einfach. Je mehr auf das Gleiche reduziert werden kann, umso mehr Differenz lässt sich herstellen.

Die besondere Wirkung des Digitalen besteht nun darin, dass eine einfache neogeo-
graphische Praxis wie die Versendung eines geocodierten Tweets aus einem beliebigen Café in einer beliebigen Großstadt mit der beliebigen Aussage, dass es sich
um ein besonders gemütliches Café handelt, drei unterschiedliche räumliche Dimensionen auf neue Weise verknüpft. Erstens beruht die Georeferenzierung des
Tweets auf der eindeutigen Identifikation eines Orts (<meta name=„geo.position"
content=„50.119524, 8.648625"/>) in einem eurozentrierten Koordinatensystem.
Ein als absolut gedachter geometrischer Raum steht am Anfang der georeferenzier-
ten Praxis (Verortung). Zweitens werden topologisch strukturierte Räume der

Konnektivität sowohl auf der Ebene digitaler Materialität wie auch auf der Ebene sozialer Beziehungen in Anwendung gebracht. Der Tweet wird in einzelne Datenpakete zerlegt, die als Signale über viele tausend Kilometer entfernte Server auf die Bildschirme jener Lesegeräte gebracht werden, die Teil meines sozialen Netzwerks sind (Vernetzung). Drittens wird durch den Tweet der Ort als abstrakte Geokoordinate symbolisch aufgeladen. Dieser Umgang mit dem Ort wird als Bedeutungszuschreibung im Archiv des Geowebs als ortsbezogenes Wissen permanent abgelegt (Verwendung).

Die zusammenhängenden Praktiken der Verortung, Vernetzung und Verwendung reduzieren die räumliche Mehrdimensionalität auf eine spezifische Form des vernetzten Umgangs mit Ort, die als „net-locality" bezeichnet wird (*Gordon & de Souza e Silva* 2011). Weil das „Lokative" sprachwissenschaftlich den Fall der Ortsangabe anzeigt, könnte man – etwas sperrig, aber durchaus zutreffend – von „vernetzter Lokativität" sprechen. Das besondere dieses digitalen Umgangs mit Orten ist, dass der Ort in seiner symbolischen Dimension immer mit dem Ort als physisch-materiellem Raum zusammenfällt und damit – zumindest theoretisch – der Geographie (wieder) ein einheitliches Raumverständnis zur Verfügung steht.

DIGITALE GEOGRAPHIE ALS (ZUKÜNFTIGE) WISSENSCHAFTLICHE PRAXIS

Wenn digitale Medien zu einer sozio-technischen Reorganisation von Raum und Ort beitragen, in welcher Weise wirkt sich Neogeographie als neuer Gegenstandsbereich dann auf die wissenschaftliche Praxis einer „Digitalen Geographie" aus?

Neogeographie geht einher mit einer bemerkenswerten Konjunktur der Kartographie. Dabei hat sich das konstruktivistische Bewusstsein durchgesetzt, dass Karten weniger Objekte als Praktiken sind und selbst ein mimetischer Kartengebrauch nicht auf eine objektive Wirklichkeit verweist. Alle Neogeographen sind heute am Kartenhandeln, an der Erfassung, Verwaltung und weiteren Vermittlungsschritten kartographisch repräsentierter Daten beteiligt.

Was aber geschieht mit Karten beim beobachtbaren Übergang von einem mimetischen zu einem überwiegend navigatorischen Gebrauch? Stellen Karten dann noch immer den zentralen Wegweiser im Netz der Dinge dar? Oder werden sie nicht eher überflüssig? Schließlich ist eine räumliche Navigation ohne kartographische Darstellung mehr als denkbar. Googles 25 selbststeuernde Autos, die in Kalifornien bislang unfallfrei knapp eine Million Kilometer zurückgelegt haben, stellen keinen Quantensprung der Anwendung künstlicher Intelligenz dar. Der Selbststeuerungsalgorithmus navigiert lediglich durch einen zuvor abgefahrenen und mit zahllosen Sensoren minutiös vermessenen Raum (*Fisher* 2013). Die Bewegungen orientieren sich an einer „Karte" im Maßstab 1:1, die auf keine kartographische Darstellung mehr angewiesen ist, sondern als digitale Materie in Serverfarmen abgelegt ist und kontextuell als digitale Technologie zur Anwendung gebracht wird.

Im Jahr 2012 wurden täglich weltweit 2,5 Exabytes ($2,5 \times 10^{18}$) produziert, ein Yottabyte (10^{24}) speichert die NSA in ihrem Rechenzentrum in Utah allein. Die drei

Trillionen Fotos, die Google noch aufnehmen möchte, um die Welt in Google Maps komplett abbilden zu können, nehmen sich dagegen vergleichsweise bescheiden aus. Big Data ist allgegenwärtig und das als Dämon und Verheißung zugleich. „Data Driven Business" gilt für Unternehmen als der wichtigste IT-Trend der kommenden Jahre. Datenhungrig wird gesammelt, was gesammelt werden kann, um mit ausgeklügelten Algorithmen nach spezifischen Informationen zu suchen und neue Informationen zu erzeugen. Insbesondere mit Blick auf die Ausbreitung von Infektionskrankheiten haben sich Big Data Systeme wie die mit „Google Correlate" erzeugten „Grippe-Trends" als durchaus nützlich erwiesen (http://www.google.org/flutrends/). Auch die visuelle Darstellung der räumlichen Verbreitung homophober und rassistischer Tweets kann hilfreiche gesellschaftliche Diskussionen anstoßen (vgl. *Abb. 4*).

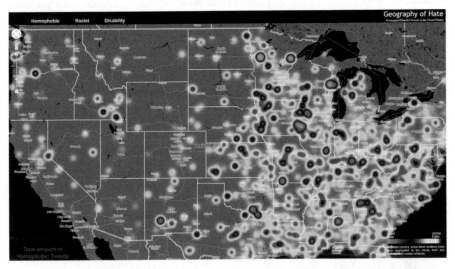

Abb. 4: Big Data-Geographien. Räumliche Verteilung homophober Tweets in den USA.
Projekt von Monica Stephens (Humboldt State University, USA; jetzt University of Buffalo)
Quelle: https://www.npr.org/sections/codeswitch/2013/05/30/187280870/haters-gonna-
hate-as-shown-on-a-map

Die Kehrseite der Daten-Sammelwut wurde im Zuge des 2013 bekannt gewordenen NSA-Abhörsystems sichtbar. Die gigantischen Ausmaße der Überwachung haben noch einmal deutlich gemacht, dass das Digitale als politische Technologie die politische Verfasstheit westlicher Demokratien verändern wird. Wie aber verändert Big Data (sozial-)wissenschaftliche Praxis? Zwei Beispiele:

Data mining. Klassische Wissenschaft im Zeitalter der Vernunft hat Hypothesen über messbare empirische Zusammenhänge in der Welt angestellt und diese anschließend methodisch überprüft. Im Zeitalter von Big Data bedarf es erstens keiner Vermutung mehr und zweitens keiner kausalen Zusammenhänge. Der Slogan „data driven" drückt die Rückkehr eines proto-naiven Positivismus hervorragend aus. Daten werden nicht mehr erhoben, sondern „abgebaut" (data mining).

Wenn sich der Zusammenhang zweier nicht zusammenhängender Merkmale als
prognostisch robust erweist, dann lassen sich mit diesen Parametern verlässliche
Aussagen über den Verlauf bestimmter Ereignisse treffen. Ein schönes Beispiel
sind hoch korrelierende Zusammenhänge bei „Google Correlate" – zum Beispiel
die Suchanfragen zu „losing weight" und „houses for rent" (vgl. *Abb. 5*).

   Temporalität. Big Data verändert unseren Zugriff auf Vergangenheit und Zukunft. Wenn zukünftig alle vergangenen digitalen Gegenwarten gespeichert und
wieder vergegenwärtigt werden können und gleichzeitig zukünftige Verläufe besser
vorhergesagt werden, dann richtet sich das Augenmerk auf die Ereignishaftigkeit
der Gegenwart. Warum sollte man weiterhin mit aggregierten Jahresdaten arbeiten
oder Interviews zu vergangenen Erlebnissen führen? In einer umfassend vernetzten
zukünftigen Gegenwart gibt es keinen Grund mehr, warum nicht auch wissenschaftliche Repräsentationen mit „real-time data" arbeiten sollten. Google und
NSA können das bereits jetzt. Außer Frage steht, dass sich die gedruckte Papier-
form dann nicht mehr als Darstellungsmedium für wissenschaftliche Arbeiten eig-
nen wird.

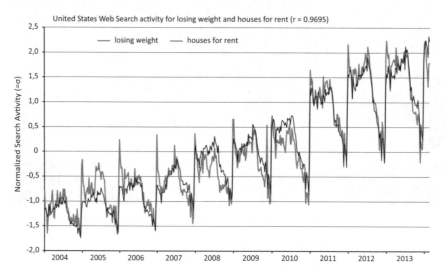

Abb. 5: Google Correlate. Robuste Zusammenhänge zwischen (nicht zusammenhängenden)
Suchanfragen („losing weight" und „houses for rent").
Quelle: www.google.com/trends/correlate/search=losing+weight&e=
houses+for+rent&t=weekly&p=us#
(Google hat den Dienst „Google Correlate" im Dezember 2019 eingestellt)

SCHLUSSBETRACHTUNG

Die Verbreitung und Anwendung mobiler Geomedien ist im Begriff Ort und Raum
als geographische Gegenstände grundlegend zu rekonfigurieren. Die neogeographi-
sche Verwandlung von Orten in „vernetzte Lokativitäten" wird überwiegend von

Kulturwissenschaften und Medientheorie bearbeitet (*Buschauer & Wills* 2013; *Döring & Thielmann* 2009). Wo der Ort der Geographie als wissenschaftliche Disziplin in diesem geomedialen Wechselverhältnis von Ortsmedien und Medienorten sein wird, bleibt abzuwarten. Die Herausforderungen sind groß. Noch größer sind aber die Potenziale dieser „Digitalen Geographie" und fast unüberschaubar bleiben die ungelösten Fragen, die sich mit jedem neuen digitalen Innovationschub auf neue Weise stellen. Ein Charakteristikum ist bereits jetzt unübersehbar. Die „neue Geographie" wird immer neu bleiben. Wie die stets unfertigen Beta-Versionen von Computersoftware befindet sie sich in einem Status permanenter Vorläufigkeit, eine Testversion von Geographie in ständiger Veränderung. Sie lässt sich bestenfalls spekulativ fassen und birgt die Gefahr, dass alles Schreiben über „neue Geographien" bereits zum Zeitpunkt der Veröffentlichung ein Text über „alte Geographien" sein wird.

LITERATUR

Buschauer, R. und *Willis, K. S.* (2013). Locative Media. Medialität und Räumlichkeit. Bielefeld: Transcript.

Döring, J. und *Thielmann, T.* (Hg.) (2009). Mediengeographie: Theorie – Analyse – Diskussion. Bielefeld: Transcript.

EU (2014). Being human in a hyperconnected era. https://digital-strategy.ec.europa.eu/en/events/being-human-hyperconnected-era. Siehe: Floridi, L. (Hg.) (2015). The Onlife Manifesto. Being Human in a Hyperconnected Era. Springer. https://www.springer.com/de/book/9783319040929, Zugriff: 14.02.2014.

Fisher, A. (2013). Googles Roadmap to Global Domination. New York Times, 11. Dezember 2013.

Gebhardt, H., Glawser, R., Radtke, U. und *Reuber, P.* (2011). Geographie: Physische Geographie und Humangeographie. Heidelberg: Spektrum Akademischer Verlag.

Goodchild, M. F. (2007). Citizens as Sensors: The World of Volunteered Geography. In: GeoJournal vol. 69 H. 4, S. 211–221.

Gordon, E. und *De Souza e Silva, A.* (2011). Net Locality: Why Location matters in a Networked World. Chichester: Wiley-Blackwell.

Günzel, S. (2013). Medienkulturgeschichte am Leitfaden des Raums. In: *Buschauer, R.* und *Willis, K. S.* (Hg.). Locative Media. Medialität und Räumlichkeit. Bielefeld: Transcript, S. 105–120.

Haklay, M. (2013). Neogeography and the Delusion of Democratisation. In: Environment and Planning A, vol. 45 H. 1, S. 55–69.

Horst, H. und *Miller, D.* (Hg.) (2012). Digital Anthropology. London: Berg.

ITU, International Telecommunication Union (2014): Measuring the Information Society. http://www.itu.int/en/ITU-D/Statistics/Pages/stat/default.aspx. Zugriff: 04.02.2014.

Latour, B. (2005). Reassembling the Social: An Introduction to Actor-Network-Theory. Oxford: Oxford University Press.

Rammert, W. (2007). Technik – Handeln – Wissen. Wiesbaden: VS.

Sony (2014). Smartband – Lifelog. http://www.sonymobile.com/de/products/smartwear/. Zugriff: 14.02.2014.

Turner, A. (2006). Introduction to Neogeography. Sebastopol, CA: O'Reilly Media.

Wardenga, U. (2010). „Man muss es mit eigenen Augen gesehen haben!" Zur Entstehung und Wirkung des Habitus von Geographen. Vortrag im Geographischen Kolloquium der Universität Mainz im Sommersemester 2010.

Wilson, M. W. und *Graham, M.* (2013). Situating Neogeography. In: Environment and Planning A 45(1), S. 3–9.

Summary: Neogeography marks the advent of a new geographic era characterised by the limitless proliferation of collaboratively produced geographic information. This paper discusses some of the challenges surrounding neogeography and proposes a disciplinary twist to approach these digitally spatialised social practices from the perspective of „digital geography".

Professor Dr. Marc Boeckler, Institut für Humangeographie, Goethe-Universität Frankfurt, Theodor-W.-Adorno-Platz 6, 60323 Frankfurt, boeckler@uni-frankfurt.de, Arbeitsgebiete/ Forschungsschwerpunkte: Wirtschaftsgeographie, Globalisierungsforschung, Science and Technology Studies

NEUE KARTOGRAFIEN, NEUE GEOGRAFIEN:
WELTBILDER IM DIGITALEN ZEITALTER

Georg Glasze

Bilder der Welt und damit nicht zuletzt Karten spielen eine wichtige Rolle bei der Herstellung und Vermittlung grundlegender Vorstellungs- und Deutungssysteme – also von Weltbildern in einem metaphorischen Sinn.[1] Der Evidenzeffekt von Karten, also ihre Augenscheinlichkeit, führt dazu, dass sie vielfach in höherem Maße als Texte als „wahr", das heißt als Abbildungen einer bestimmten Wirklichkeit, interpretiert werden.[2] Auch die Etablierung der Kartografie als Wissenschaft war eng verknüpft mit der Vorstellung, immer perfektere Abbilder der Erde schaffen zu können.

Allerdings wird die Idee einer kartografischen „Abbildung" seit den 1980er Jahren als modernistischer Mythos kritisiert. Diese sozial- und kulturwissenschaftlich orientierte, kritische Kartografieforschung betont, dass Karten immer in einem spezifischen sozio-technischen Kontext entstehen. Dieser prägt, welche Bilder der Welt hergestellt werden – und welche nicht. Karten oder, weiter gefasst, die ihnen zugrunde liegenden Geoinformationen sind also nicht einfach Abbilder der Welt, sondern (Re-)Produzenten von Weltbildern. Neuere Ansätze zeigen darüber hinaus, dass [|30] die Techniken und Praktiken der Kartografie und Geoinformation nicht nur Bilder der Welt prägen, sondern auch unsere physischen Lebenswelten mit formen.[3]

Im Folgenden werden zunächst die Ansätze einer sozial- und kulturwissenschaftlichen Auseinandersetzung mit Karten und Geoinformationen vorgestellt. Anschließend werden wichtige Bausteine der soziotechnischen Transformation der Kartografie sowie des gesamten Feldes der Geoinformation im 20. und 21. Jahrhundert beschrieben, um abschließend zu diskutieren, welche Weltbilder heute in den sogenannten Neokartografien und Neogeografien des digitalen Zeitalters entstehen.

---

1   Vgl. Christoph Markschies et al., Vorbemerkung, in: dies. (Hrsg.), Atlas der Weltbilder, Berlin 2011, S. XIV.
2   Vgl. Bruno Schelhaas / Ute Wardenga, Die Hauptresultate der Reisen vor die Augen zu bringen – oder: Wie man Welt mittels Karten sichtbar macht, in: Christian Berndt / Robert Pütz (Hrsg.), Kulturelle Geographien. Zur Beschäftigung mit Raum und Zeit nach dem Cultural Turn, Bielefeld 2007, S. 143–166.
3   Für eine deutschsprachige Einführung in die Debatte vgl. Georg Glasze, Kritische Kartographie, in: Geographische Zeitschrift, 97 (2009) 4, S. 181–191.

JENSEITS DER KARTE

Unter dem Schlagwort der „Kritischen Kartografie" hat sich seit den 1980er Jahren eine sozial- und kulturwissenschaftliche Kartografieforschung entwickelt, welche die gesellschaftliche Einbettung von Kartografie und Karten sowie die Bedeutung der Kartografie für die (Re-)Produktion bestimmter Weltbilder untersucht. In jüngerer Zeit rücken dabei vermehrt auch die Praktiken, Konventionen und Techniken des Kartenmachens und -gebrauchens ins Blickfeld sowie die Frage, wie diese „in der Welt arbeiten".

Der Geograf und Kartografie-Historiker Brian Harley bemühte sich seit den 1970er Jahren, historische Karten nicht einfach als Abbilder historischer Situationen zu interpretieren, sondern als Dokumente, die innerhalb ihres spezifischen gesellschaftlichen Kontextes verstanden werden müssen. In seinem bekanntesten Aufsatz „Deconstructing the Map"[4] von 1989 benennt Harley die Konsequenzen bestimmter sozialer Strukturen auf die Art und Weise, wie Karten produziert werden: „Monarchen, Ministerien, Institutionen des Staates, die Kirche haben alle Kartierungsprogramme für ihre eigenen Zwecke initiiert."[5] Zum anderen spricht er von der „internen Macht" des „kartografischen Prozesses".[6] Harley lässt sich hier von den Schriften der Philosophen Michel Foucault und Jacques Derrida anregen und betont, dass Karten unweigerlich immer ein bestimmtes Bild der Welt präsentieren und damit bestimmte (soziale) Wirklichkeiten herstellen – und mögliche andere „verschweigen".

Um dies zu untersuchen, schlägt Harley vor, Karten ähnlich wie Texte zu analysieren. Dabei soll von den Regelmäßigkeiten in der Gestaltung von Karten auf die impliziten Regeln der Kartografie geschlossen werden: Was wird im Zuge der Generalisierung hervorgehoben, was wird nicht dargestellt? Welche Bezeichnungen werden verwendet? Welche Grenzen werden gezogen? Welche Orte werden ins Zentrum der Karte gerückt? Wie wird die dreidimensionale Erde auf die zweidimensionale Karte projiziert, zum Beispiel winkel- oder flächentreu? Als eine Regelmäßigkeit und diskursive Regel der Kartografie identifiziert Harley beispielsweise das Prinzip der Ethnozentrizität von Karten – also die Regel, dass der Ort des Eigenen ins Zentrum von Karten gesetzt wird.[7] So ist es kein Zufall, dass die moderne europäische Kartografie genordete, eurozentrierte Karten hervorgebracht hat.

Als eine weitere Regel beschreibt Harley die „Regel der sozialen Ordnung".[8] Dabei geht er davon aus, dass Karten implizit die Prinzipien der sozialen Ordnung ihres Entstehungskontextes reproduzieren: „Häufig dokumentiert der Kartenproduzent genauso eifrig die Konturen des Feudalismus, die Umrisse der religiösen Hierarchien oder die Schritte auf den Stufen der sozialen Klasse wie eine Topografie

---

4   Vgl. Brian Harley, Deconstructing the Map, in: Cartographica, 26 (1989) 2, S. 1–20; für die folgenden Zitate aus der deutschen Fassung vgl. ders., Das Dekonstruieren der Karte, in: An Architektur, 5 (2004) 11–13, S. 4–19.
5   Ebd., S. 16.
6   Ebd., S. 17.
7   Ebd., S. 9 f.
8   Ebd., S. 10.

der physischen und menschlichen Umwelt."[9] In den Karten werden damit bestimmte soziale Wirklichkeiten (re-)präsentiert. Ganz im Sinne der Diskursforschung geht Harley davon aus, dass diese Praktiken nicht auf bewussten Entscheidungen einer Kartografin oder eines Kartografen beruhen, sondern dass diese letztlich unbewusst gesellschaftliche Selbstverständlichkeiten reproduzieren – mit anderen Worten: diskursive Regeln. [|31]

Dies lässt sich anschaulich illustrieren: So heben topografische Karten in Deutschland beispielsweise christliche Kirchen durch eine Signatur hervor. Andere religiöse Bauten werden hingegen nicht mit einer Signatur dargestellt und somit in der Regel kartografisch „verschwiegen". Für Synagogen oder Moscheen existieren in der amtlichen Kartografie in Deutschland bislang keine Signaturen – im Sinne Harleys eine Konsequenz der vorherrschenden sozialen Ordnungen, in diesem Fall der „religiösen Hierarchien".[10]

In Weiterführung der sozial- und kulturwissenschaftlichen Kartografieforschung hat sich seit Ende der 1990er Jahre zunächst in der englischsprachigen Sozial- und Kulturgeografie ein Forschungszusammenhang entwickelt, der den Blick in noch höherem Maße auf die Prozesse „vor und nach" der Karte lenkt – *beyond the map*. Diese Arbeiten beziehen viele Anregungen aus der sozialwissenschaftlichen Wissenschafts- und Technikforschung. Ins Blickfeld rücken hier die Praktiken, Konventionen und Techniken, mit denen Karten hergestellt und verwendet werden.

So hat der Wissenschaftssoziologe Bruno Latour gezeigt,[11] welche Rolle die Kartografie seit der Neuzeit bei der Produktion wissenschaftlichen Wissens und damit von Autorität in den europäischen Machtzentren spielte und wie sie genutzt wurde, um „in der Welt zu arbeiten".[12] Latour stellt dar, wie theoretische Ansätze der Kartografie sowie Kartierungstechniken und -instrumente wie Sextanten mit disziplinären Praktiken des Handels wie den standardisierten Techniken, mit denen Seefahrer räumliche Informationen erkunden, erfassen und sichern, zusammenwirkten und dadurch ermöglichten, dass systematisch Informationen von entfernten Orten gesammelt wurden. Karten schufen damit eine Voraussetzung für internationalen Handel, territoriale Expansion und globale Kolonisation.[13]

Die sozialwissenschaftliche Wissenschafts- und Technikforschung stellt also eine konzeptionelle Perspektive zur Verfügung, die es ermöglicht herauszuarbeiten, wie die Praktiken, Techniken und Konventionen sowohl der traditionellen Print-Kartografie als auch der neueren digitalen Verarbeitung und Präsentation von Geoinformationen in der Welt wirksam werden und die Welt verändern.[14]

9   Ebd.
10   Vgl. G. Glasze (Anm. 3).
11   Vgl. Bruno Latour, Science in Action, Cambridge MA 1987; ders., Die Logistik der immutable mobiles, in: Jörg Döring/Tristan Thielmann (Hrsg.), Mediengeographie, Bielefeld 2009, S. 111–144.
12   Rob Kitchin/Chris Perkins/Martin Dodge, Thinking about Maps, in: dies. (Hrsg.) Rethinking Maps, New Frontiers in Cartographic Theory, London–New York 2009, S. 14
13   Vgl. B. Latour (Anm. 11).
14   Vgl. Christian Bittner/Georg Glasze/Cate Turk, Tracing Contingencies: Analyzing the Political in Assemblages of Web 2.0 Cartographies, in: GeoJournal, 78 (2013) 6, S. 935–948.

SATELLITENGESTÜTZTE FERNERKUNDUNG

Die Bilder der Erde, die im Zuge der Entwicklung von Luftfahrt und Fotografie zunächst von Heißluft-Ballons und Flugzeugen, im 20. Jahrhundert dann auch von Satelliten aufgenommen wurden, läuteten eine neue Ära der (Re-)Präsentation der Erdoberfläche ein.[15] Obwohl räumliche (Re-)Präsentationen aus der Vogelperspektive seit vorhistorischer Zeit bekannt sind, war die kartografische Darstellung dabei in der Regel auf die Vorstellung und gegebenenfalls die mathematische Konstruktion angewiesen.

Früh nutzten die westlichen Nationalstaaten die neuen Techniken und unterstützten deren Weiterentwicklung. So setzte beispielsweise die französische Armee bereits 1794 erstmals einen Aufklärungsballon ein. Während des Ersten Weltkrieges trieb das Militär die Entwicklung der Luftbildfotografie voran, die nach wie vor für alle Arten von Spionage eingesetzt wird. Luftbildfotografie eröffnete damit neue Erkundungsmöglichkeiten für technisch hochentwickelte Staaten und stellt bis heute eine Herausforderung für die Souveränität anderer Staaten dar.[16]

Die Entwicklung der satellitengestützten Fernerkundung ab Mitte des 20. Jahrhunderts wurde zunächst vor allem von den Regierungen der USA und der Sowjetunion vorangetrieben. Der Start des ersten Satelliten „Sputnik 1" 1957 durch die Sowjetunion [|32] markierte den Beginn einer neuen Phase der Fernerkundung, da die Akzeptanz dieses Satelliten beziehungsweise das Ausbleiben eines Protests seitens der US-Regierung sowie anderer Regierungen die Grundlage für das Recht auf freien Überflug im Weltraum schuf – die sogenannte *Open Sky Doctrine*. Nur zwei Jahre später, 1959, schickte die CIA den ersten Spionagesatelliten ins Weltall.[17]

Satellitengestützte Fernerkundung hat das Bild der Erde verändert: *Erstens* fordern die Bilder der gesamten, grenzenlosen Erdoberfläche das Konzept der territorialen Souveränität heraus.[18] Ob dies als Zeitalter einer „globalen Transparenz"[19] zu feiern oder als Ära einer globalen Überwachung aller durch wenige zu fürchten ist, ist eine Frage des (geo)politischen Standpunktes. Mithilfe der Fernerkundung können sich entfernte Beobachterinnen und Beobachter teilweise mehr geografische Informationen erschließen, als Akteuren in den beobachteten Territorien vorliegen. Dies ist nicht zuletzt deshalb relevant, weil die wichtigsten Organisationen

---

15   Vgl. Wolfgang Sachs, Satellitenblick. Die Ikone vom blauen Planeten und ihre Folgen für die Wissenschaft, in: Ingo Braun / Bernward Joerges (Hrsg.), Technik ohne Grenzen, Frankfurt/M. 1994, S. 318.

16   Vgl. Denis E. Cosgrove / William L. Fox, Photography and Flight, London 2010.

17   Vgl. Walter A. McDougall, The Heavens and the Earth: A Political History of the Space Age, New York 1985.

18   Vgl. Barney Warf, Dethroning the View from Above: Toward a Critical Social Analysis of Satellite Ocularcentrism, in: Lisa Parks / James Schwoch (Hrsg.), Down to Earth: Satellite Technologies, Industries, and Cultures, New Brunswick 2012, S. 42–60.

19   Vgl. John C. Baker / Ray A. Williamson / Kevin M. O'Connell, Introduction, in: dies. (Hrsg.), Commercial Observation Satellites: At the Leading Edge of Global Transparency, Santa Monica 2001, S. 1–11.

der Satellitenindustrie bis heute in wenigen Ländern zumeist des globalen Nordens verortet sind.[20]

Zweitens hat die Verfügbarkeit von Bildern aus dem Weltall Vorstellungen der Erde als endliche, aber grenzenlose Heimstätte der Menschheit befördert: Die von Satelliten und den Mondflügen aufgenommenen Bilder eines strahlend blauen Planeten vor der Dunkelheit des unendlichen Weltraums wurden zu einem wichtigen Symbol für transnational-globale Umweltbewegungen.[21]

Nach dem Ende des Kalten Krieges beschleunigte sich die Verfügbarkeit von Satellitenbilddaten: Die Regierungen der USA und der Sowjetunion beziehungsweise Russlands gaben riesige Mengen von Daten zur zivilen Nutzung frei. Schrittweise hoben die USA Beschränkungen für die kommerzielle Satellitenindustrie auf. Es entwickelte sich eine Situation der zunehmenden wirtschaftlichen Konkurrenz der US-amerikanischen Anbieter mit Satellitenbildanbietern beispielsweise in Frankreich und Kanada.[22] Satellitenbilddaten sind heute wichtiger Teil der rasch wachsenden Verfügbarkeit digitaler, geografisch referenzierter Daten (Geodaten).

DIGITALE GEOGRAFIEN: GIS, GPS UND GEOWEB

Seit den 1960er Jahren wurde die analoge Print-Kartografie rasch und umfassend von der digitalen Kartografie und schließlich von Geografischen Informationssystemen (GIS) verdrängt. Letztere ermöglichen es, Geodaten computergestützt zu erfassen, zu speichern, zu analysieren und zu präsentieren. Die Karte steht damit nicht länger im Mittelpunkt, sie wird zu einer Präsentationsform digitaler Geoinformationen.[23] GIS wurden und werden für verschiedene Zwecke verwendet und dies zu großen Teilen außerhalb der etablierten Organisationen der staatlichen und wissenschaftlichen Kartografie: Marktforschungsunternehmen nutzen GIS für Standortanalysen, die Polizei verwendet GIS für die räumliche Analyse von Verbrechensdaten, Städte und Gemeinden setzen sie für die Flächennutzungsplanung ein. Nicht zuletzt war und ist das Militär und insbesondere das US-Militär ein wichtiger Anwender und Förderer der technischen Fortentwicklung von GIS.[24] GIS haben die alten topografischen Karten als Grundlage militärischer Operationen ab- [|33] gelöst, virtuelle 3D-Landschaftsmodelle in GIS ermöglichen die präzise Navigation militärischen Geräts.

Das US-Militär war auch ein wichtiger Akteur bei der Entwicklung einer weiteren soziotechnischen Innovation, welche grundlegend für die Transformation von

---

20   Vgl. Laura Kurgan, Close Up at a Distance: Mapping, Technology, and Politics, Cambridge MA–London 2013, S. 39 ff.

21   Vgl. Denis E. Cosgrove, Apollo's Eye. A Cartographic Genealogy of the Earth in the Western Imagination, Baltimore 2001, S. 262 ff.

22   Vgl. J. C. Baker / R. A. Williamson / K. M. O'Connell (Anm. 19).

23   Vgl. Daniel Sui / Richard Morrill, Computers and Geography. From Automated Geography to Digital Earth, in: Stanley D. Brunn / Susan L. Cutter / James W. Harrington (Hrsg.), Geography and Technology, Dordrecht 2004, S. 81–108.

24   Vgl. Neil Smith, History and Philosophy of Geography: Real Wars, Theory Wars, in: Progress in Human Geography, 16 (1992) 2, S. 257–271.

Kartografie und weiterer Geoinformation im 21. Jahrhundert sein sollte: Das Navstar Global Positional System (GPS) ermöglicht elektronischen Empfängern weltweit, ihre Position in Länge- und Breitengrad mittels Funksignalen von Satelliten zu bestimmen. Es wurde seit 1970 als sogenannte *Dual-use*-Technologie für militärische und zivile Zwecke entwickelt und ist seit 1995 funktionsfähig. Bis 2000 wurde zwischen einem Signal mit hoher Genauigkeit für das US-Militär und einem öffentlichen Signal mit limitierter Genauigkeit differenziert. Mit der Freigabe des präzisen Signals ermöglichte die US-Regierung einen Boom neuer Navigationsdienste sowie die Entwicklung weiterer sogenannter *location based services*.[25] Dabei handelt es sich um Dienstleistungen, die spezifisch für bestimmte Orte angeboten werden, wie etwa ortsbezogene Werbung über GPS- und internetfähige Smartphones sowie das Verfolgen (*tracking*) der räumlichen Mobilität von Personen oder Objekten.[26]

Auch die Ursprünge des Internet können unter anderem auf verschiedene Initiativen der US-Regierung in den 1960er Jahren zurückgeführt werden, die darauf abzielten, ein robustes und fehlerresistentes Computernetzwerk zu etablieren. Im Zuge der Zusammenführung mehrerer solcher Netzwerke und der Aufhebung von Restriktionen für deren kommerzielle Nutzung entwickelte sich in den 1990er Jahren das Netzwerk, das wir heute Internet nennen.[27] Zahlreiche Autorinnen und Autoren differenzieren eine frühe Phase des Internet als Web 1.0 und eine spätere Phase ab etwa 2004 als Web 2.0. Während für das Web 1.0 tendenziell eher statische Internetseiten sowie eine geringe Anzahl an Produzenten charakteristisch waren, zeichnet sich das Web 2.0 durch Interaktion und Kollaboration und folglich durch den Boom des *user generated content* aus.[28]

Die Techniken und Praktiken von GIS, die wachsende Verfügbarkeit digitaler Geodaten infolge der GPS-Technik und der Entwicklung einer kommerziellen Satellitenbildindustrie sowie die zunehmend einfache Nutzung des Internets über Desktop- und mobile Computer (Smartphones) sind die wichtigsten Bausteine für die Entwicklung des sogenannten Geoweb – und somit für die grundlegende Transformation von Geoinformation und kartografischer (Re-)Präsentation im digitalen Zeitalter.[29] Der Begriff „Geoweb" wird genutzt, um die wachsende Bedeutung von Geodaten für das Internet sowie den Boom neuer webbasierter Technologien, die Geodaten nutzen und vielfach produzieren, zu beschreiben. Bekannte Beispiele sind virtuelle Globen wie insbesondere Google Earth und virtuelle Kartendienste wie Google Maps, OpenStreetMap oder Here.

Zunehmend wird im Geoweb geformt, was wir über Orte und Räume der Erde wissen und wie wir in der Welt agieren. Die Entwicklung des Geoweb wurde und

25   GPS bleibt allerdings eine Einrichtung der US-Regierung. Russland und die EU haben mit GLONASS und GALILEO vergleichbare Systeme entwickelt.
26   Vgl. Hiawatha Bray, You Are Here: From the Compass to GPS, the History and Future of How We Find Ourselves, New York 2014; L. Kurgan (Anm. 20).
27   Vgl. bspw. Janet Abbate, Inventing the Internet, Cambridge MA 1999.
28   Vgl. Tim O'Reilly, What Is Web 2.0, 30.9.2005, www.oreillynet.com/pub/a/oreilly/tim/news/2005/09/30/what-is-web-20.html (11.9.2015).
29   Vgl. Georg Glasze, Sozialwissenschaftliche Kartographie-, GIS- und Geoweb-Forschung, in: Kartographische Nachrichten, 64 (2014) 3, S. 123–129.

wird in hohem Maße von Unternehmen bestimmt, die bis vor wenigen Jahren wie etwa Google oder TomTom keinen Bezug zu Geoinformation und Kartografie hatten oder noch überhaupt nicht existierten. Gleichzeitig ermöglicht der Kontext des Web 2.0 die Entwicklung von nichtkommerziellen, offenen Projekten wie OpenStreetMap und Wikimapia, in denen Tausende Freiwillige geografische Informationen erheben, organisieren und präsentieren – sogenannte *volunteered geographic information*.

Google, der sicherlich wichtigste Akteur des Geowebs, kaufte im Jahr 2004 das Start-up „Where2Technologies", das eine benutzerfreundliche Web-Oberfläche zur Präsentation geografischer Informationen geschaffen hatte. Google entwickelte die [|34] Software zu Google Maps weiter, das nach dem Start 2005 rasch zur meist genutzten digitalen Kartenplattform wurde.[30] Google Maps bietet inzwischen neben Online-Karten beispielsweise auch fotorealistische Panoramen von Straßenzügen sowie Routenplanung und Verkehrsinformationen in Echtzeit. Ebenfalls 2004 kaufte Google das Unternehmen Keyhole Inc., das auf der Basis einer Videospiel-Software einen virtuellen Globus aus Satellitendaten und Luftbildern entwickelt hatte und zeitweise von der CIA gefördert worden war. Auf der Basis der Keyhole-Software entwickelte Google den virtuellen Globus Google Earth, der 2005 online ging und einer breiten Öffentlichkeit fotorealistische Bilder der Erde in einer bis dahin unbekannten Qualität, Quantität und Abdeckung bietet.[31]

Bereits wenige Monate nach dem Start von Google Maps wurde das Programm von einem kalifornischen Informatiker gehackt und genutzt, um Immobilienangebote in Kalifornien räumlich differenziert zu präsentieren. Google erkannte, dass die Zusammenführung der Google-Basiskarte mit allen möglichen Arten weiterer georeferenzierter Daten neue Dienstleistungen ermöglicht und viele neue Nutzerinnen und Nutzer zu Google führt. Rasch schuf das Unternehmen eine Schnittstelle, die solche Zusammenführungen erleichtert und auch Menschen ohne Programmier- oder Kartografieausbildung ermöglicht, sogenannte *map mashups* zu schaffen. Auch wenn die Kartendienste von Google derzeit zumindest für die nichtkommerzielle Nutzung kostenfrei verfügbar sind, bleiben die zugrunde liegenden Geodaten allerdings nicht zugänglich und im Besitz des Unternehmens.[32]

Bei offenen Geoweb-Projekten wie dem besonders erfolgreichen OpenStreet-Map-Projekt (OSM) sind diese Daten hingegen frei verfügbar. OSM präsentiert sich auf der eigenen Webseite als „Projekt mit dem Ziel, eine freie Weltkarte zu erschaffen" – vielfach wird OSM auch als „Wikipedia der Kartografie" bezeichnet. Gestartet wurde das Projekt ebenfalls 2004 durch einen britischen Informatikstudenten, der frustriert war von der restriktiven Lizenzpolitik des staatlichen britischen Kartografiedienstleisters Ordnance Survey. Gemeinsam mit weiteren Freiwilligen der OpenData-Bewegung in London schuf er die notwendige Infrastruktur

---

30 Vgl. Craig M. Dalton, Sovereigns, Spooks, and Hackers: An Early History of Google Geo Services and Map Mashups, in: Cartographica, 48 (2013) 4, S. 261 ff.

31 Vgl. Jeremy Crampton, Keyhole, Google Earth, and 3D Worlds: An Interview with Avi Bar-Zeev, in: Cartographica, 43 (2008) 2, S. 85–93.

32 Vgl. Jeremy Crampton, Mapping. A Critical Introduction to Cartography and GIS, Malden–Oxford 2010, S. 25 ff.

zum Start des Projekts. *Mapping parties* in immer mehr Regionen brachten neue Freiwillige zu dem Projekt, die auf der Basis von selbst erhobenen GPS-*tracks* und Beobachtungen im Gelände Geodaten beitrugen.[33] Die gesamten OSM-Geodaten sind frei nutzbar und bilden die Grundlage für zahlreiche Kartendienste und andere raumbezogene Dienstleistungen. Zehn Jahre nach dem Start von OSM übertrifft die Datendichte und Aktualität der OSM-Geodaten in vielen Regionen das Angebot staatlicher und kommerzieller Anbieter.[34]

NEUE KARTOGRAFIEN – NEUE GEOGRAFIEN?

Die skizzierte Transformation von Geoinformation und kartografischer (Re-)Präsentation im digitalen Zeitalter wird vielfach mit den Begriffen „Neokartografie", *volunteered geographic information* sowie „Neogeografie" beschrieben. Welche Aspekte betonen diese Begriffe und inwiefern können sie sinnvoll voneinander unterschieden werden?

Interessanterweise ist Neokartografie dabei bislang der am wenigsten prominente Begriff. *Erstens* werden damit die Veränderung der Techniken kartografischer (Re-)Präsentation bezeichnet, insbesondere die [|35] zunehmende Dynamik in diesem Bereich: Karten werden zu einer volatilen Präsentation dynamischer Datenströme.[35] Die International Cartographic Association hat 2011 eine Kommission zur Neokartografie etabliert und betont dabei *zweitens* vor allem soziale Aspekte: So haben die „Neokartografen" in der Regel keinen traditionellen Kartografiehintergrund; bei den neokartografischen Praktiken verschwimmen die traditionellen Grenzen zwischen Kartenerstellerinnen und -nutzern.[36]

Der Begriff der *volunteered geographic information* (VGI) wurde zunächst von dem US-amerikanischen Geografen und GIS-Spezialisten Michael Frank Goodchild geprägt[37] und wird inzwischen in zahlreichen Publikationen und Forschungsprojekten aufgegriffen.[38] Goodchild betont das freiwillige Engagement vieler Bürgerinnen

---

33   Durch Schenkungen von umfangreichen Satelitenbilddaten an OSM spielt seit wenigen Jahren auch das Kartieren auf der Grundlage von Satellitenbildern eine wachsende Rolle für den Ausbau von OSM.

34   Vgl. Jamal J. Arsanjani et al., An Introduction to OpenStreetMap in Geographic Information Science: Experiences, Research, and Applications, in: dies. (Hrsg.), OpenStreetMap in GIScience. Experiences, Research, and Applications, New York–Heidelberg 2015, S. 1–15; Georg Glasze / Chris Perkins, Social and Political Dimensions of the OpenStreetMap Project: Towards a Critical Geographical Research Agenda, in: ebd., S. 143–166.

35   Vgl. Holger Faby, Von der Kartographie zur Neo-Cartography?, in: Kartographische Nachrichten, 61 (2011) 1, S. 3–9.

36   Siehe http://neocartography.icaci.org (11.9.2015).

37   Vgl. Michael F. Goodchild, Citizens as Sensors: The World of Volunteered Geography, in: GeoJournal, 69 (2011) 4, S. 211–221.

38   Vgl. bspw. Muki Haklay, How Good is Volunteered Geographical Information? A Comparative Study of OpenStreetMap and Ordnance Survey Datasets, in: Environment and Planning B, 37 (2011) 4, S. 682–703; Pascal Neis / Dennis Zielstra, Recent Developments and Future Trends in Volunteered Geographic Information Research: The Case of OpenStreetMap, in: Future

und Bürger ohne akademisch-geografische (oder kartografische) Ausbildung für die Herstellung geografischer Informationen. Der Begriff der VGI ist somit einerseits enger gefasst als Neokartografie, weil er ausschließlich auf die freiwillige (und wie einige neuere Definitionen betonen: intendierte) Sammlung und Organisation geografischer Information abhebt. Gleichzeitig ist er insofern weiter gefasst, als seine Definition über kartografische (Re-)Präsentationen hinausgeht und die komplexen Prozesse der Erstellung, Verarbeitung und Präsentation von Geodaten ins Blickfeld rückt.

Relativ weit verbreitet ist der Begriff der „Neogeografie", der allerdings sehr unterschiedlich verwendet wird. Vielfach synonym zu Neokartografie gebraucht,[39] scheint am ehesten eine weitergreifende Definition sinnvoll, wie sie von den beiden Geografen Mark Graham und Matthew Wilson vorgeschlagen wurde:[40] Sie wollen mit dem Begriff einer Neogeografie betonen, dass sich alltägliche Prozesse der (Re-) Produktion und Verwendung verschiedenster Arten geografischer Informationen im digitalen Zeitalter verändern. Ihre Definition umfasst damit nicht nur die mit VGI und Neokartografie beschriebenen Prozesse, sondern beispielsweise auch die mehr oder weniger unfreiwillige Produktion von Geodaten etwa durch das *tracking* von Smartphones oder die Georeferenzierung von Twittermeldungen sowie nichtkartografische Formen der Auswertung digitaler Geodaten. Damit bezeichnet Neogeografie das alltägliche „Geografie-Machen" im digitalen Zeitalter oder kurz: digitale Geografien.

Innerhalb der universitären Kartografie und Geografie sind die Begriffe der Neokartografie, VGI sowie Neogeografie zunächst zurückhaltend aufgegriffen worden. Aus Sorge um die Zukunft der Disziplin dominierten in der wissenschaftlichen Kartografie Abgrenzungen gegenüber neokartografischen Praktiken. In jüngerer Zeit setzt aber eine Interaktion zwischen Neokartografie und wissenschaftlicher Kartografie ein.

In der wissenschaftlichen Geografie hat sich bislang vor allem in der englischsprachigen Forschungslandschaft ein lebhafter Forschungs- und Diskussionszusammenhang entwickelt, der die soziotechnischen Hintergründe und Effekte neogeografischer Praktiken untersucht und reflektiert, und auf dem auch der folgende Ausblick aufbaut.

AUSBLICK: NEUE WELTBILDER IM DIGITALEN ZEITALTER?

Welche Weltbilder und darüber hinaus welche Geografien entstehen also im digitalen Zeitalter? Anhand zweier Spannungsfelder und einer These lassen sich grundlegende Entwicklungen skizzieren. [|36]

Internet, 6 (2014) 1, S. 76–106; Daniel Sui / Sarah Elwood / Michael F. Goodchild (Hrsg.), Crowdsourcing Geographic Knowledge. Volunteered Geographic Information (VGI) in Theory and Practice, Dordrecht–New York 2013.

39   So bspw. die frühe Definition von Andrew J. Turner, Introduction to Neogeography, Sebastopol 2006.

40   Matthew W. Wilson / Mark Graham, Guest Editorial – Situating Neogeography, in: Environment and Planning A, 45 (2013) 1, S. 3–9.

Das erste Spannungsfeld liegt zwischen den Polen „Universalisierung von Geoinformation" versus „neue Fragmentierungen". Die neuzeitliche Kartografie hat das bis heute vorherrschende Weltbild der Erde als lückenloses Mosaik politischer Territorien geprägt.[41] Zugleich wurden die westlichen Staaten zu privilegierten Akteuren der Geoinformation und kartografischen (Re-)Präsentation. Zumindest die Staaten des globalen Nordens konnten bis vor Kurzem weitgehend die Produktion von Karten und allen möglichen Formen geografischer Informationen innerhalb ihrer Grenzen kontrollieren. Mit der wachsenden Verfügbarkeit von Satellitenbilddaten sowie dem Boom neokartografischer und neogeografischer Praktiken scheinen Geoinformationen heute jedoch immer weniger durch die Nationalstaaten kontrolliert werden zu können. Neue Akteure wie Google oder OSM versprechen hingegen universell-globale Geoinformationen.

Allerdings zeigen sich auch neue Fragmentierungen. So führen die ökonomischen Interessen privatwirtschaftlicher Geoweb-Dienstleister dazu, dass in ihren Online-Karten in erster Linie kommerzielle Angebote wie etwa Pizzerien, Anwaltskanzleien oder Fitnesscenter verzeichnet werden: Die Welt wird als eine große Shopping-Mall präsentiert. Nicht zuletzt gab Google 2013 mit der Einführung einer neuen Version von Google Maps die Idee einer universellen Weltkarte auf: Je nach Suchanfrage, den besuchten Orten, dem jeweiligen individuellen Verlauf bisheriger Suchanfragen und besuchter Orte, den Spracheinstellungen und der Lokalisierung des abrufenden Computers personalisiert Google die Inhalte der Karte. Der Grund liegt im Geschäftsmodell von Google: gezielte, also möglichst personalisierte Werbung.

Neben den ökonomischen Interessen führen aber auch national differenzierte geopolitische Interessen zu neuen Fragmentierungen: So unterscheidet Google seit 2014 beispielsweise drei kartografische Präsentationen der Halbinsel Krim. Für Computer mit IP-Adressen aus der Ukraine wird die Krim als Teil der Ukraine dargestellt, für IP-Adressen aus Russland ist die Ukraine durch eine nationale Grenze von der Ukraine abgetrennt und Teil Russlands, für alle anderen Internetnutzer zeigt Google eine gestrichelte Linie im Norden der Krim als umstrittene Grenze.[42]

Das zweite Spannungsfeld liegt zwischen den Polen „Öffnung und Demokratisierung" sowie „neue Exklusionen". Insbesondere die Projekte der *volunteered geographic information* sind vielfach als „Öffnung" beziehungsweise „Demokratisierung" der Kartografie sowie der gesamten Geoinformation begrüßt worden:[43] Neue Akteure bekommen Zugang zu Geoinformationen und es eröffnen sich

41 Vgl. Jordan Branch, The Cartographic State. Maps, Territories and the Origin of Sovereignty, Cambridge 2014.

42 Vgl. Georg Glasze, Geoinformation, Cartographic (Re-)Presentation and the Nation-State: A Coconstitutive Relation and Its Transformation in the Digital Age, in: Uta Kohl (Hrsg.), Internet Jurisdiction, Cambridge 2015 (i. E.).

43 Vgl. bspw. Chris Perkins/Martin Dodge, The Potential of User-Generated Cartography. A Case-Study of the OpenStreetMap Project and Manchester Mapping Party, in: North West Geography, 8 (2008) 1; Ferjan Ormeling, From Ortelius to OpenStreetMap. Transformation of the Map into a Multifunctional Signpost, in: Georg Gartner/Felix Ortag (Hrsg.), Cartography in Central and Eastern Europe, Berlin–Heidelberg 2010, S. 1–16.

Chancen, bislang „verschwiegene" Informationen zu vermitteln. Inzwischen konnten jedoch zahlreiche Studien zeigen, dass sich auch in den VGI-Projekten Fragen von Zugang und Exklusion stellen. Vielfach werden diese Projekte von soziodemografisch sehr homogenen Gruppen geprägt: Die Teilnehmenden sind überwiegend männlich, technik-affin, jung, europäisch oder nord-amerikanisch und weiß.[44] Diese Ungleichheit prägt, welche Daten und wo Daten erhoben werden sowie die Art und Weise, wie diese Daten verarbeitet und präsentiert werden.[45]

Für den gesamten Bereich der Neokartografie und Neogeografie gilt, dass einerseits die Prozesse der Herstellung und Verarbeitung von Geoinformation in höherem [|37] Maße sichtbar werden, als dies in der traditionellen Print-Kartografie der Fall war. Andererseits wächst die Bedeutung von Code und Software für diese Prozesse und damit ein Bereich, dessen Funktionsweise und Entwicklung für die allermeisten Nutzer kaum einsichtig und verständlich ist.[46]

Letztlich stellt sich die Frage, wie sich im digitalen Zeitalter das viel diskutierte Verhältnis zwischen „Karte" und „Territorium" gestaltet, das heißt zwischen räumlicher Wirklichkeit und (Re-)Präsentation. Es zeichnet sich ab, dass eine Karte nicht länger sinnvoll als Einzelmedium konzeptualisiert werden kann.[47] Kartografische Repräsentationen sind vielmehr eingebettet in dynamische Datenströme und Prozesse. Darüber hinaus argumentieren einige Autoren, dass die Differenzierung von „Karte" und „Territorium" im digitalen Zeitalter ihre Relevanz verliert. Die Georeferenzierung immer größerer Datenmengen schafft „augmentierte Geografien".[48] Die „digitale Dimension" wird dabei „untrennbarer Bestandteil der einen räumlichen Wirklichkeit"[49] – digitale Geoinformationen werden also in sehr unmittelbarer Weise Teil der Welt.

Georg Glasze Dr. rer. nat., geb. 1969; Professor für Kulturgeographie am Institut für Geographie der Friedrich-Alexander-Universität Erlangen-Nürnberg, Wetterkreuz 15, 91058 Erlangen. georg.glasze@fau.de

44 Vgl. Mordechai Haklay, Neogeography and the Delusion of Democratisation, in: Environment and Planning A, 45 (2013) 1, S. 55–69.
45 Vgl. für ein lokales Beispiel Christian Bittner, Reproduktion sozialräumlicher Differenzierungen in OpenStreetMap: Das Beispiel Jerusalem, in: Kartographische Nachrichten, (2014) 3, S. 136–144.
46 Vgl. Rob Kitchin/Tracey P. Lauriault, Towards Critical Data Studies: Charting and Unpacking Data Assemblages and Their Work, The Programmable City Working Paper 2, Dublin 2014.
47 Vgl. Tristan Thielmann, Auf den Punkt gebracht: Das Un- und Mittelbare von Karte und Territorium, in: Inga Gryl/Tobias Nehrdich/Robert Vogler (Hrsg.), geo@web. Medium, Räumlichkeit und geographische Bildung, Wiesbaden 2013, S. 35–59.
48 Vgl. Mark Graham, Geography/Internet: Ethereal Alternate Dimensions of Cyberspace or Grounded Augmented Realities?, in: The Geographical Journal, 179 (2013) 2, S. 177–182.
49 Marc Boeckler, Neogeographie, Ortsmedien und der Ort der Geographie im digitalen Zeitalter, in: Geographische Rundschau, 66 (2014) 6, S. 7.

DIGITAL TURN, DIGITAL GEOGRAPHIES?

James Ash / Rob Kitchin / Agnieszka Leszczynski

Abstract: Geography is in the midst of a digital turn. This turn is reflected in both geographic scholarship and praxis across sub-disciplines. We advance a threefold categorization of the intensifying relationship between geography and the digital, documenting geographies produced through, produced by, and of the digital. Instead of promoting a single theoretical framework for making sense of the digital or proclaiming the advent of a separate field of 'digital geography', we conclude by suggesting conceptual, methodological and empirical questions and possible paths forward for the 'digital turn' across geography's many subdisciplines.
Keywords: computing, digital, digital geography, digital turn, geography

I INTRODUCTION

No other technological innovation in human history has affected the practice of geography in such a profound way as the computer. It has drastically transformed both geography as an acdemic discipline and the geography of the world. (Sui and Morrill, 2004: 82)

Geography, we contend, is in the midst of a digital turn. Rather than suggesting a radical rupture with extant or antecedent geographical theory and praxis, we advance the notion of the 'digital turn' to capture the ways in which there has been a demonstrably marked turn *to* the digital as both object and subject of geographical inquiry, and to signal the ways in which the digital has pervasively inflected geographic thought, scholarship, and practice.

Digital devices (computers, satellites, GPS, digital cameras, audio and video recorders, smartphones) and software packages (statistics programmes, spreadsheets, databases, GIS, qualitative analysis packages, word processing) have become indispensable to geographic practice and scholarship across sub-disciplines, [|26] regardless of conceptual approach. Current modes of generating, processing, storing, analysing and sharing data; creating and circulating texts, visualizations, maps, analytics, ideas, videos, podcasts and presentation slides; and, sharing information and engaging in public debate via mailing lists and social and mainstream media are thoroughly dependent on computational technologies (Fraser, 2007; Kitchin et al., 2013).

Moreover, as digital technologies have become pervasively quotidian, mediating tasks such as work, travel, consumption, production, and leisure, they are having increasingly profound effects on phenomena that are of immediate concern to geographers: the nature of the space economy and economic relations; the management and governance of places; the production of space, spatiality and mobilities; the processes, practices, and forms of mapping; the contours of spatial knowledge and imaginaries; and the formation and enactment of spatial knowledge politics

(Castells, 1996; Elwood and Leszczynski, 2013; Graham and Marvin, 2001; Rose et al., 2014; Wilson, 2012). Digital presences and practices are characterized by uneven geographies of underlying infrastructures, material forms, component resources, and sites of creation and disposal (Lepawsky, 2014; Zook, 2005). Similarly, there are distinct geographies of digital media such as the internet, games, the geoweb, and social, locative and spatial media (Dodge and Kitchin, 2002; Ash, 2015; Leszczynski, 2015b).

Following Lunenfeld (1999), we adopt a broad notion of 'the digital', extending beyond computational technologies to encompass ontics, aesthetics, logics and discourses. As ontics, 'the digital' designates digital systems that 'translate all inputs and outputs into binary structures of 0s and 1s, which can be stored, transferred, or manipulated at the level of numbers, or "digits" (Lunenfeld, 1999: xv). The digitally mediated material technologies we engage with have recoded multiple other technologies, media, art forms, and indeed spatialities, in ways coincident with the binary nature of computing architectures. Digitality, then, is also an aesthetics, capturing the pervasiveness of digital technologies and shaping how we understand and experience space and spatiality as always-already 'marked by circuits of digitality' that are themselves irreducible to digital systems (Murray, 2008: 40). As we adopt and ubiquitously embed networked digital technologies across physical landscapes, they come to enact progressively routine orderings of quotidian rhythms, interactions, opportunities, spatial configurations, and flows (Franklin, 2015). We use 'the digital', then, to make reference to material technologies characterized by binary computing architectures; the genre of socio-techno-cultural productions, artefacts, and orderings of everyday life that result from our spatial engagement with digital mediums; and the logics that both structure these ordering practices as well as their effects. To this we add a fourth dimension, that of digital discourses which actively promote, enable, secure, and materially sustain the increasing reach of digital technologies.

The turn to the digital in geography has, to a large degree, been thoroughly internalized and taken for granted, little acknowledged beyond some debates around epistemology and methods (e.g. critical GIS, critical data studies), and work that explicitly takes the digital as its central focus. With regard to the latter, recent conference sessions and workshops have sought to highlight what has been termed 'digital geographies',[1] which in part echo developments in other disciplines which seek to establish new fields of study, including 'digital anthropology' (Horst and Miller, 2012), 'digital sociology' (Orton-Johnson and Prior, 2013; Lupton, 2014), and 'the digital humanities' (Offen, 2013; Travis, 2015). Rather than consign the digital to a distinct disciplinary subfield, or cast all geographies reshaped or

1   For example, the 'Digital Geographies, Geographies of Digitalia' sessions at the Association of American Geographers conference, Tampa Bay, 8–12 April 2014; the 'Co-production of Digital Geography' sessions at the Royal Geographical Society conference, London, 27–29 August 2014; and the 'Digital Geography' workshop organized at the Open University, 24 March 2015. Following these sessions there have also been calls in 2016 to set up Digital Geographies study groups within the Royal Geographical Society and American Association of Geographers.

mediated by the digital as 'digital geographies', in this paper [|27] we seek to attest to the extent of the digital turn under way and argue that there is a need to more fully consider how the digital inflects geography's many subfields and mediates how geographical knowledge is produced.

We advance a threefold categorization of the relationship between geography and the digital: geographies produced *through*, produced *by*, and *of* the digital. The division between these categories is by no means mutually exclusive, with many examples overlapping between them. Nonetheless, we think it provides a useful heuristic to illustrate the scope and extent of the digital turn. In the interests of brevity, our aim is not to document all studies that involve an engagement between geography and the digital. Rather, we strive to illustrate, with selective examples, how the digital has become central to both the praxes and focus of contemporary geographical scholarship and to provide evidence of the evolving and intensifying digital turn. We conclude by suggesting conceptual, methodological and empirical questions that may aid the further development of geography's turn to the digital.

II GEOGRAPHIES *THROUGH* THE DIGITAL

The digital has long figured as a prominent site, mode, and object of/for knowledge production in human geography (Rose, 2015). By this we mean that the digital has been engaged to actualize heterodox epistemologies in the service of producing geographic knowledge and to enact knowledge politics, while simultaneously being the subject of epistemological critique.

Early approaches to engaging with digital knowledge production in human geography were rooted in the quantitative revolution and the use of computing to undertake new forms of statistical analysis and modelling (Haggett, 1966; Hagerstrand, 1967). This was accompanied by the first digital mapping projects (Balchin and Coleman, 1967; Tobler, 1959), their enrolment into national cartographic initiatives and later spatial data infrastructures, and the development of nascent geographic information systems (GIS) from the mid-1960s (Tomlinson, 1968; Foresman, 1998). Digital technologies then underwrote the development of positivist spatial science, GIS and later GIScience, as well as remote sensing and advanced photogrammetry. More recently, such quantitative geographical analysis has become more closely aligned with data science. With the rise of spatial big data and new machine learning analytics (e. g. data mining and pattern recognition, geovisualization, spatial statistics, prediction, optimization and simulation) there has been a renewed interest in developing what has been termed the computational social sciences (Lazer et al., 2009) and data-driven geography (Miller and Goodchild, 2014) to produce inherently longitudinal quantitative studies with much greater breadth, depth, scale, and timeliness (Kitchin, 2014a).

Positivist spatial science was critiqued on epistemological grounds by Marxist and humanistic geographers in the 1970s and feminist geographers beginning in the 1980s. However, the main critique of digital computation surfaced in the 1990s, particularly in debates about the role, status, and use of GIS in the discipline. The main lines of attack were drawn from emerging ideas in critical cartography, especially Harley's (1989) 'map deconstruction', and feminist critiques of science and vision. Harley emphasized that maps are never 'the territory' but rather technologies which normalize, legitimate, underwrite, and render transparent material exercises of power. As GIS became entrenched as a mainstream presence within the discipline, critical cartography likewise influenced the flourishing of critical GIS, which constituted a concerted effort at incorporating what were at the time trenchant critiques of the technology and its attendant practices (see Pickles, 1995). Critical GIS drew on feminist critiques of both science and (scientific) representation to challenge the supposed neutrality of GIS (see [|28] Leszczynski and Elwood, 2015). Feminist critiques of science were used to further challenge the inherent epistemological limitations of GIS artefacts (maps) and practices of discretization in two additional ways. First, questions were raised about exactly *whose* knowledges are being produced, *by* and *for* whom in deployments of and practices with the technology. Critiques highlighted the colonialist militarism, masculinist positivism and cartographic rationalities of GIScience that inherently produced ethnocentric, empiricist, and disembodied knowledges (Bondi and Domosh, 1992; Dixon and Jones, 1998). Second, the 'God-trick' (Haraway, 1991) of GIS – a 'view from nowhere' premised on the disembodied trope of the separation between the viewing subject (the GIS practitioner) and the object of vision (space) – was exposed as a totalizing scopic regime passed off as objective knowledge about the world (Roberts and Stein, 1995; Rocheleau, 1995).

The visual persists as an epistemological concern and entry point for engaging digital technologies in current geographic scholarship. In relation to the visual, particular emphasis is given to the epistemic rationalities imposed by the telos of digitally networked spatial platforms that continue to render the objects of representation – spaces, cities, people – 'knowable' in ways that privilege abstraction and calculability. The bulk of such approaches are most closely aligned with an aesthetic conception of the digital (Lunenfeld, 1999). Parks (2009), for instance, argues that Google Earth's vertical scopic regime encourages zooming past the geopolitical contexts of genocide (Darfur) straight into images that mobilize tropes of human misery, waste, and dispossession. At the more local end of the spectrum, Wilson (2011) demonstrates that issues of community poverty and signs of socioeconomic disenfranchisement in city neighbourhoods are reduced to superficial objects (abandoned shopping carts, refuse awaiting collection) that can be discerned by the geocoding eye. In turn, these objects can be imaged and quantified at the moment of being abstracted as digital records on location-enabled devices. Elsewhere, Rose (2016a) relates drone warfare and smart cities via a shared masculinist visuality that she terms the 'aerial view', which appears on the screens of the command-and-control centres where practices of both smart urbanism and autonomous warfare are coordinated and operationalized.

Many initial critiques of GIS sought to dismiss the technology from the discipline as the embodiment of objectionable epistemologies. However, interventions from critical GIS demonstrated precisely the opposite: digital media could be appropriated and repurposed to produce spatial knowledges that are situated, reflexive, non-masculinist, emotional/affected, inclusive and polyvocal, and flexible rather than foundational (Elwood, 2006; Kwan, 2002; Pavlovskaya, 2006; Schuurman, 2002). Feminist GIS interventions in particular repurposed quantitative method-ologies and geovisualization techniques within mixed-methods approaches that sought to effect and make subaltern and counter-hegemonic geographies visible (e. g. Kwan, 2002; Pavlovskaya, 2002). Similarly, participatory or public participa-tion GIS (P/PGIS) sought to reconfigure who performed and for whom geographic knowledge was produced by empowering groups historically on the losing side of the 'digital divide' (women, indigenous peoples, racial/ethnic minorities) to con-duct GIS analysis (Sieber, 2006).

That digital artifacts serve as objects, sites and modes of knowledge production is of course not limited to GIS. We now live in a present characterized by an abun-dant and diverse array of spatially-enabled digital devices, platforms, applications and services that have become ordinary and expected presences in our everyday lives. As a result of their pervasiveness, new spatial media are intensely bound up in the production of myriad, highly quotidian, spatial [|29] knowledges (Elwood and Leszczynski, 2013). For instance, the Surui, an indigenous Amazonian people, repurposed location-enabled Android mobile phones introduced to chronicle and geolocate instances of illegal logging and mining within their territory to document sites of cultural, historical and spiritual significance and uploaded them to Google Earth as an interactive layer for navigation and exploration (Forero, 2013).[2]

Digital technologies are also the standard media of knowledge generation and analysis in qualitative research. For example, interviews and focus groups are being captured and transcribed using digital recorders. Social interactions are being ob-served in online forums using internet ethnographies (Hine, 2000). Transcriptions are being managed and analysed using qualitative software (Hinchliffe et al., 1997). Participatory research is being conducted using digital cameras and video record-ers. Increasingly, digital methods for capturing and analysing qualitative and non-structured data, which can only be performed digitally, are being deployed (Rogers, 2013). This is particularly so with respect to the digital humanities, which seeks to use the power of computation to make sense of the vast troves of natively-digital content (e. g. radio, television, web content, etc.) as well as analogue and unstruc-tured data that has been digitized (e. g. millions of books, documents, newspapers, photographs, art works, material objects, etc.). Digital humanities research is aided by new tools of data curation, management, and analysis capable of handling mas-sive numbers of data objects. Rather than concentrating on a handful of novels or photographs, it becomes possible to search, connect and analyse across a large number of related works and use key techniques such as mapping and geovisualiza-tion to reveal spatial patterns and processes (Travis, 2015).

The proliferation and public accessibility of digital platforms for geographic knowledge production '[poses] epistemological challenges to the dominant theory of truth, in particular advancing a shift away from the correspondence model of

---

2   The Surui cultural map Google Earth layer (.kmz) may be downloaded at https://www.google. co.uk/earth/outreach/stories/surui.html

truth towards consensus and performative interpretations' (Warf and Sui, 2010: 197). As such, the politics of geographical knowledge production with the digital – which involves questions of how particular knowledges come to be considered legitimate (Elwood and Leszczynski, 2013) – remains influenced and marked by hegemonic social relations of, amongst others, race, class, and gender, as well as global digital divisions of labour (Graham and Foster, 2016). Moreover, they increasingly reflect the interests of the corporate entities that own and exert control over dominant digital spatial platforms by, for example, managing the use of APIs (application programming interfaces) to which they may revoke access, without explanation, at any time (Leszczynski, 2012).

The necessity for geographers to continue to move between enrolling the digital within critical geographic praxis whilst simultaneously engaging digitally-mediated knowledges is imperative in a present characterized by the diversification, rampant commercialization, and pervasiveness of locative media (Leszczynski, 2015b; Wilson, 2012) and the rollout of digital archives and repositories (Offen, 2013). As digital platforms simultaneously deterritorialize labour practices and reentrench spatially uneven patterns of the precarious positioning of workers in content and commodity chains that reflect global coreperipheries (Graham et al., 2014, 2015), we need to attend to the geographies produced *by* the digital.

III GEOGRAPHIES *PRODUCED BY* THE DIGITAL

Since the early 1990s, there have been a series of studies that have examined how the digital is mediating and augmenting the production of space and transforming socio-spatial relations. [|30] Initially, this work concentrated on how ICTs, and the internet in particular, were transforming economic, cultural, social, and political geographies. Some work took a technologically determinist position, declaring that networked ICTs flattened distance and rendered geography irrelevant by overcoming space with time through the instantaneous transfer of information (Cairncross, 1997; Friedman, 2005). Others, however, argued that while ICTs produced space-time compression and distanciation, geography remained critical.

Examined from a political economy perspective, it was clear that the new information economy was leading to changes in how companies and employment patterns were spatially structured through processes of concentration and dispersal, inducing significant urban-regional restructuring and the creation of a postindustrial landscape (Castells, 1996; Graham and Marvin, 2001). Geographical research highlighted how urban hierarchies were being reinforced through the concentration of command and control, and the agglomeration of information-rich business into key places (Moss, 1986). Consequently, many cities sought to pro-actively 'wire' themselves to attract inward investment and position themselves in the global informational economy (Warf, 1995). At the same time, many office activities, business services and production centres were decentralized to the suburbs, more peripheral cities or other countries to take advantage of cheaper rent and labour costs (Breathnach, 2000).

Simultaneously cities were starting to become much more reliant on digital systems with respect to their planning, management and governance, and digital infrastructures and devices were starting to be routinely embedded into the spatial fabric of cities themselves (Mitchell, 1996). Although city managers had been experimenting with using computer models and management systems to inform policy and govern cities since the early 1970s (Flood, 2011), it was only from the mid-1980s onwards that GIS and other land-use, planning and architecture software packages became common tools for urban management, along with updated urban control rooms for utility and transport infrastructures. From the 1990s onwards, cities became increasingly computational with traditional infrastructures augmented with networked sensors, transponders, and actuators, enabling new forms of real-time operational governance. For Graham and Marvin (2001), these new digital tools and mediated infrastructures were key components of the emerging neoliberal city, becoming increasingly privatized but also important means for enacting governance and control and creating particular power geometries. This generated what they termed splintering urbanism, a planning logic characterized by uneven development through the creation of differential and fragmentary infrastructures and services that are organized as much, if not more, for profit than public good.

Related research, also rooted in political economy, noted that far from flattening social and economic divides, digital social inequalities have only intensified along lines of access to ICTs. For Castells (1996), the social and spatial polarization inherent in the digital divide was characterized by a separation between what he termed the 'space of flows' (well-connected, mobile and more opportunities) and the 'space of places' (poorly connected, fixed, and isolated). This digital divide takes many forms, including divisions between classes, urban locations and nations (Dodge and Kitchin, 2002). This continues to be an ongoing issue, both with respect to access to digital technologies and infrastructures but also the content of the internet, which is decidedly skewed in its focus (Graham et al., 2014).

The 'digital divide' has more recently been complicated by the proliferation of digital technologies and content (data) in the spaces and practices of everyday life – such as the growth [|31] of smartphones – as well as the now entirely quotidian nature of information communication technologies (ICTs) around the world (Graham, 2011; Kleine, 2013). Questions of how digitally-mediated knowledge is produced, by whom, and in whose interests continue to attract attention. For instance, Graham and collaborators (Graham et al., 2015a, 2015b) have demonstrated that increased connectivity in Africa, in the form of expanded telecommunications infrastructures, has not translated into direct increases in individual participation on digital platforms or resulted in a seamless, proportionate incorporation of African economies into global technology and information sectors.

Over the past decade, much of the work on the relationship between the digital and the urban has focused on smart cities. Some of this research is informed by a political economic framework for documenting how the underlying discourses and rollout of smart city technologies are rooted in a neoliberal ethos of market-led and technocratic solutions to city governance and development that reinforce existing power geometries and social and spatial inequalities rather than eroding or reconfiguring them (e. g. Greenfield, 2013; Datta, 2015; Shelton et al., 2015). Smart cities have also been approached from a more positivistic stance that utilizes and promotes a computational social science approach. Here, research is principally concerned with utilizing urban big data to computationally model and simulate urban processes and with producing new tools and apps, such as urban dashboards, that reshape how cities are planned and how people navigate and interact with urban spaces (Batty, 2014).

Elsewhere, research draws from poststructural theory to consider the ways in which the digital production of space and mobilities is mediating new forms of governmentality. At the turn of the millennium, Amin and Thrift (2002: 125) noted that '[n]early every urban practice is becoming mediated by code'. Dodge and Kitchin (2005) argued that such was the importance of software to the production of space that in many cases code and space were mutually constituted as 'code/space': if the software failed, the space could not be produced as intended. However, they asserted that the relationship between code and space is neither deterministic nor universal. Rather code/space emerges in contingent, relational, contextdependent and imperfect forms.

One of the key ways in which code/spaces are enacted is in the regulation and control of space and the reproduction of regimes of governmentality. The policing of areas is increasingly being undertaken through networked surveillance and security apparatuses, and how populations are managed is mediated by information systems and databases. Such technologies on the one hand enforce new forms of (self) disciplining (Foucault, 1977), and on the other enact new forms of control (Deleuze, 1992; Sadowski and Pasquale, 2015). With respect to the latter, expressions of power are not visible and threatening, as with sovereign or disciplinary regimes. Rather, power is exerted subtly through distributed protocols that define and regulate access to resources and spaces and reshape behaviour. One manifestation of such control is socio-spatial sorting, whereby people are evaluated via algorithms that calculate and enforce differential access with respect to perceived worth (e. g. customer, credit and crime profiling) (Graham, 2005). Discipline and control are increasingly being dispensed through forms of automated management wherein governmentality is enacted through automated, automatic and autonomous systems (Amoore and Hall, 2009; Kitchin and Dodge, 2011).

Over the last decade, research has focused not only on the wiring of the networked smart city, but also on how to theoretically and empirically engage the technologies themselves. Specifically, this swathe of research has attended to the rollout and effects of new spatial and [|32] locative technologies, such as online mapping tools with accompanying APIs that enable the easy production of map mashups, usergenerated spatial databases and mapping systems (e. g. OpenStreetMap and WikiMapia), and locative media and augmented reality (e. g. satnavs and location-based social networking). Collectively, these were initially engaged as constituting what was termed 'the geoweb' – the aggregate of spatial technologies and geo-referenced information organized and transmitted through the internet and accessed through spatial media. These spatial media are having profound effects on the production of space/spatiality, mobility and knowledge politics. As geographic spaces are being evermore complemented with various kinds of georeferenced and real-time data (Gordon and De Souza e Silva, 2011; Graham and Zook, 2013), spatial media is creating new spatial practices enabling individuals to check into locations, create personalized georeferenced data, navigate routes, and locate friends and services. As such, spaces are being increasingly mediated and experienced through digital interfaces, in turn transforming the 'social production of space and the spatial production of society' (Sutko and De Souza e Silva, 2010: 812) and generating new spatialities that have variously been termed code/spaces, hybrid spaces (De Souza e Silva, 2006), digiplace (Zook and Graham, 2007), net locality (Gordon and De Souza e Silva, 2011), augmented reality (Graham et al., 2013), and mediated spatiality (Leszczynski, 2015b). Spatial media mediate social encounters within spaces and provide different ways to know and navigate locales, enabling on-the-fly scheduling of meetings and serendipitous encounters. Importantly, they do so in situ, on-the-move and in real-time, augmenting a whole series of activities such as shopping, wayfinding, sightseeing, and protesting, They also alter the traditional basis of knowledge politics because they transform the nature of expertise in terms of who can generate spatial data, and open up different epistemological strategies for asserting 'truth' (Elwood and Leszczynski, 2013).

IV GEOGRAPHIES *OF* THE DIGITAL

While work in contemporary human geography is attentive to the pervasiveness of digital, networked spatial media in the spaces and practices of everyday life, geographers' early engagements with charting the geographies *of* the digital took the form of a theoretical and empirical exploration of the digital as a particular geographical domain with its own logics and structures. These studies sought to apply pre-existing geographical ideas and methodologies to study what it considered to be a new material, spatial and technical realm of communication and interaction (the internet/cyberspace, virtual worlds, digital games) and their associated sociotechnical assemblages of production.

Initially, geographies of the digital conceptualized digitally-mediated experience as a form of cyber or virtual space (Crang et al., 1999; Fisher and Unwin, 2003). Cyberspace served as a kind of metaphor for understanding the worlds accessed by digital technologies, such as webpages, forums, multi-user dungeons and online video games, and how those worlds are constructed through sets of ICTs (Dodge and Kitchin, 2002). Here, cyberspace was understood as the outcome of a set of connected material objects (screens, routers, servers), working in relation to a human body (Zook et al., 2004; Kinsley, 2013b). As Hillis (1999) has helpfully shown, this metaphor of cyberspace operated around a predominantly visual understanding of space in which various computer-generated environments were accessed via a screen. Cyberspace was something to be surveyed, made sense of, and experienced by the eye. In doing so, spatial experience was primarily understood as the co-production between a cognitively imbued human body, a set of objects that made up an environment, and the mind [|33] which operated to unify this set of disjunctive entities into a holistically experienced world. As a kind of spatial landscape, it appeared logical to map cyberspace as one would any new terrain: as a set of material infrastructures and a space for shared experience (Shields, 2003). However, as Kinsley (2013b) and Graham (2013) have argued, the terms cyberspace and virtual space are problematic because they create a distinction between two supposedly different realms (digital and analogue, or virtual and actual), covering over the complex processes through which they are entwined. Extending earlier work by economic geographers interested in the distribution and concentration of internet infrastructure (e.g. Malecki, 2002; Zook, 2005), Blum (2012) and Starosielski (2015), amongst others, have grounded metaphors, such as those of 'cyberspace' and 'the cloud', by tracing the actual spatialities of internet infrastructures at both local and global scales. These spatialities include the instantiation of digital networks as internet exchanges, data centres, fibre optic cables and their landing sites, as well as the contentious economic, social, political, and historical contexts of their geographies.

Another body of work has charted the spatialities of video games and social media. What unites these areas of research is a concern for theorizing the relationship between body and screen and how engaging and communicating through screens alters the spatial understandings, embodied knowledge, political awareness and social relationships of users. In the case of video games, Ash (2009, 2010, 2012) has suggested that engaging with game environments cultivates new modes of spatial awareness organized around ethologies of action that guide players without thinking in order to capture and hold their attention. Shaw and Warf (2009) suggest these digital environments can also influence geopolitical understandings by shaping how users imagine other people and places around the world.

Working from a feminist perspective, geographers have explored how digital technologies transform social reproduction. For example, Longhurst (2013) has argued that the visual nature of digital technologies, such as Skype video calls, re-orients bodily relations between family members and create feelings of connection that are absent when communicating through telephone or email. Others note how digital technologies reorganize socio-spatial relations between different activities such as work, rest and mobility and between different family members, such as adults and children (Chan, 2008; Larsen, 2006; Valentine and Holloway, 2002). These studies highlight how digital technologies challenge notions of place-based identity as defined by a shared location and how pre-existing social relations are not extinguished, but rather transformed.

Distinct from this approach, a related body of work has plotted the material geographies of ubiquitous computing (digital objects and processes embedded into the environment, such as RFID tags and sensors) (Galloway, 2004). Here, digital geographies are figured as sets of technologies that go beyond an engagement with an interface or screen as a virtual geography (Kinsley, 2013a), or as an infrastructure whose primary aim is to enable this virtual geography (Graham, 1998). Instead, the focus is on the 'actual geographies that evolve on the surface of the earth in the information age: the changes in and among places resulting from the increased ability to store, transmit and manipulate vast amounts of information, and the new patterns of geographical differentiation, privilege and disadvantage that these changes are bringing about' (Sheppard et al., 1999: 798). As Galloway (2004: 387) argues, ubiquitous computing 'did not seek to transcend the flesh and privilege the technological'. Instead, 'ubiquitous computing was positioned to bring computers to "our world" (domesticating them), rather than us having to adapt to the "computer world" (domesticating us)'. Geographies of [|34] ubiquitous computing have thus examined the insertion and uptake of digital objects and markers into environments, such as place tagged podcasts (Arikawa et al., 2007), barometric pressure sensors (Retscher, 2007) and WiFi routers (Köbben, 2007).

Most recently, an emerging body of work has begun to trace the generation and flows of big data and algorithms. While geographies of the digital have understood data to be key to all digital communication, big data refers to a quantitative and qualitative shift in the amount, velocity, variety, resolution and flexibility of data that is now collected and analysed by a range of devices (Kitchin, 2014a). Geographers have explored the spaces of big data, including volunteered geographic information, in a variety of ways. Crampton et al. (2013) have detailed how geotagged data from services such as Twitter can be used to understand socio-spatial processes such as riots and response to natural disasters. They also recognize the limitations of such an approach, suggesting geotagged data is often non-representative given that it is generated by a relatively small number of people within any population. Further, analysts are typically working with secondary data 'fumes' visible to users of locative social media services, rather than full data sets, as these data sets remain commercially confidential and inaccessible to researchers (Arribas-Bel, 2014; Thatcher, 2014). Elsewhere, Graham and Shelton (2013) argue that any spatial big data necessarily create large data shadows, where groups who are considered valuable are increasingly data mined, while other populations are excluded from analysis. DeLyser and Sui (2013) thus argue that analysing the spatiality of big data requires novel methodological approaches that cross between qualitative and quantitative methods because big data alone cannot offer a comprehensive geography of the digital.

Emerging research has also identified glaring inequalities in the geographies of big data production. Graham et al. (2015) in particular evidence stark global North–South polarities in the geographies of information that reflect and reproduce global economic coreperipheries. For example, there have been more Wikipedia articles written about the uninhabited continent of Antarctica than all of the countries of Africa combined (Graham, 2009). The production of geocoded content about places furthermore exhibits a form of informational magnetism, whereby individuals in digitally underrepresented parts of the world, such as the Middle East–North Africa (MENA) region, are more likely to contribute content and edits to Wikipedia about places in the global North (the 'core') than they are about the places in which they themselves live (global informational peripheries) (Graham et al., 2015). These uneven contours of geographic content also manifest locally. For example, in the aftermath of Hurricane Sandy, the wealthy New York borough of Manhattan cast a far larger Twitter 'data shadow' than the most severely affected, more socioeconomically deprived areas of the tristate coastline, giving the impression that Manhattan was more deserving of an earlier and/or more concentrated emergency response than was merited (Shelton et al., 2014).

As Kwan (2016) has recently contended, however, much of what geographers have to date been engaging as 'big data' is actually the effect of algorithms, i. e. not unfiltered big data but the result of algorithmic processing of datasets. In human geography, this turn to algorithms as an object/subject of research is reflected in increasing interest in algorithmic governance and governmentality (Kitchin and Dodge, 2011; Amoore and Poitukh, 2015; Leszczynski, 2016), as well as the spatialities of algorithms themselves, i. e. the geographies of their coding, circulation, and appropriation (Amoore, 2016).

V DIGITAL TURN, DIGITAL GEOGRAPHY?

If the definition of a 'turn' is a concerted reorientation of focus of attention and approach, [|35] then it is fair to say that over the past two decades geography has experienced a 'digital turn'. Across all sub-disciplines, there has been a recognition of how the digital is reshaping the production and experience of space, place, nature, landscape, mobility, and environment. This recognition is underpinned by a turn *to* the digital as subject/object of geographical scholarship, and a profound inflection of geographic theory and praxis by the digital, whether understood as ontics, aesthetics, logics, or discourse, or an assemblage thereof.

In this paper, we have strived to evidence the digital turn by charting the intensifying history of geography's engagement with the digital, with an emphasis on contemporary theoretical and empirical interventions that we have approached through the tripartite heuristic of geographies *through* the digital, geographies *produced by* the digital, and geographies *of* the digital. Given the scope and extent of the digital turn, we have had space to focus on only a small sample of such work. Our choice to profile work concerned with the relationship between the digital and the urban, for example, is not to the exclusion of non-urban research, such as that investigating the negative regional impacts of the lack of broadband infrastructure in rural areas or the use of software-enabled technologies in farming, or the robust body of work on ewaste and digital dumping grounds, which are disproportionately located in impoverished regions of the Global South. Indeed, there are countless other interventions we could have discussed that trace, either explicitly or more obliquely, how digital technologies recast economic, political, social, cultural, health, and other geographies.

These exclusions notwithstanding, the epistemologies and methodologies of geographical scholarship and research are now thoroughly mediated by digital technologies. These technologies alter, in all kinds of explicit and subtle ways, the kinds of questions that are asked, how they are asked and answered, the ways in which knowledges are constructed, communicated and debated, as well as the material spatialities and geographies of their production, transmission, and appropriation. For us, these considerations capture the extent, emphases, and effects of geography's 'digital turn' and not the imperative towards designating a field of 'digital geography' that should or could be established within the discipline. Similar attempts have been underway in anthropology and sociology for a number of years. In both cases, the focus is broad, encompassing the anthropology and sociology 'of', 'produced by', and 'produced through' the digital. The consequence, we believe, is to recast nearly all anthropology and sociology as 'digital anthropology' and 'digital sociology' to some degree, especially given the reliance on digital technologies in knowledge production. But if everything becomes 'digital' then 'digital' becomes an empty signifier and unworthy of distinct denotation. While we do maintain that there is a need to think critically about the relationship between geography and the digital, we contend that rather than cast all of those geographies as 'digital geography' it is more meaningful to think about how the digital reshapes many geographies, mediates the production of geographic knowledge, and itself has many geographies.

By framing the digital in this way, we avoid the decontextualization of digital approaches, methodologies, and research studies from their subdisciplinary domains such as urban geography or geographies of development. Instead, the emphasis remains on how an engagement with the digital develops our collective understandings of cities and development, as well as health, politics, economy, society, culture, and the environment, amongst others. It also allows for 'the digital' to function as a site and mode for intersectional research that cuts across research foci and leverages methodologies from multiple geographical subdisciplines. Attending to the geography of rare metals used in the production of digital technologies, for instance, [|36] raises questions in the fields of resource and development geographies, postcolonial studies, as well as geopolitics. This enables the differences the digital makes to research, epistemology and knowledge production to be contextualized within a broader knowledge base and history of theory, concepts, models and empirical findings within and across geographic sub-domains. For example, we feel it makes sense to frame smart city developments within debates around the long history of urbanization and urbanism, rather than to set them within a field of digital geography.

Disciplinary engagement with the digital is a rapidly developing field with many aspects of the intensifying relationship between geography and digitality deserving of further conceptual, methodological, and empirical attention. As a preliminary prospectus for future work, we argue that there is much to be gained from identifying synergies with the theory and praxis of disciplines that focus more substantially on the specifically technical aspects of the digital, such as science and technology studies, software studies, cybernetics, critical data studies, game studies, platform studies, (new) media studies, informatics, and human-computer interaction. We believe this is critically important because if we are to identify and meaningfully influence the effects and outcomes of digital technologies, then it is imperative that we understand the nature and operationalization of technology infrastructures and protocols. As Nadine Schuurman (2000) argued with reference to GIS and its critics in the late 1990s, epistemological quandaries of the technology arise from its material architectures.

We believe geographers are uniquely placed to interrogate the materialities of digital computation in innovative ways. Geographers' theorizations of space, time and relationality can be fruitfully developed to consider how digital computation and its associated objects are both singular things, with particular capacities, that also create shared space times for both other technical objects and the humans who use those objects. This calls for further attention to be given to the work that non-human infrastructures perform that always exceeds the technical parameters of their design. Tim Schwanen (2015) develops three potential strategies for studying digital computation in this way. In relation to smartphone apps, he suggests that researchers begin with the app itself rather than 'the human individual, her needs, preferences, valuations or even the social practices she is enrolled in' (Schwanen, 2015: 682). Practically, this can take the form of understanding the script design of the app and then understanding how users engage with the script design, for example. Schwanen also suggests that we consider how engagements between the objects of digital computation and humans create new objects: in terms of apps, this might be affective senses of reward or competition (also see Cockayne, forthcoming). Finally, we can understand how the disjunction between design and use shapes broader practices with these technologies.

A substantive empirical examination and theorization of the political economies of spatial big data, algorithms and geolocation technologies remains underdeveloped. While work in this area has begun (see Leszczynski, 2012, 2014; Wilson, 2012), to date, there has been little engagement with the ways in which 'disruptive' activities of the sharing/platform/gig economies are completely contingent on geologistics as a business model (e. g. Uber as a business model; app-driven services of the 'last mile' economy; accommodation platforms such as Airbnb). There is a need to further connect empirical research in this vein to burgeoning geographical analyses of the reconfiguration of labour in the gig economy, the rise of digital labour, and the uneven global geographies of microwork.

Rather than advocating for a single focus on political economic concerns, we encourage geographers to critically reflect upon the wider [|37] *dispositif* or assemblage of the digital. Foucault's (1977) concept of the *dispositif* refers to a 'thoroughly heterogeneous ensemble consisting of discourses, institutions, architectural forms, regulatory decisions, laws, administrative measures, scientific statements, [and] philosophical, moral and philanthropic propositions' (in Gordon, 1980: 194), which enhance and maintain the exercise of power within society. Unpacking a digital *dispositif* involves charting the wider discursive and material practices that interact in relational, contingent and contextual ways to shape the design, deployment, normalization and use of digital technologies in ways that serve and sustain particular kinds of interests (the economy, social capital) in society, consolidating and channelling the exercise of power. Kitchin (2014a) sets out a similar notion with respect to mapping out what he has termed data assemblages, arguing for the need to examine digital objects and infrastructures comprehensively, critically engaging their interlocking technical stack (platform, operating system, code, data, interface) and the epistemological, political economic, institutional, legal, and governmentalized contexts of their production, circulation, and operationalization in society.

Such a focus on data assemblages is one approach to tackle Crampton et al.'s (2013) imperative for empirically and methodologically going 'beyond the geotag', but work remains to be done in identifying and addressing the exclusions and inclusions of digital connectivity and discourse. As a first prerogative, there is a pressing need to destabilize the dominance of the Global North as a universal placeholder and de facto field site for geographical research about the digital. The recent expansion of digital infrastructures into parts of the world that have been historically disconnected allows for empirical assessment of the relationships between connectivity, digital inclusion, and economic integration in ways that are not possible in the already connected Global North.

There is also further need to attend to questions around the ways in which big data economies, algorithms, digital technology design, and utopian narratives are informed by the persistence of colonialism and masculinism. Western-centric prototypes of the 'smart city' cannot – and should not – be transplanted onto megacities of the Global South with no consideration of a city's unique history, infrastructure, or context (Datta, 2015). Similarly, as Gillian Rose (2016b) has recently argued, visions and discourses of the city are characteristically devoid of the presences of women; when they do appear, it is almost exclusively as the victims of violence.

Continuing to think beyond the geotag, geographers need to be increasingly attuned to the ways in which algorithms and spatial big data – namely, personal locational traces – participate within epistemologies that equate data with definitive evidence of spatial presence, movement and behaviour in what Crawford (2014) terms 'data-driven regime[s] of truth'. As a function of the relationality of big data phenomena, data indicative of spatial presence, movement and behaviour are being used to infer social, political and religious affiliations about individuals, as well as their involvement and complicity in events and occurrences such as protests and their predisposition or likelihood towards participation in particular kinds of activities (see Leszczynski, 2015a). Such datadriven correlations are deeply informed by, and reproduce, longstanding socio-economic inequalities, which must continue to be made visible. Related to this, there is much work to be undertaken in mapping out the politics and ethics of spatial big data, open data initiatives, algorithms, and the impetus towards smart cities. This includes the need to examine the ownership and control of data; the integration of data within urban operating systems, control rooms, and data markets; data security and integrity; data protection and privacy; data quality and provenance and dataveillance. [|38]

It is clear that ideals such as the OECD's (1980) Fair Information Practice Principles concerning notice, choice, consent, security, integrity, access, use and accountability are treated as redundant, with data being generated without consent and repurposed in the service of datadriven urbanism and the 'data-security assemblage' (Aradau and Blanke, 2015, Kitchin, 2014a; Shelton et al., 2015). As Datta (2015), Greenfield (2013), Kitchin (2014a) and Leszczynski (2016) note, there is a strong neoliberal ethos underpinning such appropriations of data, with the technological solutionism deployed and the corporatization of city services designed to buttress inequalities and enforce securitized regimes of law and order. Geographers are ideally placed to map the socio-spatial materialities of these various data regimes and to chart the promises and perils, socio-spatial processes and political economies of data-driven urbanism. At the same time, geographers are well positioned to undertake normative analyses to investigate what a more fair, equitable and ethical smart city might look like. This is important because discourses of equitability are currently controlled by corporations who own these data and their platitudes regarding 'citizen-centric design' should not be taken at face value.

The digital has reshaped how geographical research is conducted, becoming a central focus across geography's various subdisciplines. In this paper, we have traced the multiple diverse epistemological and methodological frameworks through which the digital has been engaged in geography over the last half-century. With a particular emphasis on contemporary human geographies, we have intentionally abstained from promoting particular methods and/or theoretical approaches above others. Rather, we believe that the rapidly diversifying and burgeoning universe of networked digital content, presences, praxes, phenomena and technical protocols is deserving of a parallel multiplicity of epistemologies, political projects, and methodologies. As the proliferation, commercialization, and popularization of geolocation technologies is itself engendering the flourishing of spatial ontologies and epistemologies, we encourage geographers to adopt and embrace an epistemological, ontological, and methodological openness in their engagements with the digital.

Acknowledgements

The authors wish to thank the organizers of the 7th Annual Doreen Massey Event, the theme of which was 'Digital Geographies'. It was held at the Open University on 24 March 2015. This paper comes out of fruitful discussions had by the authors as they participated together on the opening panel of the event. The authors would also like to thank the two editors and four referees for their helpful comments on an earlier version of this paper.

Declaration of conflicting interests

The author(s) declared no potential conflicts of interest with respect to the research, authorship, and/or publication of this article.

Funding

The author(s) disclosed receipt of the following financial support for the research, authorship, and/or publication of this article: Rob Kitchin's contribution to this paper was supported by a European Research Council Advanced Investigator Award, 'The Programmable City' (ERC-2012-AdG-323636). [|39]

REFERENCES

Amin A and Thrift N (2002) *Cities: Reimagining the Urban*. London: Polity.

Amoore L (2016) *History, algorithms, ethics*. Paper presented at the Association of American Geographers Annual Meeting, San Francisco, 29 March–3 April.

Amoore L and Hall A (2009) Taking people apart: Digitised dissection and the body at the border. *Environment and Planning D: Society and Space* 27: 444–464.

Amoore L and Piotukh V (2015) *Algorithmic Life: Calculative Devices in the Age of Big Data*. London: Routledge.

Aradau CT and Blanke T (2015) The (big) data-security assemblage: Knowledge and critique. *Big Data & Society* 2(2). DOI: 10.1177/2053951715609066.

Arikawa M, Tsuruoka K, Fujita H and Ome A (2007) Place-tagged podcasts with synchronized maps on mobile media players. *Cartography and Geographic Information Science* 34: 293–303.

Arribas-Bel D (2014) Accidental, open and everywhere: Emerging data sources for the understanding of cities. *Applied Geography* 49: 45–53.

Ash J (2009) Emerging spatialities of the screen: Video games and the reconfiguration of spatial awareness. *Environment and Planning A* 41: 2105–2124.

Ash J (2010) Teleplastic technologies: Charting practices of orientation and navigation in videogaming. *Transactions of the Institute of British Geographers* 35: 414–430.

Ash J (2012) Attention, videogames and the retentional economies of affective amplification. *Theory, Culture and Society* 29: 3–26.

Ash J (2015) *The Interface Envelope: Gaming, Technology, Power*. London: Bloomsbury.

Balchin W and Coleman A (1967) Cartography and computers. *The Cartographer* 4: 120–127.

Batty M (2014) *The New Science of Cities*. Cambridge, MA: MIT Press.

Blum A (2012) Tubes: A Journey to the Center of the Internet. New York: HarperCollins.

Bondi L and Domosh M (1992) Other figures in other places: On feminism, postmodernism, and geography. *Environment and Planning D: Society and Space* 10: 199–213.

Breathnach P (2000) Globalisation, information technology and the emergence of niche transnational cities: The growth of the call centre sector in Dublin. *Geoforum* 31: 477–485.

Cairncross F (1997) *The Death of Distance*. London: Orion Business Books.

Castells M (1996) *The Rise of the Network Society*. Oxford: Blackwell.

Chan AH-N (2008) 'Life in Happy Land': Using virtual space and doing motherhood in Hong Kong. *Gender, Place and Culture* 15: 169–188.

Cockayne D (forthcoming) Affect and value in critical examinations of the production and 'prosumption' of big data. *Big Data & Society*.

Crang M, Crang P and May J (1999) *Virtual Geographies: Bodies, Space and Relations*. London: Psychology Press.

Crampton JW, Graham M, Poorthius A, Shelton T, Stephens M, Wilson MW and Zook M (2013) Beyond the geotag: Situating 'big data' and leveraging the potential of the geoweb. *Cartography and Geographic Information Science* 40: 130–139.

Crawford K (2014) When FitBit is the expert witness. *The Atlantic*, 19 November. Available at: http://www.theatlantic.com/technology/archive/2014/11/when-fitbit-is-the-expert-witness/382936/ (accessed 5 August 2016).

Datta A (2015) New urban utopias of postcolonial India: 'Entrepreneurial urbanization' in Dholera smart city, Gujarat. *Dialogues in Human Geography* 5: 3–22.

De Souza and Silva A (2006) From cyber to hybrid: Mobile technologies as interfaces of hybrid spaces. *Space and Culture* 9: 261–278.

Deleuze G (1992) Postscript on the societies of control. *October* 59: 3–7.

DeLyser D and Sui D (2013) Crossing the qualitative-quantitative divide II: Inventive approaches to big data, mobile methods, and rhythmanalysis. *Progress in Human Geography* 37: 293–305.

Dixon DP and Jones JP III (1998) My dinner with Derrida, *or* spatial analysis and poststructuralism do lunch. *Environment and Planning A* 30: 247–260.

Dodge M and Kitchin R (2002) *Mapping Cyberspace*. London: Routledge.

Dodge M and Kitchin R (2005) Code and the transduction of space. *Annals of the Association of American Geographers* 95: 162–180.

Elwood S (2006) Beyond cooptation or resistance: Urban spatial politics, community organizations, and GISbased spatial narratives. *Annals of the Association of American Geographers* 96: 323–341.

Elwood S and Leszczynski A (2013) New spatial media, new knowledge politics. *Transactions of the Institute of British Geographers* 38: 544–559.

Fisher P and Unwin D (2003) *Virtual Reality in Geography*. Boca Raton: CRC Press. [|40]

Flood J (2011) The Fires: *How a Computer Formula, Big Ideas, and the Best of Intentions Burned Down New York City – and Determined the Future of Cities*. New York: Riverhead.

Forero J (2013) From the stone age to the digital age in one big leap. *NPR.org*, 28 March. Available at: http://www. npr.org/2013/03/28/175580980/from-the-stone-age-to-the-digital-age-in-one-big-leap (accessed 5 August 2016).

Foresman TW (1998) *The History of Geographic Information Systems: Perspectives from the Pioneers*. Upper Saddle River: Prentice Hall.

Foucault M (1980 [1977]) The confession of the flesh. In: Gordon C (ed.) *Power/Knowledge*. New York: Pantheon Books, 194–228.

Foucault M (1977) *Discipline and Punish: The Birth of the Prison*. London: Vintage Books.

Franklin S (2015) *Control: Digitality as Cultural Logic*. Cambridge, MA: MIT Press.

Fraser A (2007) Coded spatialities of fieldwork. *Area* 39: 242–245.

Friedman T (2005) *The World Is Flat: A Brief History of the Twenty-First Century*. New York: Farrar, Straus and Giroux.

Galloway A (2004) Intimations of everyday life: Ubiquitous computing and the city. *Cultural Studies* 18: 384–408.

Gordon C (1980) *Power/Knowledge: Selected Interviews and Other Writings of Michel Foucault*. New York: Pantheon Books.

Gordon E and De Souza e Silva A (2011) *Net Locality: Why Location Matters in a Networked World*. Chichester: Wiley-Blackwell.

Graham M (2009) Wikipedia's known unknowns. *The Guardian*, 2 December. Available at: http://www.theguardian.com/technology/2009/dec/02/wikipedia-known-unknowns-geotagging-knowledge (accessed 5 August 2016).

Graham M (2011) Time machines and virtual portals: The spatialities of the digital divide. *Progress in Development Studies* 11: 211–227.

Graham M (2013) Geography/internet: Ethereal alternate dimensions of cyberspace or grounded augmented realities? *The Geographical Journal* 179: 177–182.

Graham M and Foster C (2016) Geographies of information inequality in sub-Saharan Africa. *The African Technopolitan* 5: 78–85.

Graham M and Zook M (2013) Augmented realities and uneven geographies: Exploring the geo-linguistic contours of the web. *Environment and Planning A* 45: 77–99.

Graham M, Andersen C and Mann L (2015) Geographical imagination and technological connectivity in East Africa. *Transactions of the Institute of British Geographers* 40(3): 334–349.

Graham M, Hogan B, Straumann RK and Medhat A (2014) Uneven geographies of user-generated information: Patterns of increasing informational poverty. *Annals of the Association of American Geographers* 104: 746–764.

Graham M, Straumann M and Hogan B (2015) Digital divisions of labour and informational magnetism: Mapping participation in Wikipedia. *Annals of the Association of American Geographers* 105: 1158–1178.

Graham M, Zook M and Boulton A (2013) Augmented reality in urban places: Contested content and the duplicity of code. *Transactions of the Institute of British Geographers* 38: 464–479.

Graham S (1998) The end of geography or the explosion of place? Conceptualising space, place and information technology. *Progress in Human Geography* 22: 165–185.

Graham SDN (2005) Software-sorted geographies, *Progress in Human Geography* 29: 562–580.

Graham S and Marvin S (2001) *Splintering Urbanism: Networked Infrastructures, Technological Mobilities and the Urban Condition*. London: Routledge.

Greenfield A (2013) *Against the Smart City*. New York: Do Publications.

Hagerstrand T (1967) The computer and the geographer. *Transactions of the Institute of British Geographers* 42: 1–20.

Haggett P (1966) *Locational Analysis in Human Geography*. New York: St Martin's Press.

Haraway DJ (1991) *Simians, Cyborgs, and Women: The Reinvention of Nature*. New York: Routledge.

Harley JB (1989) Deconstructing the map. *Cartographica* 26: 1–20.

Hillis K (1999) *Digital Sensations: Space, Identity, and Embodiment in Virtual Reality*. Minneapolis: University of Minnesota Press.

Hinchliffe S, Crang M, Reimer S and Hudson A (1997) Software for qualitative research 2: Some thoughts on 'aiding' analysis. *Environment and Planning A* 29: 1109–1124.

Hine C (2000) *Virtual Ethnography*. London: SAGE.

Horst H and Miller D (2012) *Digital Anthropology*. New York: Bloomsbury. [|41]

Kinsley S (2013a) Beyond the screen: Methods for investigating geographies of life 'online'. *Geography Compass* 7: 540–555.

Kinsley S (2013b) The matter of 'virtual' geographies. *Progress in Human Geography* 36: 364–384.

Kitchin R (2014a) *The Data Revolution: Big Data, Open Data, Data Infrastructures and Their Consequences*. London: SAGE.

Kitchin R (2014b) The real-time city? Big data and smart urbanism. *GeoJournal* 79: 1–14.

Kitchin R and Dodge M (2011) *Code/Space: Software and Everyday Life*. Cambridge, MA: MIT Press.

Kitchin R, Linehan D, O'Callaghan C and Lawton P (2013) Public geographies and social media. *Dialogues in Human Geography* 3: 56–72.

Kleine D (2013) *Technologies of Choice? ICTs, Development, and the Capabilities Approach*. Cambridge, MA: MIT Press.

Köbben B (2007) Wireless campus LBS: A testbed for WiFi positioning and location based services. *Cartography and Geographic Information Science* 34: 285–292.

Kwan MP (2002) Feminist visualization: Re-envisioning GIS as a method in feminist geographic research. *Annals of the Association of American Geographers* 92: 645–661.

Kwan MP (2016) Algorithmic geographies: Big data, algorithmic uncertainty, and the production of geographic knowledge. *Annals of the American Association of Geographers* 106: 274–282.

Larsen J, Axhausen KW and Urry J (2006) Geographies of social networks: Meetings, travel and communications. *Mobilities* 1: 261–283.

Lazer D, Pentland A, Adamic L, Aral S, Barabási A-L, Brewer D, Christakis N, Contractor N, Fowler J, Gutmann M, Jebara T, King G, Macy M, Roy D and Van Alstyne M (2009) Computational social science. *Science* 323: 721–723.

Lepawsky J (2014) The changing geography of global trade in electronic discards: Time to rethink the e-waste problem. *The Geographical Journal* 181: 147–159.

Leszczynski A (2012) Situating the geoweb in political economy. *Progress in Human Geography* 36: 72–89.

Leszczynski A (2014) On the neo in neogeography. *Annals of the Association of American Geographers* 104: 60–79.

Leszczynski A (2015a) Spatial big data and anxieties of control. *Environment and Planning D: Society and Space* 33: 965–984.

Leszczynski A (2015b) Spatial media/tion. *Progress in Human Geography* 39: 729–751.

Leszczynski A (2016) Speculative futures: Cities, data, and governance beyond smart urbanism. *Environment and Planning A*. Epub ahead of print 22 May 2016. DOI: 10.1177/0308518X16651445.

Leszczynski A and Elwood S (2015) Feminist geographies of new spatial media. *The Canadian Geographer* 59: 12–28.

Longhurst R (2013) Using Skype to mother: Bodies, emotions, visuality, and screens. *Environment and Planning D: Society and Space* 31: 664–679.

Lunenfeld P (1999) Screen grabs: The digital dialectic and new media. In: Lunenfeld P (ed.) *The Digital Dialectic: New Essays on New Media*. Cambridge, MA: MIT Press, xiv–xxi.

Lupton D (2014) *Digital Sociology*. London: Routledge.

Malecki EJ (2002) The economic geography of the internet's infrastructure. *Economic Geography* 78: 399–424.

Miller HJ and Goodchild M (2015) Data-driven geography. *GeoJournal* 80: 449–461.

Mitchell WJ (1996) *City of Bits: Space, Place and the Infobahn*. Cambridge, MA: MIT Press.

Moss M (1986) Telecommunications, world cities and urban policy *Urban Studies* 24: 534–546.

Murray S (2008) Cybernated aesthetics: Lee Bull and the body transfigured. *Performing Arts Journal* 30: 38–65.

OECD (1980) *OECD Guidelines on the Protection of Privacy and Transborder Flows of Personal Data*. Available at: https://www.oecd.org/sti/ieconomy/oecdguidelinesontheprotectionofprivacy andtransborderflowsofpersonaldata.htm (accessed 5 August 2016).

Offen K (2013) Historical geography II: Digital imaginations. *Progress in Human Geography* 37: 564–577.

Orton-Johnson K and Prior N (2013) *Digital Sociology: Critical Perspectives*. Basingstoke: Palgrave.

Parks L (2009) Digging into Google Earth: An analysis of 'crisis in Darfur'. *Geoforum* 40: 535–545.

Pavlovskaya M (2002) Mapping urban change and changing GIS: Other views of economic restructuring. *Gender, Place and Culture* 9: 281–290.

Pavlovskaya M (2006) Theorizing with GIS: A tool for critical geographies? *Environment and Planning A* 38: 2003–2020.

Pickles J (1995) *Ground Truth: The Social Implications of Geographic Information Systems*. New York: Guildford Press. [|42]

Retscher G (2007) Augmentation of indoor positioning systems with a barometric pressure sensor for direct altitude determination in a multi-storey building. *Cartography and Geographic Information Science* 34: 305–310.

Roberts S and Stein R (1995) Earth shattering: Global imagery and GIS. In: Pickles J (ed.) *Ground Truth: The Social Implications of Geographic Information Systems*. New York: Guilford Press, 171–195.

Rocheleau D (1995) Maps, numbers, text, and context: Mixing methods in feminist political ecology. *The Professional Geographer* 47: 458–466.

Rogers R (2013) *Digital Methods*. Cambridge, MA: MIT Press.

Rose G (2015) Rethinking the geographies of cultural 'objects' through digital technologies: Interface, network and friction. *Progress in Human Geography*. Epub ahead of print 19 April 2015. DOI: 10.1177/0309132515580493.

Rose G (2016a) Smart cities and drone warfare: A shared visuality? *Visual/Method/Culture*, 18 February. Available at: https://visualmethodculture.wordpress.com/2016/02/18/smart-cities-and-drone-warfare-a-shared-visuality/ (accessed 5 August 2016).

Rose G (2016b) So what would a smart city designed for women look like? (and why that's not the only question to ask). *Visual/Method/Culture*, 22 April. Available at: https://visualmethodculture.wordpress.com/2016/04/22/so-what-would-a-smart-city-designed-for-women-be-like-and-why-thats-not-the-only-question-to-ask/ (accessed 5 August 2016).

Rose G, Degen M and Melhuish C (2014) Networks, interfaces, and computer-generated images: Learning from digital visualisations of urban redevelopment projects. *Environment and Planning A* 32: 386–403

Sadowski J and Pasquale F (2015) *The spectrum of control: A social theory of the smart city.* First Monday. Available at: http://firstmonday.org/ojs/index.php/fm/article/view/5903/4660 (accessed 5 August 2016).

Schuurman N (2002) Women and technology in geography: A cyborg manifesto for GIS. *The Canadian Geographer* 46: 258–265.

Schwanen T (2015) Beyond the instrument: Smartphone app and sustainable mobility. *EJTIR* 15: 675–690.

Shelton T, Poorthius A and Zook M (2015) Social media and the city: Rethinking urban socio-spatial inequality using user-generated geographic information. *Landscape and Urban Planning* 142: 198–211.

Shelton T, Poorthius A, Graham M and Zook M (2014) Mapping the data shadows of Hurricane Sandy: Uncovering the socio-spatial dimensions of 'big data'. *Geoforum* 52: 167–179.

Shelton T, Zook M and Wiig A (2015) The 'actually existing smart city'. *Cambridge Journal of Regions, Economy and Society* 8: 13–25.

Sheppard E, Couclelis H, Graham S, Harrington JW and Onsrud H (1999) Geographies of the information society. *International Journal of Geographical Information Science* 13: 797–823.

Shields R (2003) *The Virtual.* London: Routledge.

Sieber R (2006) Public participation geographic information systems: A literature review and framework. *Annals of the Association of American Geographers* 96: 491–507.

Starosielski N (2015) *The Undersea Network.* Durham: Duke University Press.

Sui D and Morrill R (2004) Computers and geography: From automated geography to digital earth. In: Brunn SD, Cutter SL and Harrington JW (eds) *Geography and Technology.* New York: Springer, 81–108.

Sutko DM and De Souza e Silva A (2010) Location-aware mobile media and urban sociability. *New Media & Society* 13: 807–823.

Thatcher J (2014) Living on fumes: Digital footprints, data fumes, and the limitations of spatial big data. *International Journal of Communication* 8: 1765–1783.

Tobler WR (1959) Automation and cartography. *The Geographical Review* XLIX: 526–534

Tomlinson RF (1968) A geographic information system for regional planning. In: Stewart GA (ed.) *Land Evaluation.* Melbourne: Macmillan, 200–210.

Travis C (2015) *Abstract Machine: Humanities GIS.* Redlands: ESRI Press.

Valentine G and Holloway SL (2002) Cyberkids? Exploring children's identities and social networks in on-line and off-line worlds. *Annals of the Association of American Geographers* 92: 302–319.

Warf B (1995) Telecommunications and the changing geographies of knowledge transmission in the late 20th century. *Urban Studies* 32: 361–378.

Warf B and Sui D (2010) From GIS to neogeography: Ontological implications and theories of truth. *Annals of GIS* 16: 197–209.

Wilson MW (2011) 'Training the eye': Formation of the geocoding subject. *Social & Cultural Geography* 12: 357–376. [|43]

Wilson MW (2012) Location-based services, conspicuous mobility, and the location-aware future. *Geoforum* 43: 1266–1275.

Zook MA (2005) *The Geography of the Internet Industry: Venture Capital, Dot-coms and Local Knowledge.* Oxford: Blackwell.

Zook MA and Graham M (2007) Mapping DigiPlace: Geocoded internet data and the representation of place. *Environment and Planning B: Planning and Design* 34: 466–482.

Zook M, Dodge M, Aoyama Y and Townsend A (2004) New digital geographies: Information, communication, and place. In: Brunn SD, Cutter SL and Harrington Jr SL (eds) *Geography and Technology*. Dordrecht: Kluwer, 155–178.

James Ash is a geographer and senior lecturer in media and cultural studies at Newcastle University, UK. His research is concerned with the cultures, economies and politics of digital interfaces. He is author of The Interface Envelope: Gaming, Technology, Power (Bloomsbury) and has published a range of papers on technology, interface design and gaming in journals including Theory, Culture and Society and Environment and Planning D: Society and Space.

Rob Kitchin is a professor and ERC Advanced Investigator in the National Institute of Regional and Spatial Analysis at Maynooth University. He is currently a principal investigator on the Programmable City project, the Digital Repository of Ireland, the All-Island Research Observatory and the Dublin Dashboard.

Agnieszka Leszczynski is a lecturer in the School of Environment at the University of Auckland, New Zealand. Her work is situated at the subdisciplinary interfaces of GIScience and human geography and examines issues around geospatial technologies and critical GIScience.

II WIE SICH DAS DIGITALE UND GESELLSCHAFTSPROZESSE VERSCHRÄNKEN

FEMINIST GEOGRAPHIES OF DIGITAL WORK

Lizzie Richardson

Abstract: Feminist thought challenges essentialist and normative categorizations of 'work'. Therefore, feminism provides a critical lens on 'working space' as a theoretical and empirical focus for digital geographies. Digital technologies extend and intensify working activity, rendering the boundaries of the workplace emergent. Such emergence heightens the ambivalence of working experience: the possibilities for affirmation and/or negation through work. A digital geography is put forward through feminist theorizations of the ambivalence of intimacy. The emergent properties of working with digital technologies create space through the intimacies of postwork places where bodies and machines feel the possibilities of being 'at' work.
Keywords: digital geographies, economy, feminist theory, intimacy, work

I INTRODUCTION

The digital has been a focus for enquiry in geography from a variety of perspectives. Scholarship has considered interactions between software and space including GIS and spatial knowledge; digital divides and development; code/spaces; robots; big data and forms of governance (Del Casino Jnr, 2015; Graham, 2011; Kitchin and Dodge, 2011; Kitchin, 2013, 2014a; Kleine, 2013; Wilson, 2014). The differentiated materialities of digital technologies and their relationship with lived experience have also been addressed (Ash, 2013; Kinsley, 2014; Kirsch, 2014; Leszczynski, 2014; Rose, 2015; Wilson, 2011, 2012). This article contributes to these cultural perspectives by building on feminist geographical analyses of the digital (e. g. Elwood and Leszczynski, 2011; Kwan, 2002, 2007; Leszczynski and Elwood, 2015) as a means through which geographers can 'more assertively' contribute to debates on 'the digital economy' (Kinsley, 2014: 378).

From the (limited) perspective of the 'Global North',[1] I contend feminist critique provides an important analytical lens for understanding the role of digital technologies in geographies of work. These technologies enact an *extension* of the activities that count as work, together with an *intensification* of working practices, rendering the boundaries of the workplace *emergent*. In addition to making digital products, digital work is understood to include broader practices that extend and intensify working activities through digital technologies: spatial processes that should be of significant [|245] interest to geographers. Such (technological) changes to work are *ambivalent*: they provide opportunities for *affirmation* and *negation* (as long noted, e. g. Beck, 2000; Hardt, 1999; Negri, 1989; Sennett, 1998).

---

1   The differing practices that might constitute 'digital work' are highly uneven globally; however, it is not within the scope of this article to examine these.

Affirmatively, work offers a basis for utopian demands and might be experienced as creative fulfilment. Negatively, work reduction underpins claims for work/life balance, particularly when excessive work is experienced as exploitation. Therefore my aim is not to delimit 'the digital', claiming it marks a 'phase shift' in understandings of technology and work. Neither is it to suggest that feminist critique is the only way to approach such work. Rather I demonstrate the richness and complexity of feminist thought[2] for critical perspectives on the emergent properties of work with and through digital technologies. Feminist critiques of work have been articulated through such extensions and intensifications of working activity.

A motivating feminist issue has been the emergence of different working experiences beyond *and* within what counts as 'work' (Cameron and Gibson-Graham, 2003). Therefore, much feminist critique has operated by challenging the spatio-temporal boundaries of work, for example through extending work to include 'social reproduction', as well as by highlighting the ways working activity is intensified through forms of 'emotional labour' (Hochschild, 1983; McDowell, 1991; Pratt, 2004). In questioning what counts as work by foregrounding different and differentiated working practices, feminists have shown how the 'workplace' is ambivalent; it might be a space for affirmation *and* for negation. The 'home' has served as a paradigmatic site in which working experience can be both affirming, for example through the pleasures derived from various labours of love, but also negating, because such work is unremunerated and therefore potentially exploitative (Boris and Parreñas, 2010; Pratt, 2012). Thus feminist demands have both sought affirmative recognition of the category of 'women's work' but have also shown how the differentiated nature of work negates any straightforward categorization. Therefore, feminist critique offers a way of considering the ambivalent emergence of digital work. Feminist approaches aid understanding of how the emergent properties of the digital workplace might result in a fulfilling and inclusive experience but also an exploitative and isolating one. Specifically, *intimacy* is put forward as a frame for a geography of the digital that develops through work's ambivalence. Feminist approaches to intimacy emphasize its ambivalent potential to be productive and destructive, corresponding to the possibilities for digital work to be fulfilling yet exploitative (Pratt and Rosner, 2012). Intimacy occurs through a contradictory spatial sense as private and proximate, public and distant (Berlant, 1998). Thus I argue for a geography of the digital in the extensive and intensive emergence of working space through intimacy.

To know the digital workplace through intimacy is to attune to the ambiguity of its actualization where the establishment, displacement or transcendence of working limits occurs through what it feels like to be doing work. By emphasizing how

---

2   My emphasis is on 'feminism' as a form of political thought extending beyond but arising from (and in turn shaping) experiences of 'women'. This illustrates that what counts as 'femininity', and thus 'women's work', is open to debate. Nonetheless, there is fantastically valuable scholarship by geographers highlighting the continued inequalities, for example in pay, for 'women' in the workplace (e. g. Larner, 1991; Molloy and Larner, 2010; Cox, 1997, 2007; Reimer, 2016; Epstein and Kalleberg, 2004).

knowledge unfolds through experience, an intimate geography of the digital combines theory and practice to put knowing 'working space' together with changing 'the workplace'. The article first substantiates the claim that the geographies of digital work are ambivalent. It shows how digital technologies result in extensions and intensifications of work that hold possibilities for working affirmation and negation. Second, it considers how the direction of feminist critique follows two 'moves' that are articulated through such ambivalences of work extension and intensification. The 'anti-essentialist' move challenges singular locations of work through forms of extension, whilst the 'anti-normative' move questions singular performances of work through modes of intensification. Third, it develops from strands of feminist thought an [|246] intimate geography of digital work. The emergent properties of digital working create space through the intimacies of postwork places where bodies and machines feel the possibilities of being 'at' work.

II GEOGRAPHIES OF DIGITAL WORK

The 'saturation' of space with forms of software, computational systems and devices is increasingly well documented by geographers (see Kinsley, 2014). Economic questions are examined in this scholarship, for example the value of geo-locational data; the productivity of 'smart cities' and the provision of digital infrastructure (e. g. Graham et al., 2012; Kitchin, 2014b). However, the embodied processes by which such digital economies take place receive less investigation. People, and how they live and make a living with and through these technologies, deserve greater attention. This section puts forward the geographies of work as one solution to this problem of the 'relatively few empirical examinations of contemporary digital geographies' (Kinsley, 2014: 368). I outline first how digital technologies extend work beyond the 'firm' workplace and, second, how this results in intensified worker practices to realize working location. I show that such working extensions and intensifications through digital technologies are ambivalent: work can be both a form of self-exploitation and self-fulfilment. These emergent – and thus uncertain – spatial experiences of work through digital technology deserve attention because they are *normal*, not exceptional, as noted by those challenging the novelty of labour precarity (Neilson and Rossiter, 2008) or the 'cybertariat' (Huws, 2003). The stability of the 'standard employment relationship' that reduces acts of making work into the institutions of finding a job have been forms of economic security extended to a 'relatively privileged group of disproportionately White, male workers in the global North' (De Peuter, 2011: 419). Understanding how digital technologies both produce and mitigate such 'precarious' working geographies by folding in life 'beyond' work is therefore vital for approaching economic futures.

1 Extending work

Digital technologies enable work to extend or take place beyond the firm or formal workplaces. In the following I draw out three modes of digital work extension (that are not exhaustive): intermediation, co-creation and multilocation work. First, digital technologies extend the possibilities of 'outsourcing' work through mediating labour markets. This occurs through the parcelling up and apportioning of 'jobs' through digital platforms. These platforms expand on the role of 'labour market intermediaries' in shaping both the supply of workers and the legal conditions of employment within 'subcontracted capitalism' (Wills, 2009; Coe and Jordhus-Lier, 2011; Coe, 2013). Crowdsourcing labour platforms, of which Amazon's Mechanical Turk is paradigmatic, enable companies to obtain labour to undertake 'tasks that could alternatively be performed internally by employees' (Bergvall-Kareborn and Howcroft, 2014: 214). Such leveraging of the crowd can both build on forms of outsourcing, where companies externalize (menial) 'human intelligence tasks' (Mullings, 1999; James and Vira, 2010), but also depart from them where platforms operate peer-to-peer (e. g. the platform TaskRabbit), such that client and service-provider are (theoretically) interchangeable (Bauer and Gegenhuber, 2015). These forms of distributed production heightened by digital technologies constitute the 'openness paradigm' (Ettlinger, 2014: 100) in which a 'new tier in the division of labour facilitated by new communications networks' entails 'unregulated freelance work'. Second, digital technologies extend work through opening up possibilities for forms of 'co-creation'. Rather than work taking place 'in-house', 'value' is created in partnership [|247] between producers and consumers, further blurring any distinction between them. This has long been noted in forms of creative media activity (Banks and Deuze, 2009; Roig et al., 2014) where terms such as 'free labour', 'fan labour' and the 'audience commodity' (Manzerolle, 2010; Terranova, 2000; Scholz, 2013) have been evoked to indicate how 'time spent on Facebook and other corporate platforms is not simply consumption or leisure time, but productive time that generates economic value' (Fuchs, 2014: 98). In addition to digital content production, co-creation might also include activities like online (garment) customization, internet shopping, and self-service check-outs, which mean that consumers might be understood as workers, or as 'working customers' (Cova and Dalli, 2009; Gabriel et al., 2015).

Third, digital technologies enable forms of 'multi-location' work that takes place beyond the 'workplace'. This builds on the growth in teleworking and the home office (Avery and Baker, 2002; Greenhill and Wilson, 2006; Johnson et al., 2007; Steward, 2000; Vartiainen and Hyrkkänen, 2010; Vilhelmson and Thulin, 2001). The possibilities for networked connection mean 'white-collar' workplaces are interstitial, producing 'plural workscapes' (Felstead et al., 2005) that might affect how work is done. Sites such as airports, hotel lobbies and cafes exhibit different 'task-space relationships' in which the character of the space and feature of the work task interact. 'Mobile' and 'global' work (Cohen, 2010; Jones, 2008) illustrate how mobilities can produce highly localized experiences of space (Cresswell, 2011; Hannam et al., 2006; Laurier, 2004) in which workers deploy 'micro-practices' to allow them 'to adapt the spaces […] to their particular work needs and in doing so create a temporary workspace' (Hislop and Axtell, 2009: 72). Taken together, these extensions of work beyond 'firm' location result in greater intensities of activity so that 'space' can be transformed into a 'workplace', examined below.

2 Intensifying work

This section considers the intensifications of work, showing how they can be understood both quantitatively and qualitatively, before emphasizing the ambivalence of emergent working practices wrought by digital technologies. Without static location, a greater number of tasks are required so that space can be 'fixed' for work. For example, such intensification might involve the considerable effort invested by 'multi-location workers' to produce a temporary workplace (Hislop and Axtell, 2009), including the management of spatio-temporal arrangements through the 'perpetual coordination' (Larsen et al., 2008) of interactions with colleagues and clients enabled by digital technologies. Work-life balance policies might be framed as a response to such long-running problems of 'overflow' and intensification of work (Jarvis and Pratt, 2006) that are exaggerated by the logic of continuous connectivity of handheld mobile technologies (Wilson, 2014). These policies have aimed to validate paid work in combination with other activities, including leisure. However, Perrons et al. (2006: 75) claim that 'life' activities are more often constituted by forms of care work, such that balance leads 'to long overall working days rather than the implied harmonious equilibrium'. This problem of 'balance' is further entrenched through digital technologies, where intensification occurs not only through calculable dimensions but through the qualities of 'network time' in which the 'unpredictable' and 'volatile […] temporality of the network [is carried] into almost every aspect of our lives' (Hassan, 2003: 236). Such qualities of intensity that allow the office to be 'always on' (Wainwright, 2010) produce 'variable geometries of connection' (Crang et al., 2006) that challenge the possibilities for precise balance, meaning that if any equilibrium is to be found, it is a chance 'teeter on the brink' (Hassan, 2003: 239). [|248]

The 'network economy and network worker' (Hassan, 2003) then occur through forms of work intensification that involve not only a greater quantity of tasks to 'delimit' the workplace, but also tasks of a different quality. When the workplace might 'crop up' unaccounted for, (potential) workers must 'start engaging in *interpretive (imaginative)* labour' (Graeber, 2015: 101, my emphasis). Rather than 'brought into being by institutionalised frames of action' (p. 99), the workplace is performed by self-organized workers who decide where and when to 'clock off', and also therefore by imagining what might (not) constitute 'working activity'. Such a relationship between worker independence and 'mental labours' (Ross, 2000) of conjuring working possibilities beyond institutional architectures has long been associated with forms of artistic work. The expansion of these *auteur* practices of the artist to a much wider section of the workforce through 'creative work' means:

people increasingly have to become their own micro-structures, they have to do the work of
the structures by themselves, which in turn requires intensive practices of self-monitoring or
'reflexivity'. (McRobbie, 2002: 518)

So the self-organization of work enabled by digital technologies extends the intensive artistic work of sustaining (and promoting) oneself alongside one's art. Digital
technologies deepen the importance of 'netWORKing' (Nardi et al., 2002), and
'professional cool', in the 'foundationless suspense' and 'perpetual anxiety' of
work in the 'knowledge' and 'creative' economies (Liu, 2004: 19). Such 'networking is an additional form of labour that is required to demonstrate ongoing employ-
ability' (Gregg, 2011: 11), when permanent employment is unlikely or undesirable.
This performance of 'cool' amidst insecurities is enhanced by online social networking (such as Facebook and LinkedIn), 'virtualizing' the possibilities for entrepreneurial promotion of the self (Flisfeder, 2015; Gregg, 2009).

These emergent properties of digital work are ambivalent. The extension of
work beyond the formal/institutionalized workplace requires greater intensities of
work that might be both affirming and negating. The possibilities for worker freedoms through digital technologies simultaneously require greater attachments to
work to secure working conditions of possibility – uncertainties that emerge through
the contradictions of 'self-exploitation in response to the gift of autonomy, and
dispensability in exchange for flexibility' (Ross, 2008: 34). Thus in the forms of
co-creative digital work outlined above, a 'Marxist' framing of the exploitative extraction of surplus value through such 'free labour' can also be read with notions of
'affective consumer labour', and the potential affirmation and enjoyment derived
by engaging in such activities (Beverungen et al., 2015; Jarrett, 2016). Equally, the
drive to balance or limit working hours through work-life balance policies sits
alongside the possibility that the 'work world offers a range of consolations when
one's private life may demand more effort and less reward than […] paid pursuits'
(Gregg, 2011: 5). So the creative, affirming possibilities of (self-organized) work
through digital technologies that might 'make-over the world of work into something closer to a life of enthusiasm and enjoyment' (McRobbie, 2002: 521) simultaneously result in working conditions that are 'less just and equal in […] provision
of guarantees' (Ross, 2008: 35). This section has outlined a geography of the digital
economy through a focus on the ambivalent extensions and intensifications of work.
A critical vocabulary for approaching these emergent properties of work, in which
theories for changing work might occur through working practices, is traced in
feminist geographies below.

III FEMINIST GEOGRAPHIES OF WORK

Feminist critique has used the extensions and intensifications of work to outline a
politics of [|249] the emergent properties of working activity by emphasizing their
ambivalence. Recognizing the diversity of feminisms, I draw on the ways such
thought and practice have involved a complex politics of demanding in which
claims for inclusion within the category of 'work' have the potential to both affirm

and negate working activity. Feminist politics has highlighted extensions of work to claim recognition within existing modes of value (namely 'capitalist waged labour'), but also has shown how the intensities of working activity demand alternative understandings beyond the wage relation. Regarding the digital extensions and intensifications outlined above then, a key question explored in feminist thought concerns where and when work that is not recognized as such becomes a space for affirmation or negation. This holds in tension the possibilities and limits of a politics of recognition when work is emergent. Therefore, whilst drawing on feminist traditions within (economic) geography (e. g. Gibson-Graham, 1996, 2006; McDowell, 1991, 1997, 2015; Rose, 1993, 1997; Pollard, 2013), the aim is not 'to reduce feminism down to the identity politics of gender' (Wright, 2010: 60). As should already be apparent, feminist thought might take the situated and embodied struggles of 'women's work' as a starting point, but the orientation of these struggles does not necessarily follow any singularly 'gendered' figure. To illustrate this, I outline two critical 'moves' of feminist politics that are useful for understanding the emergent properties of work. Neither move is intended to speak for the diversity of feminist projects, nor is each without intersections with other forms of politics, as will be indicated. The aim in using these broad brushstrokes is to signal some of the direction of action opened by feminist thought for approaching work.

1 The anti-essentialist move

The anti-essentialist move is one that questions unitary, coherent identity. McDowell (1993: 157) demonstrates this feminist approach in the aim 'to challenge the very nature and construction of that body of knowledge that is designated academic geography'. For some feminists interested in economy, this has meant questioning both the essence of the category of work as well as refusing 'capitalism a planetary identity as the most coherent, powerful force on earth' (Wright, 2010: 61). The reading of feminist anti-essentialism presented here foregrounds the ways this move deals with extensions of work beyond waged labour, or 'organized' production. Drawing on an interpretation of Marx, this might be understood as the way capital attempts to exploit 'absolutely', for example by extending the working day, rather than 'relatively', for example through raising the intensity of labour (Witheford, 1994: 89). To do this, feminist scholarship has questioned the absolute nature of 'production' as the site of value extraction by 'capital'. Accepting the Marxist critique of production as the grounds for exploitation, feminist thought has challenged the assumption of family or domestic labour as 'non-productive' and therefore not a site of exploitation (Dalla Costa and James, 1972). In foregrounding such seemingly non-productive activity, feminists have highlighted the 'relationship between the production of value "at work" and the social reproduction of labour-power' (Mitchell et al., 2003: 415; Bakker, 2007). Thus through child-raising and housework, women's role in the reproduction of labour power is 'a form of exploitation simultaneously distinct from, and complementary to, that of wage labour in production' (Witheford, 1994: 98).

One way of questioning any essential understanding of work has been through strategies that 'read for difference' (Gibson-Graham, 2008), undermining the dominance of conventional framings of labour. An example of this strategy is the tactic of 'renaming' work, such as through the term 'emotional labour' (Hochschild, 1983). Whilst deployed differently, this [|250] term has broadly been used to extend what counts as productive activity, with geographers often focusing on the role of emotions in constituting both remunerated and unremunerated care work (Boyer et al., 2013; Dyer et al., 2008; Huang and Yeoh, 2007). Such an antiessentialist politics is ambivalent: it manifests as both a negation and an affirmation of work. On the one hand, the demand is for forms of (women's) 'nonwork' to count as work such that this activity can be understood as (double) exploitation. Work must be recognized for it to be negated, understood as oppressive and therefore a site of struggle. On the other hand, the demand is (for women) to be counted as workers in order to realize agency in political struggle. Work is affirmed as a means through which to enact potentially subversive politics. Therefore, the anti-essentialist move exposes the 'performance' of work as a category with seeming coherence but without essential (affirming or negating) substance. Work is evoked and occurs with some stability, yet simultaneously extends beyond its boundaries, evading absolute categorization. This raises the question of why, if without essence, such 'performances' of work occur as they do. There is a strange historicity to the act of work that despite its diverse possibilities means that 'the act that one does, the act that one performs, is, in a sense, an act that has been going on before one arrived on the scene' (Butler, 2004: 160).

2 The anti-normative move

Through exposing and challenging 'social norms', the anti-normative move can be understood as a response to this historicity to the act of work. This move foregrounds the 'relation of individuals to productive work' (Donzelot, 1991: 251) when the 'contemporary blurring of work and nonwork is accepted and understood as normal or even positive' (Mitchell et al., 2003: 429). The reading of antinormativity presented here shows how work is intensified (in and beyond the 'workplace') to include all kinds of embodied efforts that might cultivate and be cultivated according to certain standards. This intensification results from the ways these standards, or norms, simultaneously require *and* produce 'work'. Such simultaneous 'production' and 'consumption' occurs because, for Butler (1993), norms are not stable but nonetheless continue to direct work despite the very production (i. e. instability) of their normative nature. In Butler's (1993: 13) words, a norm shapes the (working) subject 'to the extent that it is "cited" as such a norm, but it also derives its power through the citations that it compels'. Thus, in addition to work as an extended relation to organized production, work also becomes an intensive relation to the self through the embodiment and (re)production of norms. The ongoing shaping of the (working) subject occurs through 'work' on the self, such that norms do not 'simply exist [...] "out there"' but instead are involved in 'how bodies *work* and are *worked* upon' (Ahmed, 2004: 145, my emphasis). This normative 'work on the self' as the primary relation of the worker to productive work, similar to the processes of individuation McRobbie (2002) notes as characteristic of creative work, has been recognized as part of wider intensifications of 'work' in 'life'. Such intensification is not simply a quantitative concentration of 'work' tasks that undermines the possibilities of enjoyable 'social' time, as the political claim underpinning demands for work-life balance might suggest (Burchell, 2006).

Intensification is also a qualitative shift in which the conceptual distinction between the two is undone, 'making work itself the territory of the social, the privileged space for the satisfaction of social need' (Donzelot, 1991: 251). This 'socialized worker' (Negri, 1989) is 'socialized' by 'her participation in a far more ramified and expansive system of value creation' in which 'a central element in her work involves *communicative* and *coordinative* tasks' [|251] (Witheford, 1994: 95). Witheford's reading of Negri's position is symptomatic of the way this intensification of work through embodiment of norms can be understood to take on 'feminine' qualities, typified in the sorts of emotive and connective practices required to balance both caregiving and running the household. In this 'homework economy' outside 'the home':

> Work is being redefined as both literally female and feminised, whether performed by men or women. To be feminised means to be made extremely vulnerable; able to be disassembled, reassembled, exploited as a reserve labour force; seen less as workers than servers; subjected to time arrangements on and off the paid job. (Haraway, 1991: 166)

Thus, whilst not necessarily undertaken by women, the intensification of work as the territory of the social normatively constructs work as feminine (and therefore to some degree disempowered, in this extract from Haraway). The 'microstructures' and forms of 'self-exploitation' (McRobbie, 2002) that constitute the normative 'work on the self' required to be a 'socialized worker' are continuations of forms of 'women's work'. Therefore, in considering how norms construct the relationship between working subjects and work, the anti-normative move illustrates the challenges to feminist demands for recognition of and through work. First, as with the anti-essentialist move, the anti-normative move illustrates work's ambivalence: its paradoxical negations and affirmations. Work is shown to be normative, negatively constraining working bodies through forms of self-management that might be oppressive.

Yet the anti-normative moves also illustrate the instability of normative *work on bodies* and therefore the affirmative possibility for shaping *bodies at work* beyond these constraints. This posits 'work' as an activity in which women 'participate in their own representation of themselves while simultaneously rejecting the representation of themselves as exemplars of Woman' (Wright, 1997: 278). Second, in thus focusing on work's emergent intensities, the anti-normative move emphasizes a politics of the *qualities* rather than *quantities* of work. Instead of 'counting in' and 'adding on' (Cameron and Gibson-Graham, 2003) working activities so that they can be recognized as such, the focus is on how work is internally differentiated: it is undertaken through diverse styles and performances that evade absolute

categorization and recognition. So if work is 'a kind of imitation for which there is
no original' (Butler, 1991: 21), the feminist politics lies in dismissing any 'original'
essence of work that could be empowered through recognition, but also in showing
how such appearance of internal essence continues to be differentially produced.
Thus this post-structuralist move to challenge any 'quest for completeness' in work
(Cameron and Gibson-Graham, 2003: 152) might be taken up as a negation (i.e.
normative constraints) or an affirmation (i.e. alternative scripting) of working experience. This is the distinction between the negations of the subject that can be
identified in Butler's reading of Derrida's *différance*, and the affirmations of the
subject to be found in Gibson-Graham's (2008: 614) transformative politics that
might 'bring new worlds into being'. To conclude this section examining how feminist geographies have approached the emergent properties of work, I emphasize
the significance of focusing on the qualities of work. The emergence of the digital
workplace means that what counts within the category of 'work' is open, necessitating a politics located in knowledge of the digital workplace as it unfolds, understood through the geographies of intimacy, as I argue next.

IV INTIMACIES OF DIGITAL WORK

I have outlined the emergent properties of digital work, and how feminist critique
[|252] operates through a politics of work's emergence: its extensions and intensifi-
cations. This section combines these threads to consider how feminist thought can
approach the emergence of work through digital technologies. I foreground feminist
theorizations of intimacy as a critical lens for the geographies of digital work. Feminist approaches to intimacy emphasize its ambivalent potential to be both affirming
and negating. Thus knowing working space through intimacy means paying close
attention to what it feels like to be 'at' work amidst the emergent properties of the
digital workplace. First, intimacy is situated within a wider sense of ambivalence towards work in contemporary feminist thought. In part, this stems from the contradictions of a politics of recognition, where the promise of working 'equality' for women
either fails to be realized or proves insufficient when faced with the excesses of life
beyond straightforward gendered work identities. Instead of trying to balance 'the
work ethic' and 'family values', I illustrate how Weeks' (2011) 'postwork imaginary'
builds on the anti-essentialist move to propose a politics suited to the extensive properties of digital work. Second, I indicate how feminist understandings of 'sexual difference', rather than categories of gendered identity, build on the anti-normative move
to respond to these ambivalences of recognition through work. In emphasizing the
unstable 'matter' of embodiment, such 'sexual difference theories' (e.g. Grosz, 2005)
illustrate the complications of categories of maker and made in working relationships
with machines, necessitating an understanding of digital work intensity as both quantification and qualification of the self. Third, I outline how such experiences of digital
work occur through the spatial sensibilities of intimacy (Berlant, 1998). The intimacies of the digital workplace occur through what it feels like to be at work when
working space is disruptive and mobile.

1 Postwork places

> The home, workplace, market, public arena, the body itself – all can be dispersed and inter-
> faced in nearly infinite polymorphous ways, with large consequences for women and others.
> (Haraway, 1991: 163)

To understand how the intimate might become the critical location of digital work requires focusing on work extension beyond the formal 'workplace'. This builds on the anti-essentialist feminist move in which a politics is constructed through the extensions of 'work' as both part of but also separate from 'life'. Here I consider how the 'feminist insistence on expanding the concept of labour beyond its waged form' (Weeks, 2011: 122) provides a frame suited to the extensive properties of digital work. This means acknowledging the contradictory possibilities of work as a demand *on* time but also, simultaneously, as a demand *for* time. 'Work time' can itself be demanded as a space for desires, wishes and wants, in what Weeks (2011) terms a 'postwork' imaginary. For all that might be understood as monotonous, routine and binding about work, alternative possibilities exist and can be illuminated through critical scrutiny of present working ideals and realities. This postwork project of critique is an affirmative one that both 'refuses the existing world of work that is given to us and also demands alternatives' (2011: 233). This means that demands for and through work can 'be understood as an invocation of the possibility of freedom; [...] as the time and space for invention' (Weeks, 2011: 145). Therefore, Weeks' imaginary is one suited to the emergent properties of digital work because – if 'work' occurs anytime and anywhere – the opportunity arises for a (feminist) politics that cannot be subsumed into the (often competing) discourses of family and work values.

Such a 'postwork' condition extending 'work' beyond the formal 'workplace' can be understood in (at least) two ways in relation to [|253] digital technologies. One is to find opportunity in the 'deskilling' and unemployment that some worry will result from digital technologies as the latest wave of 'automation'. Rather than a source of job losses, automation becomes the grounds for demands for full unemployment and therefore an apparently 'postwork' condition in the temporal sense. This would be enabled, according to Srnicek and Williams (2015), by some form of Universal Basic Income[3] and, perhaps unsurprisingly, would not mean no 'work' per se, but rather a revaluation of work based on what it involves. Boring and unattractive work would have to be better paid, whereas attractive work that people want to do would need less remuneration. Another 'postwork' expression through digital technologies is perhaps more nuanced. Rather than seeking full unemployment, digital technologies open up possibilities for work time to be more creative. The potential for 'smart machines' to 'increase the intellectual content of work'

---

3   Other terms are used, but the principle is that work does not form the main mechanism for
    distribution of income in society. For example see Hardt and Negri's (2009: 380) call for a
    'basic income sufficient for the necessities of a productive dignified existence' and Ferguson's
    (2015) discussion of a politics of distribution (as opposed to production) in relation to social
    welfare programmes in southern Africa.

(Zuboff, 1988: 243) enables distributed decision-making that goes beyond 'responsible autonomy' (Friedman, 1977) of workers within organizations. Instead, work
occurs independently through semi-institutionalized attachments to 'entrepreneurial' practices of starting-up (Cockayne, 2015) and the broader informalities of
'maker cultures' utilizing small-scale digital ('additive') manufacturing technologies (Richardson, 2016; Rosner and Turner, 2015) that have evolving logics of circulation (Birtchnell and Urry, 2013a, 2013b). Here, the digital is held to enable a
democratization of desirable creative work, although still by no means universally
available.

    Thus elements of the 'postwork' imaginary might be far-fetched, but this utopi-
anism is deliberate. For Weeks' feminist politics, the aim is to move beyond the
constraints of existing categorizations of working hours to enable people 'to imagine and explore alternatives to the dominant ideals of family form, function and
division of labour' (Weeks, 2011: 170). This seeks to pick up from forms of feminist
utopianism that declined from the 1970s and gave way, according to Weeks (2011:
184), to a more limited politics of recognition:

> the aspiration to move beyond gender as we know it was supplanted by efforts to secure the
> recognition and equal treatment of a wider variety of the genders we now inhabit; the project
> of 'smashing the family' and seeking alternatives was largely abandoned in favour of achieving
> a more inclusive version of the still privatised model; postwork militancy was eclipsed by the
> defence of the equal right to work balanced with family.

To rectify this, a (feminist) politics of work needs to be *more*, not less, demanding,
redirecting 'our attention and energies toward an open future' (2011: 206). Utopian
demands for work beyond the regularities of the workplace can feed playful im-
provisation and performance of differently desirable working practices that might
turn into wider realities. In such postwork possibilities enabled by digital technologies the distinction between own time and owned time, between the personal and
the professional, is necessarily blurred. So, if digital technologies mean work can
become a creative, lively practice that extends the 'workplace', it is possible to see
how this working geography might become an 'intimate' one. The postwork imaginary offers a framework for the ambivalence of the extensive properties of digital
work that, whilst potentially resulting in more 'work' time, might also mean more
desirable activity, pointing to the need to examine the differing intensities of such
working experiences, taken up below.

2 Bodies and machines 'at' work

> Ambivalence towards the disrupted unities mediated by high-tech culture requires not sorting
> consciousness into categories [...] but subtle understanding of emerging pleasures, experi-
> ences, and powers with serious potential for changing the rules of the game. (Haraway, 1991:
> 172–3) [|254]

As well as extending the location of 'work' into 'life', intimate geographies of the
digital also intensify individual working experience. This picks up the anti-normative feminist move connecting the ways work is categorized (and 'represented')

with differently embodied working experiences. In these relationships between work on bodies and bodies at work through digital technologies, the category of 'women's work' holds an ambivalent position. Technological advancements erase divisions between male and female work, as hypothesized by Marx and Engels (1967: 88): 'the more modern industry becomes developed, the more is the labour of men superseded by that of women. Differences of age and sex have no longer any distinctive validity for the working class.'

Equally, though, by opening possibilities for the self-organization of work digital technologies encourage flexible, relational practices – negotiations and compromises that are forms of 'soft mastery' (Turkle, 1997: 56) often associated with normative working styles of women. Turkle (1997) has argued that whilst such 'soft' skills with and through technologies are not unique to either gender, they are nonetheless styles 'to which many women are drawn'. Given these ambivalences of the category of 'women's work' in relation to digital technologies, I suggest that experiences of digital work might be understood through the intensities of 'sexual difference'. Rather than framing the experience of work as a priori 'male' or 'female', foregrounding notions of 'sexual difference' means focusing on how working processes are indeterminate, exploratory, often transforming senses of gendered (and/or working) identities. In this understanding, gender identity is an effect, not a cause, of working experiences, similar to Butler's model of performativity.

However, in contrast to Butler's theorization of the production of gender, sexual difference theories emphasize the biological and/or nonhuman constitution of 'matter', characteristic of what has been termed 'new materialist' feminism (see Barad, 2003; Colls, 2012; Hird, 2009; Fannin et al., 2014; Jagger, 2015; Kirby and Wilson, 2011). Broadly, the contention is that Butler's account of materiality overemphasizes 'culture/discourse' and therefore does not upset gender enough; it does not focus on the (biological) instability of 'matter/bodies'.[4] Instead, through emphasizing the changing materiality of bodies,[5] sexual difference unsettles the boundaries between the human and nonhuman, organism and machine, in ways that are useful for conceptualizing the experience of digital work.

By foregrounding the role of technology in working experiences, digital work magnifies how the worker is both creator and created through relationships with machines. In forms of 'artistic' work such as music-making and photography, (digital) technology might be held to function in a mimetic capacity, operating as a

---

4   For an insight into these debates see Ahmed's (2008) discussion of 'new materialist' claims that feminism is 'anti-biology'. She argues (in an unapologetically provocative manner) that the 'new materialist' scholarship risks turning 'matter' into a 'fetish object'. Instead she suggests that there is a need to at least partially let go of these theoretical 'objects' to find out what happens when (theoretical) things fall apart, given that what counts as biology and materiality within the sciences is itself disputed.

5   Noting that there are significant philosophical disparities between the 'materialisms' under the banner of this 'new' feminist scholarship, e.g. see Hein's (2016) discussion of immanence in Deleuze (as taken up in Grosz's vitalist approach) set against the transcendence of Barad's performativity of matter.

'mirror' of the observing subject in the world. In less 'creative' work, (digital) technologies objectify the worker, often through processes of observation and quantification now heightened through 'wearables' and other selftracking devices (Moore
and Robinson, 2015). So in these human-nonhuman relations, the difference between the maker and the made becomes uncertain, meaning that 'what people are
experiencing is not transparently clear' (Haraway, 1991: 173). As well as potential
tools of control, the 'enchanting' and 'romantic automatism' (Sussman, 1999;
Turner, 2008) of working technologies result in 'close, sensuous and relational'
(Turkle, 1997: 62) engagements with the worker. Thus, digital work necessitates
careful attention to the co-constitutive experiences of worker with technology, com-
plicating 'simple models of subject and object' (Crang, 1997: 366).

In focusing on processes that exceed categorization, sexual difference emphasizes a qualitative approach to the intensities of work experience, holding open
working subjectivity [|255] through these worker-technology relations. Grosz
(2005: 160) argues that:

> sexual difference is not a measurable difference between two given, discernible, different
> things – men and women, for example – but an incalculable process, not something produced
> but something in the process of production.

The provisional nature of the sexed and/or working body that appears in this account requires understandings of the 'intensity' working space that combine the
objective and the subjective, appropriate for digital technologies. As well as being
intensive – comprising dense spatio-temporal concentrations of tasks – digital work
is also *intense* – comprising degrees of directed attention through and towards
tasks. Digital working space is objectively intensive through high volumes of jobs
and subjectively intense through the variegated experiences of their undertaking.
This has implications for how 'quantification' in digital work is understood. Whilst
there may be a proliferation of (wearable) tools that can package, monitor and direct
forms of embodied skill, often through seemingly relentless 'notifications', such
work also retains elements of the 'unpredictable and unforeseeable' (Grosz, 2005:
189). As Swan (2013) argues, the 'quantified self' is also qualitative, combining
objective data metrics with the subjective experiences of the impact of this data.
Such quantifications are entwined with qualifications of the self as workers sense
and interpret their environment differently, making adjustments to their incorpora-
tion of working tasks. So by creating more tasks that are further proliferated through
the variations in how they take place, accounts of digital work intensities must re-
tain the qualitative 'exploration of difference' (Grosz, 2005: 190) in interaction with
processes of quantification. I now turn to the spatial senses of intimacy through
which these working experiences unfold.

3 Feeling working space

> There is no drive in cyborgs to produce total theory, but there is an intimate experience of
> boundaries, their construction and deconstruction. (Haraway, 1991: 181)

As both personal and professional, private and public, these intimacies of being
'at' work with digital technologies require an approach to 'collectively building
effective theories' of space that emphasize 'what it means to be embodied in high-
tech worlds' (Haraway, 1991: 173). I outline here how these intimate geographies
of the digital occur through what it feels like to be at work. Digital technologies
exaggerate the potential inconsistencies between being 'at' work and 'doing' work,
meaning working space is constituted by combinations of 'objective' fixed working location and the 'subjective' senses of work taking place. To label this 'workspace' intimate is to emphasize the ambivalent feelings of proximity and distance,
connection and disconnection, that constitute experiences of working with and
through digital technologies. Gregg (2011: 1) discusses such uncertainties of
'work's intimacies' through her notion of 'presence bleed' in which digital technologies challenge conventions of availability when 'firm boundaries between
personal and professional identities no longer apply'. In such digital work she
shows how there are:

> opportunities for connection, community and solidarity, generating relationships that compli-
> cate what we mean by the notion of friendship. At the same time however, technology has
> played on feelings of instability, threat and fear among workers facing an unstable employment
> landscape and the death of the linear career path. (2011: 8)

Here, the intimate appears as both a location beyond the formal workplace and a
style of working experience in connection with technology. The 'intimate' is a more
or less fixed [|256] spatial *object*, a sphere in which personal roles are negotiated
and reconfigured. Equally, 'intimacy' constitutes *subjective* processes of work taking place through the forming of attachments beyond the self, including with/
through technologies.

Geographies of intimacy[6] might then trace the parameters and follow the movements of this 'subject object' (Suchman, 2011), this disruptive and mobile working
space, to understand the ambivalences of digital work. Thinking through this, a first
step takes the disruptions of the intimate sphere to consider how digital working
space is not a bounded, contained location. Feminist approaches to the 'intimate
sphere' highlight its constitution through forms of external penetration that challenge senses of proximity as quantifiably nearby. The closeness of intimacy can be
packed with explosive feelings of, and 'unbearable' compulsions to, distance

---

6   These complexities of intimacy have been of interest to geographers looking to connect the
    encounters through which embodied experiences of race, sexuality and so on are played out
    singularly, yet connect with broader societal 'aggregations' and inequalities cut along such
    'cultural' lines (e. g. Nayak, 2011; Saldanha, 2010; Valentine, 2008). This 'intimate turn' in-
    volves a complex 'politics of proximities' that highlights 'potential shifts in the boundaries of
    what counts as political subject matter' (Price, 2013: 578), including through a reorientation of
    the geopolitical (Pain, 2015).

(Berlant and Edelman, 2014). Thus occupying and being occupied 'in' the intimate sphere occurs through complex senses of shared working space through digital technologies that might involve, for example, experiences of regulation and exploitation. As a biopolitical project (Oswin and Olund, 2010), the intimate sphere is governed to shape the working subject, such as through the wearable technologies mentioned above that seek to optimize embodied performances of tasks in and beyond the 'workplace'. As a commercial project (Boris and Parreñas, 2010), the intimate sphere is exploited, undermining the authentic (working) subject, for example through the performances of domesticity and hospitality required by AirBnB hosts (Molz, 2012). In these cases, working space is felt through disruption of the boundaries of the intimate sphere through digital technologies.

A second step examines the movements of intimacy to understand how working space (re)configures feelings of continuous connection to and through digital work. Connecting with work – understood as knowledge of working activity built through 'close' association and repeat 'encounter' – takes on a mobility through the emergent properties of digital work. The attachments to and encounters with work through digital technologies constitute working connections that travel, mingling with practices that make other places familiar, such as cafes, bedrooms and parks. These mobile connections with work possess an amorphous, spreading presence so that working space has provisional dimensions that evade leading 'to a stabilising something, something institutional' (Berlant, 1998: 287). Thus through these working connections, digital work's intimacy has relational coordinates that resist a static geometry, meaning that working space shapes and is shaped by the wider environment. Intimacy fails to take fixed parameters, to solidify, and instead is 'portable, unattached to a concrete space', and is in fact a 'drive that creates spaces around it' (Berlant, 1998: 284). This moving and enveloping intimacy with work through digital technologies might be sensed through an 'aesthetic of attachment' but without 'inevitable forms or feelings' (1998: 285).

Together then, these disruptive and mobile intimacies of digital working space might be both overwhelming and desirable (Berlant, 2016), requiring close attention to what it feels like to be working. Being bound to work through technology is simultaneously an attachment to working freedoms; to the promise that these digital tools will bring working activities and forms of connection that are better for the self beyond work. As well as disrupting boundaries, this intimate geography involves processes of orientation and disorientation (Ahmed, 2006), a 'feeling out' (Berlant, 2011: 17) of what work might become and what might become work through these extensions and intensifications. Social media operates through such intimate orientations, where personal 'status' and 'feeds' are forms of production and consumption that can simultaneously direct [|257] work (e.g. through self-promotion and 'networking') but also send working activity astray (e.g. 'scrolling' as shifting registers of distracted interaction). In such attachments to diffuse updates and notifications through embodied engagements with technologies, intimacies are where work might be shared and where sharing occurs through work. So as well as focusing on disruptive 'spheres' of work/life interaction, these intimate geographies of the digital also highlight mobile quasi-institutional spatial forms, felt for example in co-working spaces,[7] that operate as shared 'micro-structures' (McRobbie, 2002) sustaining attachments to and through working futures. Working with and through digital technologies then occurs through the intimacies of work taking place that might produce experiences of fulfilment, exploitation, or both simultaneously.

V CONCLUSION

Feminist critique provides an analytical means to situate work as an empirical focus for digital geographies. Geographers can contribute to debates on the digital economy by examining how working activities take place through and with digital technologies. These technologies extend and intensify work, rendering the boundaries of the workplace emergent. This heightens the ambivalence of work, and the possibilities for working experience to be that of affirmation and/or negation. Building on anti-essentialist and anti-normative politics, the utility of feminist thought for approaching such extensions and intensifications of work through digital technologies has been shown through the lens of intimacy. The uncertainties over when and with what implications working space becomes workplace, or open connection becomes rule-governed network, requires focusing on what it feels like to be doing work. Knowledge of this intimate geography of the digital combines theory and practice to produce descriptions of the workplace that might themselves become culturally creative acts, to go 'ahead of social practices in order to open a field for them' (De Certeau, 1984: 125). I point to three areas for the continuation of this project of knowing and changing the 'digital workplace' through intimacy.

First, 'postwork' places require further examination to understand the relationship between workplace and worker. Whether 'flexible' or 'insecure', digital workplace and worker are engaged in nuanced mimetic relations. The complex adaptive functions of mimicry, producing contrary effects of travesty, camouflage and intimidation (Lacan, 1979: 99), demand a focus on what workplace representation for/of the worker might mean with digital technologies. The mobility of the workplace and the mix of differently orientated workers that occupy it undermine existing attempts to regulate and manage the quantities and intensities of work, for example through work-life balance policies. Examining the implications of the extending digital workplace for the worker would contribute to debates concerning the working conditions and worker wellbeing at different ends of the income spectrum. Equally, for those interested in measures of 'productivity' and their investigation, there are likely implications for the quality and quantity of work undertaken through these postwork places.

Second, the prevalence of different combinations of bodies and machines 'at' work through digital technologies necessitates greater understanding of evolving forms of worker embodiment. People engage in a range of intensities of interaction with, supervision over and direction by technologies that might make work appear

---

7   For London examples of these spaces see the GLA's 2014 report 'Supporting Places of Work: Incubators, Accelerators and Co-working Spaces'.

closer or further away. These ambiguous proximities with work through machines, and the implications of work disappearing at a distance (Suchman, 1995), raise questions concerning understandings of skill. If skill is something of the activity of doing work, digital technologies result in bodily elaborations in close connection [|258] with machines that might limit or displace past working 'skills', whilst proliferating the informational content of work (Berardi, 2009). This shifts understandings of human labour with machines from 'the mechanistically automated to the electronically autonomous' (Stacey and Suchman, 2012: 28), perhaps resulting in job losses but also in new styles of working. The question of what constitutes skill in these experiences of proximity through digital work, and how workers train for and improvise through such working environments, is an area ripe for geographical investigation.

Third, the idea and practice of 'feeling working space' opens up further questions concerning politics. The immanent feeling out of digital work provides opportunities for working differently but is also problematic. Employment becomes a nebulous category as 'entrepreneurship' and 'self-employment' expand, and the 'old word unemployment' is no longer recognizable 'in the scene that word named for so long' (Derrida, 1994: 101). Digital technologies might therefore exaggerate the discrepancies between what it feels like to be at work and the distribution of income in society. Greater understanding of the emergent properties of the 'digital workplace' therefore seems vital for those seeking a politics that challenges contemporary and future relationships between work and income.

Acknowledgements

I thank the editors and reviewers, whose attentive comments have been very helpful for my (still sometimes failed) attempts to clarify the argument and contribution, together with Tara Cookson, Sarah Radcliffe and Carmen Teeple Hopkins who provided valuable feedback on an early draft. I am very grateful to Daniel Cockayne for a critical but generous reading of some nascent muddled thoughts. I also acknowledge and thank David Bissell for sharing with me a currently unpublished paper on skills for future environments, which informed my thinking in the revision of the essay. [|259]

Declaration of conflicting interests

The author(s) declared no potential conflicts of interest with respect to the research, authorship, and/or publication of this article.

Funding

The author(s) received no financial support for the research, authorship, and/or publication of this article.

REFERENCES

Ahmed S (2004) *The Cultural Politics of Emotion*. Edinburgh: Edinburgh University Press.

Ahmed S (2006) *Queer Phenomenology: Orientations, Objects, Others*. Durham: Duke University Press.

Ahmed S (2008) Open forum imaginary prohibitions: Some preliminary remarks on the founding gestures of the 'New Materialism'. *European Journal of Women's Studies* 15: 23–39.

Ash J (2013) Rethinking affective atmospheres: Technology, perturbation and space times of the non-human. *Geoforum* 49: 20–28.

Avery G and Baker E (2002) Reframing the infomated household-workplace. *Information and Organization* 12: 109–134.

Bakker I (2007) Social reproduction and the constitution of a gendered political economy. *New Political Economy* 12: 541–556.

Banks J and Deuze M (2009) Co-creative labour. *International Journal of Cultural Studies* 12: 419–431.

Barad K (2003) Posthumanist performativity: Toward an understanding of how matter comes to matter. *Signs* 28: 801–831.

Bauer RM and Gegenhuber T (2015) Crowdsourcing: Global search and the twisted roles of consumers and producers. *Organization* 22: 661–681.

Beck U (2000) *The Brave New World of Work*. Cambridge: Polity Press.

Berardi F (2009) *The Soul at Work: From Alienation to Autonomy*. Los Angeles: Semiotext(e).

Bergvall-Kareborn B and Howcroft D (2014) Amazon Mechanical Turk and the commodification of labour. *New Technology, Work and Employment* 29: 213–223.

Berlant L (1998) Intimacy: A special issue. *Critical Inquiry* 24: 281–288.

Berlant L (2011) *Cruel Optimism*. Durham: Duke University Press.

Berlant L (2016) The commons: Infrastructures for troubling times. *Environment and Planning D: Society and Space* 34: 393–419.

Berlant L and Edelman L (2014) *Sex, or the Unbearable*. Durham: Duke University Press.

Beverungen A, Bohm S and Land C (2015) Free labour, social media, management: Challenging Marxist organization studies. *Organization Studies* 36: 473–489.

Birtchnell T and Urry J (2013a) 3D, SF and the future. *Futures* 50: 25–34.

Birtchnell T and Urry J (2013b) Fabricating futures and the movement of objects. *Mobilities* 8: 388–405.

Boris E and Parreñas R (eds) (2010) *Intimate Labors: Cultures, Technologies, and the Politics of Care*. Stanford: Stanford University Press.

Boyer K, Reimer S and Irvine L (2013) The nursery workspace, emotional labour and contested understandings of commoditised childcare in the contemporary UK. *Social & Cultural Geography* 14: 517–540.

Burchell B (2006) Work intensification in the UK. In: Perrons D, Fagan C, McDowell L, Ray K and Ward K (eds) *Gender Divisions and Working Time in the New Economy: Changing Patterns of Work, Care and Public Policy in Europe and North America*. Northampton, MA: Edward Elgar, 21–34.

Butler J (1991) Imitation and gender subordination. In: Fuss D (ed.) *Inside/Out: Lesbian Theories, Gay Theories*. New York: Routledge, 13–30.

Butler J (1993) *Bodies that Matter*. London: Routledge.

Butler J (2004) Performative acts and gender constitution: An essay in phenomenology and feminist theory. In: Bial H (ed.) *The Performance Studies Reader*. London: Routledge.

Cameron J and Gibson-Graham JK (2003) Feminising the economy: Metaphors, strategies, politics. *Gender, Place & Culture* 10: 145–157.

Cockayne D (2015) Entrepreneurial affect: Attachment to work practice in San Francisco's digital media sector. *Environment and Planning D: Society and Space*. DOI:10.1177/0263775815618399.

Coe NM (2013) Geographies of production III: Making space for labour. *Progress in Human Geography* 37: 271–284. [|260]

Coe NM and Jordhus-Lier D (2011) Constrained agency? Re-evaluating the geographies of labour. *Progress in Human Geography* 35: 211–233.

Cohen R (2010) Rethinking 'mobile work': Boundaries of space, time and social relation in the working lives of mobile hairstylists. *Work, Employment & Society* 24: 65–84.

Colls R (2012) Feminism, bodily difference and non-representational geographies: Feminism, bodily difference and non-representational geographies. *Transactions of the Institute of British Geographers* 37: 430–445.

Cova B and Dalli D (2009) Working consumers: The next step in marketing theory? *Marketing Theory* 9: 315–339.

Cox R (1997) Invisible labour: Perceptions of paid domestic work in London. *Journal of Occupational Science* 4: 62–67.

Cox R (2007) The au pair body: Sex object, sister or student? *European Journal of Women's Studies* 14: 281–296.

Crang M (1997) Picturing practices: Research through the tourist gaze. *Progress in Human Geography* 21: 359–373.

Crang M, Crosbie T and Graham S (2006) Variable geometries of connection: Urban digital divides and the uses of information technology. *Urban Studies* 43: 2551–2570.

Cresswell T (2011) Mobilities I: Catching up. *Progress in Human Geography* 35: 550–558.

Dalla Costa M and James S (eds) (1972) *The Power of Women and the Subversion of the Community*. Bristol: Falling Wall Press.

De Certeau M (1984) *The Practice of Everyday Life*. Berkeley: University of California Press.

Del Casino VJ (2015) Social geographies II: Robots. *Progress in Human Geography*. DOI: 10.1177/0309132515618807.

De Peuter G (2011) Creative economy and labor precarity: A contested convergence. *Journal of Communication Inquiry* 35: 417–425.

Derrida J (1994) *Spectres of Marx*. London: Routledge.

Donzelot J (1991) Pleasure in work. In: Burchell G, Gordon C and Miller P (eds) *The Foucault Effect*. Chicago: University of Chicago Press, 251–280.

Dyer S, McDowell L and Batnitzky A (2008) Emotional labour/body work: The caring labours of migrants in the UK's National Health Service. *Geoforum* 39: 2030–2038.

Elwood S and Leszczynski A (2011) Privacy, reconsidered: New representations, data practices, and the geoweb. *Geoforum* 42: 6–15.

Epstein C and Kalleberg A (2004) *Fighting for Time: Shifting Boundaries of Work and Social Life*. New York: Russell Sage Foundation.

Ettlinger N (2014) The openness paradigm. *New Left Review* 89: 89–100.

Fannin M, MacLeavy J, Larner W and Wang W (2014) Work, life, bodies: New materialisms and feminisms. *Feminist Theory* 15: 261–268.

Felstead A, Jewson N and Walters S (2005) *Changing Places of Work*. Basingstoke: Palgrave Macmillan.

Ferguson J (2015) Give a Man a Fish: Reflections on the New Politics of Distribution. Durham: Duke University Press.

Flisfeder M (2015) The entrepreneurial subject and the objectivization of the self in social media. *South Atlantic Quarterly* 114: 553–570.

Friedman A (1977) *Industry and Labour*. Basingstoke: Macmillan.

Fuchs C (2014) Digital prosumption labour on social media in the context of the capitalist regime of time. *Time & Society* 23: 97–123.

Gabriel Y, Korczynski M and Rieder K (2015) Organizations and their consumers: Bridging work and consumption. *Organization* 22: 629–643.

Gibson-Graham JK (1996) *The End of Capitalism (As We Knew It): A Feminist Critique of Political Economy*. Minneapolis: University of Minnesota Press.

Gibson-Graham JK (2006) *A Postcapitalist Politics*. Minneapolis: University of Minnesota Press.

Gibson-Graham JK (2008) Diverse economies: Performative practices for 'other worlds'. *Progress in Human Geography* 32: 613–632.

GLA (Greater London Authority) (2014) *Supporting places of work: Incubators, accelerators and coworking spaces*. Available at: http://www.london.gov.uk/sites/default/files/supporting_places_of_work_-_iacs.pdf (accessed 18 October 2016).

Graeber D (2015) *The Utopia of Rules: On Technology, Stupidity, and the Secret Joys of Bureaucracy*. London: Melville House.

Graham M (2011) Time machines and virtual portals: The spatialities of the digital divide. *Progress in Development Studies* 11: 211–227.

Graham M, Hale S and Stephens M (2012) Featured graphic: Digital divide: The geography of internet access. *Environment and Planning A* 44: 1009–1010. [|261]

Greenhill A and Wilson M (2006) Haven or hell? Telework, flexibility and family in the e-society: A Marxist analysis. *European Journal of Information Systems* 15: 379–388.

Gregg M (2009) Banal Bohemia: Blogging from the ivory tower hot-desk. *Convergence: The International Journal of Research into New Media Technologies* 15: 470–483.

Gregg M (2011) *Work's Intimacy*. Cambridge: Polity.

Grosz E (2005) *Time Travels: Feminism, Nature, Power*. Durham: Duke University Press.

Hannam K, Sheller M and Urry J (2006) Editorial: Mobilities, immobilities and moorings. *Mobilities* 1: 1–22.

Haraway D (1991) *Simians, Cyborgs and Women: The Reinvention of Nature*. New York: Free Association Books.

Hardt M (1999) Affective labor. *Boundary* 2 26: 89–100.

Hardt M and Negri A (2009) *Commonwealth*. Cambridge, MA: Belknap.

Hassan R (2003) Network time and the new knowledge epoch. *Time & Society* 12: 225–241.

Hein S (2016) The new materialism in qualitative inquiry: How compatible are the philosophies of Barad and Deleuze? *Cultural Studies Critical Methodologies* 16: 132–140.

Hird M (2009) Feminist engagements with matter. *Feminist Studies* 35: 329–346.

Hislop D and Axtell C (2009) To infinity and beyond?: Workspace and the multi-location worker. *New Technology, Work and Employment* 24: 60–75.

Hochschild A (1983) *The Managed Heart: The Commercialization of Human Feeling*. Berkeley: University of California Press.

Huang S and Yeoh B (2007) Emotional labour and transnational domestic work: The moving geographies of 'maid abuse' in Singapore. *Mobilities* 2: 195–217.

Huws U (2003) *The Making of a Cybertariat: Virtual Work in a Real World*. New York: Monthly Review Press.

Jagger G (2015) The new materialism and sexual difference. *Signs* 40: 321–342.

James A and Vira B (2010) 'Unionising' the new spaces of the new economy? Alternative labour organising in India's IT enabled services–business process outsourcing industry. *Geoforum* 41: 364–376.

Jarrett K (2016) *Feminism, Labour and Digital Media: The Digital Housewife*. London: Routledge.

Jarvis H and Pratt A (2006) Bringing it all back home: The extensification and 'overflowing' of work. *Geoforum* 37: 331–339.

Johnson L, Andrey J and Shaw S (2007) Mr. Dithers comes to dinner: Telework and the merging of women's work and home domains in Canada. *Gender, Place & Culture* 14: 141–161.

Jones A (2008) The rise of global work. *Transactions of the Institute of British Geographers* 33: 12–26.

Kinsley S (2014) The matter of 'virtual' geographies. *Progress in Human Geography* 38: 364–384.

Kirby V and Wilson E (2011) Feminist conversations with Vicki Kirby and Elizabeth A. Wilson. *Feminist Theory* 12: 227–234.

Kirsch S (2014) Cultural geography II: Cultures of nature (and technology). *Progress in Human Geography* 38: 691–702.

Kitchin R (2013) Big data and human geography: Opportunities, challenges and risks. *Dialogues in Human Geography* 3: 262–267.

Kitchin R (2014a) Big data, new epistemologies and paradigm shifts. *Big Data & Society* 1. DOI: 10.1177/2053951714528481.

Kitchin R (2014b) The real-time city? Big data and smart urbanism. *GeoJournal* 79: 1–14.

Kitchin R and Dodge M (2011) *Code/Space: Software and Everyday Life*. Boston, MA: MIT Press.

Kleine D (2013) *Technologies of Choice? ICTs, Development, and the Capabilities Approach*. Boston, MA: MIT Press.

Kwan M-P (2002) Feminist visualization: Re-envisioning GIS as a method in feminist geographic research. *Annals of the Association of American Geographers* 92: 645–661.

Kwan M-P (2007) Affecting geospatial technologies: Toward a feminist politics of emotion. *The Professional Geographer* 59: 22–34.

Lacan J (1979) *The Four Fundamental Concepts of Psycho-analysis*. London: Penguin.

Larner W (1991) Labour migration and female labour: Samoan women in New Zealand. *Journal of Sociology* 27: 19–33.

Larsen J, Urry J and Axhausen K (2008) Coordinating face-to-face meetings in mobile network societies. *Information, Communication & Society* 11: 640–658.

Laurier E (2004) Doing office work on the motorway. *Theory, Culture & Society* 21: 261–277.

Leszczynski A (2014) Spatial media/tion. *Progress in Human Geography*. DOI: 10.1177/0309132514558443.

Leszczynski A and Elwood S (2015) Feminist geographies of new spatial media: Feminist geographies of new [|262] spatial media. *The Canadian Geographer / Le Géographe canadien* 59: 12–28.

Liu A (2004) *The Laws of Cool: Knowledge Work and the Culture of Information*. Chicago: University of Chicago Press.

Manzerolle V (2010) Mobilising the audience commodity: Digital labour in a wireless world. *Ephemera: theory and politics in organization* 10: 455–469.

Marx K and Engels F (1967) *The Communist Manifesto, trans. Moore S*. London: Penguin Books.

McDowell L (1991) Life without father and Ford: The new gender order of post-Fordism. *Transactions of the Institute of British Geographers* 16: 400–419.

McDowell L (1993) Space, place and gender relations: Part I. Feminist empiricism and the geography of social relations. *Progress in Human Geography* 17: 157–179.

McDowell L (1997) *Capital Culture: Gender at Work in the City*. Oxford: Blackwell.

McDowell L (2015) Poepke Lecture in Economic Geography: The lives of others: Body work, the production of difference, and labor geographies. *Economic Geography* 91: 1–23.

McRobbie A (2002) Clubs to companies: Notes on the decline of political culture in speeded up creative worlds. *Cultural Studies* 16: 516–531.

Mitchell K, Marston S and Katz C (2003) Life's work: An introduction, review and critique. *Antipode* 35: 415–442.

Molloy M and Larner W (2010) Who needs cultural intermediaries indeed?: Gendered networks in the designer fashion industry. *Journal of Cultural Economy* 3: 361–377.

Molz J (2012) CouchSurfing and network hospitality: 'It's not just about the furniture'. *Hospitality & Society* 1: 215–225.

Moore P and Robinson A (2015) The quantified self: What counts in the neoliberal workplace. *New Media and Society*. DOI: 10.1177/1461444815604328.

Mullings B (1999) Sides of the same coin?: Coping and resistance among Jamaican data-entry operators. *Annals of the Association of American Geographers* 89: 290–311.

Nardi B, Whittaker S and Schwarz H (2002) NetWORKers and their activity in intensional networks. *Computer Supported Cooperative Work* 11: 205–242.

Nayak A (2011) Geography, race and emotions: Social and cultural intersections. *Social & Cultural Geography* 12: 548–562.

Negri A (1989) *The Politics of Subversion: A Manifesto for the Twenty-First Century*. Cambridge: Polity.

Neilson B and Rossiter N (2008) Precarity as a political concept, or, Fordism as exception. *Theory, Culture & Society* 25: 51–72.

Oswin N and Olund E (2010) Governing intimacy. *Environment and Planning D: Society and Space* 28: 60–67.

Pain R (2015) Intimate war. *Political Geography* 44: 64–73.

Perrons D, Fagan C, McDowell L, Ray K and Ward K (eds) (2006) *Gender Divisions and Working Time in the New Economy: Changing Patterns of Work, Care and Public Policy in Europe and North America.* Northampton, MA: Edward Elgar.

Pollard J (2013) Gendering capital: Financial crisis, financialization and (an agenda for) economic geography. *Progress in Human Geography* 37: 403–423.

Pratt G (2004) *Working Feminism.* Philadelphia: Temple University Press.

Pratt G (2012) *Families Apart: Migrant Mothers and the Conflicts of Labor and Love.* Minneapolis: University of Minnesota Press.

Pratt G and Rosner V (eds) (2012) *The Global and the Intimate: Feminism in Our Time.* New York: Columbia University Press.

Price PL (2013) Race and ethnicity II: Skin and other intimacies. *Progress in Human Geography* 37: 578–586.

Reimer S (2016) 'It's just a very male industry': Gender and work in UK design agencies. *Gender, Place and Culture* 23: 1033–1046.

Richardson M (2016) Pre-hacked: Open design and the democratisation of product development. *New Media and Society* 18: 653–666.

Roig A, San Cornelio G, Sanchez-Navarro J and Ardevol E (2014) 'The fruits of my own labor': A case study on clashing models of co-creativity in the new media landscape. *International Journal of Cultural Studies* 17: 637–653.

Rose G (1993) *Feminism and Geography: The Limits of Geographical Knowledge.* Cambridge: Polity.

Rose G (1997) Situating knowledges: Positionality, reflexivities and other tactics. *Progress in Human Geography* 21: 305–320.

Rose G (2015) Rethinking the geographies of cultural 'objects' through digital technologies: Interface, network and friction. *Progress in Human Geography.* DOI: 10.1177/0309132515580493.

Rosner D and Turner F (2015) Theaters of alternative industry: Hobbyist repair collectives and the legacy [|263] of the 1960s American counterculture. In: Plattner H, Meinel C and Leifer L (eds) *Design Thinking Research.* New York: Springer International, 59–69.

Ross A (2000) The mental labour problem. *Social Text* 63: 1–31.

Ross A (2008) The new geography of work: Power to the precarious? *Theory, Culture & Society* 25: 31–49.

Saldanha A (2010) Skin, affect, aggregation: Guattarian variations on Fanon. *Environment and Planning A* 42: 2410–2427.

Scholz T (ed.) (2013) *Digital Labor: The Internet as Playground and Factory.* New York: Routledge.

Sennett R (1998) *The Corrosion of Character: The Personal Consequences of Work in the New Capitalism.* New York: Norton.

Srnicek N and Williams A (2015) *Inventing the Future: Postcapitalism and a World without Work.* Brooklyn, NY: Verso Books.

Stacey J and Suchman L (2012) Animation and automation: The liveliness and labours of bodies and machines. *Body and Society* 18: 1–46.

Steward B (2000) Changing times: The meaning, measurement and use of time in teleworking. *Time & Society* 9: 57–74.

Suchman L (1995) Making work visible. *Communications of the ACM* 38: 56–64.

Suchman L (2011) Subject objects. *Feminist Theory* 12: 119–145.

Sussman M (1999) Performing the intelligent machine: Deception and enchantment in the life of the automaton chess player. *TDR/The Drama Review* 43: 81–96.

Swan M (2013) The quantified self: Fundamental disruption in big data science and biological discovery. *Big Data* 1: 85–99.

Terranova T (2000) Free labour: Producing culture for the digital economy. *Social Text* 2: 33–58.

Turkle S (1997) *Life on the Screen: Identity in the Age of the Internet.* New York: Simon & Schuster.

Turner F (2008) Romantic automatism: Art, technology, and collaborative labor in Cold War America. *Journal of Visual Culture* 7: 5–26.

Valentine G (2008) Living with difference: Reflections on geographies of encounter. *Progress in
Human Geography* 32: 323–337.

Vartiainen M and Hyrkkänen U (2010) Changing requirements and mental workload factors in mobile multi-locational work. *New Technology, Work and Employment* 25: 117–135.

Vilhelmson B and Thulin E (2001) Is regular work at fixed places fading away? The development of
ICT-based and travel-based modes of work in Sweden. *Environment and Planning A* 33: 1015–
1029.

Wainwright E (2010) The office is always on: DEGW, Lefebvre and the wireless city. *The Journal
of Architecture* 15: 209–218.

Weeks K (2011) *The Problem with Work: Feminism, Marxism, Antiwork Politics, and Postwork
Imaginaries.* Durham: Duke University Press.

Wills J (2009) Subcontracted employment and its challenge to labor. *Labor Studies Journal* 34:
441–60.

Wilson A (2016) The infrastructure of intimacy. *Signs* 41: 247–280.

Wilson M (2011) Data matter(s): Legitimacy, coding, and qualifications-of-life. *Environment and
Planning D: Society and Space* 29: 857–872.

Wilson M (2012) Location-based services, conspicuous mobility, and the location-aware future.
Geoforum 43: 1266–1275.

Wilson M (2014) Continuous connectivity, handheld computers, and mobile spatial knowledge.
Environment and Planning D: Society and Space 32: 535–555.

Witheford N (1994) Autonomist Marxism and the information society. *Capital & Class* 18: 85–125.

Wright MW (1997) Crossing the factory frontier: Gender, place, and power in the Mexican maqui-
ladora. *Antipode* 29: 278–302.

Wright MW (2009) Gender and geography: Knowledge and activism across the intimately global.
Progress in Human Geography 33: 379–386.

Wright MW (2010) Gender and geography II: Bridging the gap – feminist, queer, and the geographical imaginary. *Progress in Human Geography* 34: 56–66.

Zuboff S (1988) *In the Age of the Smart Machine.* New York: Basic Books.

Lizzie Richardson research explores urban life in the UK, drawing on approaches to culture, economy and politics from human geography and beyond. After completing her PhD in
the Department of Geography at Durham University in 2014, Lizzie worked as a Lecturer in
Human Geography at the University of Cambridge, before returning to Durham in 2016 as a
Leverhulme Early Career Research Fellow.

THE DIGITAL TURN IN POSTCOLONIAL URBANISM: SMART CITIZENSHIP IN THE MAKING OF INDIA'S 100 SMART CITIES

Ayona Datta

The smart city as a "digital turn" in critical urban geography has gone largely unnoticed in postcolonial urbanism. This paper seeks to address this gap by examining the emergence of new forms of postcolonial citizenship at the intersection of digital and urban publics. In particular, I investigate the production of a "smart citizen" in India's 100 smart cities challenge – a state-run inter-urban competition that seeks to transform 100 existing cities through ICT-driven urbanism. By examining the publicly available documents and online citizen consultations as well as observations of stakeholder workshops in four of the proposed smart cities, I illustrate how a global technocratic imaginary of "smart citizenship" exists alongside its vernacular translation of a "*chatur* citizen" – a politically engaged citizen rooted in multiple publics and spatialities. This takes place through three key processes – enumerations, performances and breaches. Enumerations are coercions by the state of an urban population that has so far been largely hidden from analogue technologies of governance and governmentality. Articulations are the performances of smart citizenship across digital and material domains that ironically extend historic social inequalities from the urban to the digital realm. Finally, breaches are the ruptures of the impenetrable technocratic walls around the global smart city, which provides a window into alternative and possible futures of postcolonial citizenship in India. Through these three processes, I argue that subaltern citizenship in the postcolony exists not in opposition, but across urban and digital citizenships. I conclude by offering the potential of a future postcolonial citizen who opens up entangled performances of compliance and connivance, authority and insecurity, visibility and indiscernibility across political, social, urban and digital publics.
Keywords: digital citizenship, India, postcolonial urbanism, smart cities, subaltern citizen

Funding information

The Arts and Humanities Research Council, Grant/Award Number: AH/N007395/1

1 INTRODUCTION

The smart city as an emerging area of scholarship indicates a clear "digital turn" (Ash, Kitchin, & Leszczynski, 2016) in urban geography. Although the idea of "networked urbanism" (Graham & Marvin, 2001) has been around for the last decade or so, as Glasmeier and Christopherson argue "what is new about the contemporary smart city narrative is the emphasis on places transformed by the application of technologies" (2015, p. 4). The contemporary smart city takes earlier ideas of a [|2]*

---

* Die Seitenzahlen beziehen sich auf die Onlineversion, nicht auf die Printversion.

digital or networked city beyond mere connectivity to a new regime of speculative futures that combine big data, algorithmic governance and automated urban management (Leszczynski, 2016). Recent scholarship on smart cities has argued that it represents a techno-utopian fantasy (Datta, 2015; Watson, 2014), a mode of governmentality (Kitchin, 2015; Vanolo, 2013) driven by corporate interests that in western contexts has become a smokescreen for implementing a range of "cost containment measures and supporting the shift to pro-innovation public expenditures" (Pollio, 2016, p. 514). Often presented by state–corporate partnerships as "non-ideological, commonsensical and pragmatic", Kitchin (2015, p. 131) notes that smart cities bring together two problematic neoliberal urban visions – first that the use of ICT will drive economic growth and urban prosperity; second that the use of ICT can make urban governance more efficient, manageable, transparent and hence equitable.

It is now accepted that the processes through which smart cities are conceived, implemented and received are diverse, contextual and often contradictory. Yet scholarship on smart cities is dominated by a "one-size-fits-all" critique where broader theoretical arguments are seen to stand for and "reveal the discursive and material realities of actually existing smart city developments" (Kitchin, 2015, p. 131). As Luque-Ayala and Marvin argue, there is therefore a need to

> explore the contradictions of smart urbanism, its differential expressions across global North and South, and the potential this creates to develop more oppositional, contested forms of knowledge and subjectivity that emerge from these contexts. (2015, p. 2113)

Moreover, while there now exists rich scholarship on smart cities in the global North, research on this theme is only just emerging in the global South (Datta, 2015; Odendaal, 2011; Shin, 2017). This is surprising, given the take up of smart cities in the global South has been at a faster rate than in the West, with countries like India, China, Korea, Saudi Arabia and others being the largest "consumers" of the global smart city market (McKinsey, 2011). More importantly, while smart cities seek to transform the social, economic and political life of cities through digital technologies, it is surprising that postcolonial urban theory has largely side-stepped this "digital turn" in the global South. This is a significant gap which this paper seeks to address using the case of India's national 100 smart cities programme.

In 2014, the newly elected ruling party in India announced that they will build 100 smart cities through a national programme to "leapfrog" India towards a digital urban future. Seeking to produce the "smart city" and the "smart citizen" as two sides of this future, this programme seeks to apply a range of digital technologies from e-governance to smart utility networks to produce ubiquitously networked cities. Doing this requires a number of manoeuvres (Figure 1). First, since 80% of its citizens are currently outside the digital divide, they need to be drawn into digital space in order to produce a "user base" for the smart city services. Second, these new digital subjects have to be shown how to perform as "smart citizens" in order to contribute to the "success" of the smart city. In this context, this paper asks: How are smart citizenships envisaged in the postcolonial city? What are the transformations required in state-citizen relationships in order to produce smart citizenship? What potential does smart citizenship offer for articulating new emergent forms of subaltern citizenships across digital and urban realms of the smart city? In this paper, I argue that the production of smart citizens in the future Indian city has become synonymous with the production of a postcolonial technocratic subjectivity, which needs to be critically investigated. To do so, this paper understands postcolonial urbanism as a historiography of knowledge-power systems that produces the trajectories of urban futures in the global South and the rendering of postcolonial citizenship therein as an amorphous set of rights claims across digital and material domains of the postcolonial city. Drawing on and connecting this to the scholarship on "subaltern urbanism" (Roy, 2011a, p. 223), the paper seeks to situate the "smart citizen" within a critical postcolonial urbanism which reveals the trajectories of postcolonial subjecthood across digital, social and material "publics" of the city and across the political, social and performative transformation of citizenship in a new digital urban age.

2 THE "DIGITAL TURN" IN POSTCOLONIAL URBANISM

Making deep genealogical connections with colonial histories of planning and governance, postcolonial urbanism has provided us with a rich critique of the rhetoric of "disorder" in colonial and postcolonial cities which produces violently exclusionary practices against subaltern groups in the name of modernity and development (Dupont, 2008; Legg, 2007; Tarlo, 2003). Postcolonial urban scholarship has critiqued the "north-centrism" (Robinson, 2003, p. 273) of urban studies, of the "geographies of theory" (Roy, 2009, p. 819) that flow from the global North to global South, and of the processes of planetary gentrification (Ghertner, 2014) and urbanisation which support and even perpetuate the exclusionary ways in which cities and citizenship in the global South are imagined and conceptualised. Recent literature on postcolonial urbanism [|3] argues for a cosmopolitan vision of cities (Robinson, 2005) as experiments in "worlding" (Roy & Ong, 2011) and as local "mutations" (Rapoport, 2014) of global urban visions, knowledge and policies. A substantial feature of this literature has been to understand varied models of "homegrown neoliberalism" (Roy, 2011b) emerging in postcolonial cities through the new power–knowledge nexus of state–corporates–experts. This has produced "thick" descriptions of everyday life around the struggles of marginal groups living under a violence of law (Datta, 2012), threat of evictions (Bhan, 2009) or a "rule by aesthetics" (Ghertner, 2011a). While this literature has firmly established postcolonial urban theory as a body of knowledge in its own right, there has been a surprising lack of engagement with the "digital" as a simultaneously transformative space of the city.

Yet postcolonial governance in the last decade or so has been radically transformed through a variety of programmes and initiatives for e-governance. As Hoelscher notes, urban governance in India

since the mid-2000s has been characterized by a "silent revolution" in the ICT and telecoms sector, intertwined with the propagation of mobile-based technology in the realm of urban policy. (2016, p. 32)

E-governance, in its provision, maintenance and accountability of municipal urban basic services and infrastructures of water, waste, energy and transport, has made ICT-driven urbanism both acceptable and commonplace in the last decade. This has not done away with the bureaucratisation of urban governance evident in form-filling, surveying and record keeping that as Gupta (2012) notes has been the hallmark of uneven and arbitrary state governmentality for the last few decades. Rather e-governance in the 2000s lay the foundations for a more ubiquitous form of technocratic governance that would expand and consolidate state rule in India's future cities. In this context, smart cities can no longer be cast solely as the mobility of global visions of urban futures, rather as the extension of state aspirations of governmentality, modernity and the control of "disorder" into a digital urban age.

To be sure, the relationship between the postcolonial city and digital technologies has been debated before. Writing over a decade ago, Graham (2000) noted that in cities of the global South there is emerging an uneven array of

> premium networked spaces: new or retrofitted transport, telecommunications, power or water infrastructures that are customized precisely to the needs of powerful users and spaces, whilst bypassing less powerful users and spaces. (2000, p. 185)

He suggested that urban development and planning in global South cities is largely driven by middle-class privilege, and therefore networked spaces are unevenly distributed and coincide with elite and middle-class access, leading to a socially and physically "splintered urbanism". Sundaram noted that this splintering "has become a significant theatre of elite engagement with claims of globalisation" (2004, p. 64) in India. This is also evident in South Africa, where Odendaal (2011) noted that digital access has been spatialised across existing investment patterns, which consolidates social and economic inequalities. The Indian smart city emerges in this moment to rewire historical–material–social inequalities in postcolonial cities through new power–knowledge networks of ubiquitous connectivity.

An examination of smart cities in India helps us reframe postcolonial urban theory and its central concerns with the genealogies of power, knowledge and politics in substantially different ways. First, behind the emergence of smart cities in the global South lurks a "rhetorics of urgency" (Datta, 2015, p. 5) to address the crises represented in the "disorders" of population explosion, uncontrolled urbanisation and climate change. Alongside a hubris of "India's Urban Awakening" (McKinsey, 2010), the smart city presents the postcolonial city as the new frontier of opportunities in the midst of crises, and digital technologies as a tool for eliminating its pre-existing disorders. Second, and in order to achieve this, digital technologies prise open territories and populations that were earlier invisible, unaccountable, unrecordable and thus ungovernable in the postcolonial bureaucratic state by extracting them into formalised and visible structures of data, governance and management. Finally, the above produce smart cities as new hybrids of power–knowledge networks that simultaneously traverse the discursive, material and digital realms of the postcolonial city. They produce new urban spatialities that are always already mediated through digital technologies and infrastructures of urban governance and encourage new technologies of visibility (Sundaram, 2004) in the postcolonial digital urban age.

Such a conceptualisation of postcolonial urbanism tests the limits of "worlding" cities (Roy & Ong, 2011) or "homegrown neoliberalism" (Roy, 2011b) or the "propriety of property" (Ghertner, 2011b) in unpacking the heterogeneous and uneven reach of algorithmic governmentality over urban territories and populations. It calls on us to expand the horizon of postcolonial governmentality beyond "regimes of dispossession" (Levien, 2013) to new regimes of digital incorporation in [|4] the smart city. Examining citizenship in the smart city means asking what postcolonial urbanism demands of the digital urban age in order to establish new regimes of control. Seeing the city through the lens of a postcolonial "digital turn" means seeing urban space as the "test-bed" for state–corporate partnerships in governmentality, achieving what Hardt and Negri call "new and complex regimes of differentiation and homogenisation, deterritorialization and reterritorialization" (2001, p. xiii). This means understanding how smart cities as a new form of postcolonial urbanism rearrange power–knowledge networks across territories and populations beyond the material. It means understanding how postcolonial citizenship must now directly address the subaltern "other" within and across new forms of power vested in the digital.

3 SMART CITIZEN AS POSTCOLONIAL SUBJECT

The figure of the smart citizen in India reinforces historical and contemporary paradoxes of identity and belonging. This is because the notion of citizenship is a relatively new concept in India. Under colonial rule, bodies and identities of Indians existed as "lesser" subjects, and were "scientifically" categorised along caste, religious and ethnic markers (Chatterjee, 2004) in order to assist colonial rule. Civil society and citizenship within an equal public sphere was largely absent, and any claims to citizenship by colonial subjects was primarily conceived through civil disobedience of the colonial state and ultimately freedom from it. On independence in 1947, the Indian Constitution explicitly provided universal rights to all citizens while simultaneously evoking rights to positive discrimination in social, economic and political spheres (such as education, employment, housing and so on) for those historically marginalised by caste, religious or gender ideologies. Indian citizenship is now caught between the search for a uniform civil society under the rubric of universalised rights and the demand for differentiated rights for those who have historically faced social and economic marginalisation in society (Jayal, 2013). The smart citizen – a ubiquitously connected subject, participating equally in digital space – remains poised between these two rights-based demands.

The paradoxes of Indian citizenship and its particular forms of universal and differentiated rights produces what Chatterjee notes as the differentiation between civil and political society. For Chatterjee (2004), while civil society is the realm of the privileged, it is through political society – the actually existing realm of subaltern politics – that marginal groups make moral rights claims on the basis of caste, class, religion, gender and thus transform their bare life conditions to citizenship. This distinction, however, is not supported in the case of digital space as Sundaram (2009) notes how media piracy in Delhi opened new spaces of disorder across political and civil society. Sundaram's work suggests that conventional forms of power located across civil and political society are making way for more multi-sited amorphous forms of power and subjectivity across digital and material spheres of the city. This amorphousness is crucial in smart citizenship, which can no longer be considered in the same way as subaltern citizens living in slums and informal settlements make moral rights claims through political society. While exclusion from the "paradigms of propertied citizenship" (Roy, 2003, p. 463) does not exclude access to digital space, lack of digital access often overlaps with the absence of property rights. This is paradoxical, since digital access offers the illusion of a neutral, transparent and equal digital civil society unattached to material socio-economic markers, and yet is actually deeply connected to it. As Jayal notes, this has redefined the notion of "citizenship obtained through struggle, to citizenship as a gift of the state" (2013, p. 3).

Smart citizenship thus emerges at the moment in the large-scale transformation of India's economy when a new individualist and consumerist identity has begun to take hold across different social groups. In the newly formed Indian republic in 1947, the "quest for citizenship … [was] conducted in tandem with a quest for democratic modernity" (Jayal, 2013, p. 274), but in India's new urban age, the quest for a smart citizen has been given shape by a quest for technocratic urbanism. This has transformed moral rights claims to more universalised demands among subaltern groups to have a share in the economic prosperity of the free market. Following Chatterjee (2004), this move from subjecthood to citizenship is now one that makes claims on governmental authorities over services and benefit such as digital access. Thus, Jayal notes that "new forms of individuated citizenship are being enabled by electronic media and social networking, which offer unusual opportunities for expressive citizenship" (2013, p. 280) for these groups. This resonates with Sadoway and Shekhar's (2014) conceptualisation of smart citizenship as bridging the binaries of analogue and virtual modes of civic engagement which they claim with is "deep democracy" (Appadurai, 2001).

I argue that smart citizenship in India cannot be seen through this binary lens of subject/citizen, but rather as an amorphous and dialectic identity across three overlapping vectors. First, as *enumeration* through citizen consultations of a digital population that had so far been hidden from analogue technologies of measurement (such as the Census). This is a call by the state that evokes the moral imperatives of *all* citizens to engage in citizen consultations, but that paradoxically imparts subjecthood to subaltern groups. As Isin and Ruppert (2015) suggest, the digital citizen now becomes both "subject to [|5] power" as well as "subject of power", simultaneously controlling and controlled by the structures and networks of digital technologies and information flows in the smart city.

Second, *articulation* of how to become smart citizens is key to establishing and reinforcing social and political power over both territorial and digital spaces of the city. This emerges in the performative acts of citizenship across multiple publics of the smart city – in the "digital acts" of a population participating online and making claims to the future smart city, in the corporeal performances of a population through marathons, poster, essay and poetry competitions in the public sphere, and in middle-class vigilante performances over subaltern populations.

Finally, *breaches* to particular forms of authority and power across digital and urban publics open up "opportunities, possibilities and beginnings" (Isin & Ruppert, 2015) from within the spaces of the smart city. Breaches open up the space for a "pirate modernity" (Sundaram, 2004, p. 66) where the subaltern citizen refuses to interact with the brand-driven economy of big data, open data and data governance. This, I argue, is the process of translation where the smart citizen as a ubiquitously connected subject emerges as a "*chatur* citizen" – a postcolonial subject who shares the same space with a globally recognisable smart citizen. The *chatur* citizen, as I argue in this paper, is a cultural rather than literal translation, which loses its neutrality and acquires heightened political meaning in the process. Through rhetorical confusion and reinterpretation of the "smart city", the *chatur* citizen offers hope for the vernacularisation of smart citizenship through an embodiment of situated and discursive politics across digital and subaltern publics.

4 INVESTIGATING SMART CITIZENSHIP IN THE FUTURE SMART CITY

How does one investigate the digital turn in postcolonial cities before the arrival of the smart city in its material form? How does one investigate smart citizenship in the absence of subaltern rights to the city? How does one investigate citizenship in the future smart city? These and other questions guided our work on a UK Research Council funded development grant titled "Learning from the Utopian City". Crucial to our project were two scales of inquiry. First, a horizon scanning of publicly available smart city documents and applications. A key source was the federal Indian Ministry of Urban Development (MoUD) website, which hosts vast resources of online citizen consultations and smart city applications submitted by different urban municipalities towards the 100 smart cities challenge. A large part of the analysis in this paper is driven by a discursive analysis of the documents, applications and citizen consultations in terms of their references to and descriptions of the smart citizen.

The second scale of inquiry is related to the space of a series of four stakeholder workshops held in India from January to June 2016 in four cities bidding for the challenge – Varanasi, Chandigarh, Navi Mumbai and Nashik. The workshops brought together a range of participants (between 20 and 30) from government departments, private developers, ICT companies, NGOs, residents' welfare associations, slum dwellers' associations and so on. For the first time, these stakeholders were meeting around a table to debate the future Indian city. For the first time, stakeholders in government, third sector and subaltern groups realised that the smart city held very different meanings across the table. This to us was the defining moment of emergent citizenship, where we observed the rhetorical confusion and breach of the imagined smart city at different scales.

We made in-depth observations of the formal presentations, arguments, debates and discussions alongside video and audio recordings. In doing so, the four city workshops were taken as ethnographic sites for observing the unfolding of enumerations, articulations and breaches of smart citizenship. Our role as facilitators of the workshop directed these interactions quite specifically for the stakeholders. On the one hand, we asked all stakeholders to articulate their aspirations for the future city. On the other hand, we also requested stakeholders to articulate details of and their specific role in the future smart city (as policymakers, planners, consultants, activists or citizen participants). We observed the blueprints of smart urbanism presented by government agencies which valorised technology as a means to control the perceived disorder of existing Indian cities. We also observed the performative citizenships of subaltern actors in the workshops where they described themselves as "dutiful subjects" by law, and made claims to wider inclusion in the economic and entrepreneurial successes of the future city. Finally, we also observed how through these discussions, a new rhetorical and paradoxical smart citizen became indigenous to the historical struggles of subaltern citizenship. Further to these workshops we contacted participants in Navi Mumbai and Nashik to discuss some of the issues highlighted in face-to-face or Skype interviews. These were unrecorded, but these discussions have contributed to the analysis of citizen consultations.

These two scales of inquiry brought out the complex nature of smart cities, citizenship and its contested translation in postcolonial contexts. These translations are both discursive and material in that these are both embedded in policy documents and embodied in everyday life struggles. Following Jazeel's (2014, p. 89) call to "dwell in the domain of representation and to trouble over the mechanics of representation at work in geographical knowledge production", my approach [|6] complicates the distinction between discursive and material in the future city. My examination therefore aims at simultaneity rather than causality in a digital urban age – seeing the discursive representations of smart citizenship in policies, its material politics of enumeration and articulation and the "speech acts" of ordinary citizens, all as co-constitutive of transformations in postcolonial citizenship. The methodological approach then is to unpack how data, policy, representations and discourses of smart citizenship together rework the everyday struggles of ordinary citizenship. In the sections that follow, I present three simultaneous processes of smart citizenship in India – enumerations, articulations and breaches. Through these processes, I outline an emergent postcolonial subject who is simultaneously transgressive and compliant in their practices and performances of citizenship in the new smart city. I will conclude with the potential of this new postcolonial citizen in India to embody the inherent conflicts and contradictions of a digital urban age.

5 "CITYZENS BECOME NETIZENS"

In May 2014, a new ruling party came to power in India on the basis of their promises of good governance and economic growth. The creation of 100 smart cities formed the backbone of these promises, which sought to "leapfrog" India into a digital urban age of innovation, entrepreneurialism and endless prosperity. The overlaying of a smart urbanism onto Indian cities represents India's route to "planned" cities by claiming to make governance transparent, open and accountable to citizens in the new smart city. The key feature of the smart cities agenda in India is a national competition between 100 existing cities nominated by regional states. The federal state allocated a total budget of ₹ 100 crore (approx. £11 million) to be equally divided across the 100 cities irrespective of their metropolitan size, demographic base and aggregate revenue. Each regional state was expected to make similar financial commitments. Each nominated city would then make a bid to receive the allocated money, as well as raise matching funds from its own revenue sources. The Indian definition of the "smart city", however, significantly played down the global imaginary of a ubiquitously connected Internet of Things (IoT) and Big Data to emphasise the importance of local situatedness. Indeed, the Smart City guidelines explicitly made the case that

> there is no universally accepted definition of a Smart City. It means different things to different people. The conceptualisation of Smart City, therefore, varies from city to city and country to country, depending on the level of development, willingness to change and reform, resources and aspirations of the city residents. (GoI, 2015, p. 5)

Despite its locally situated rhetoric, India's smart cities challenge encompassed a neoliberal logic that made private sector involvement obligatory. From assisting in the preparation of proposals to the implementation of smart city projects, each city authority had to appoint consultants chosen exclusively from a list prepared by the federal Ministry of Urban Development (MoUD), and an external "hand-holding" agency (such as DfID, World Bank, USAID, ADB). The final deployment of smart technologies was to be tendered to international ICT companies such as IBM, Cisco, Siemens, Hewlett Packard and others who were named as potential partners. This meticulous attention to each process and stakeholder in the planning and apparent democratisation of smart cities claimed to reverse some of the current planning thinking on Indian cities, which argues that India is not able to "plan" its cities since it is itself an "informalized entity, engaging in deregulation, ambiguity and exception" (Roy, 2009, p. 76).

Crucial to the seeming democratisation and formalisation of the smart city, the guidelines highlighted the "citizen" as central to its ethos. Very early on, the draft concept note stated that the intention was

> to promote cities that provide core infrastructure and give a decent quality of life to its citizens, a clean and sustainable environment and application of "Smart" Solutions. (GoI, 2015, p. 5)

Indeed, it emphasised that quality of life in smart cities would be on a par with European standards, indicating that despite its situated rhetoric, the Indian smart cities mission had set for itself the parameters of western modernity and urbanism. This was legitimised through citizen consultations. Compared with other evaluation categories in the smart cities challenge, "citizen participation" had the third highest weightage of 16% after "feasibility and cost-effectiveness of proposals" (30%) and "result orientation" (20%). Other criteria such as "smartness of proposal" (10%), "strategic plan" (10%), "vision and goals" (5%), "evidence-based city profiling and key performance indicators" (5%), and "processes followed" (4%) came much lower in the priority list. Not only that, digital presence was a key [|7] criteria of citizen participation as evident in the assessment questions: "How much of social media, community, mobile governance have been used during citizen consultation?" and "How well does the Vision articulate the use of information and communication technologies to improve public service delivery and improve the quality of life of local citizens?" (GoI, 2016). This then was not just a digital turn in urbanism but an urbanism for and by digital technology.

In February 2016, as the first set of 20 winning cities was revealed, MoUD announced that the 100 smart cities challenge had kick-started the process of "cityzens becoming netizens" (Naidu, 2016). The Ministry reported that a total of 15.2 million citizens participated in the preparation of smart city plans at various stages. The national smart cities mission had tapped into the digital sphere to produce millions of digital citizens projected to become smart citizens in the future. Although these statistics were at a scale not achieved or attempted earlier, it still constituted a small share of the total urban populations (about 12% of the total population of the participating cities). But who were these smart citizens of the future? How were they discursively and performatively produced? What was their role in the future smart city?

5.1 Enumeration: Fast-tracked citizens

The first indications of the production of "netizens" became apparent in the draft preceding the final smart city guidelines in 2014. The "Smart City Mission document", as it was called, went through a number of iterations from concept notes to draft guidelines to a final public document. Crucially, the "netizen" also underwent a set of transformations. The draft concept note identified the smart city as

> not only described by the level of qualification or education of the citizens but also by the quality of social interactions regarding integration and public life and the openness towards the "outer" world. (MoUD, 2014, p. 9)

It further noted that it was important to have urban citizens in the digital sphere since online participation will lead to "Social pressure on other citizens [that] can often remove resistance and facilitate a greater degree of civic discipline" (MoUD, 2014, p. 17). These explicit links between digital participation and digital surveillance were dropped in the final Smart City mission guidelines, to describe a more active citizen.

> The Proposal development will lead to creation of a smart citizenry. The proposal will be citizen-driven from the beginning, achieved through citizen consultations, including active participation of groups of people, such as Residents Welfare Associations, Tax Payers Associations, Senior Citizens and Slum Dwellers Associations. During consultations, issues, needs and priorities of citizens and groups of people will be identified and citizen-driven solutions generated. (GoI, 2015, p. 22)

This focus on "smart citizenry" reflects the general transformation in processes of deliberative democracy which Kitchin (2015, p. 133) observes have a "discursive emphasis" on "inclusivity and citizen empowerment" without any real impacts on democratic participation. The process of enumerating this "smart citizenry" relied on what Dan Hill (2012, location 744, kindle edition) has called a "push button democracy" where online endorsement is seen to stand for democratic deliberation in the public sphere. With the announcement of the nominated 100 smart cities, overnight each city began their own "enumeration" process through which an urban population was incorporated into the logics of big data and digital governance. This enumeration occurred across digital and material publics, with municipalities summoning its urban residents to participate in online "citizen consultations" to produce what I will call here "fast-tracked digital citizens". These consultations were both online and face-to-face, virtual and analogue. Each city carried out extensive public mobilisation events (smart city walks and marathons, essay, poster and logo competitions). They conducted surveys in middle-class neighbourhoods as well as informal settlements, in government-run schools and colleges and in marginal city wards. The public were invited to meetings and workshops with local councillors and civil servants through newspaper announcements, through neighbourhood residential associations, through slum dwellers' organisations, traders' associations and so on.

One of the key platforms where these processes came together was the MoUD website, which hosted the smart city proposals and citizen consultations with over 2 million online comments received across 98 cities. Despite being a small share of the total urban population of these 100 cities, this nevertheless revealed the huge shift towards digital enumeration in urban governance. For example, by the end of 2015, Chandigarh Municipal Corporation reported that there had been about [|8]

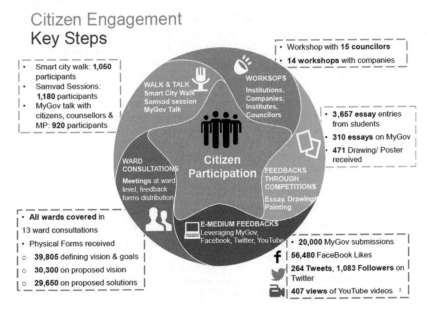

Fig. 1 Chandigarh Smart city consultation graphic
Source: MoUD (2015)

20,000 MyGov submissions with over 56,000 Facebook likes, 260 tweets and 400 views of YouTube videos of their smart cities proposal. The language of consultations on the digital platforms was English, although responses were sometimes in Hindi or other regional languages. The platform to interact and input suggestions also required skills that were clearly absent among those with lower education and digital capabilities.

Capturing online citizens and enumerating the data on consultations served the MoUD in highlighting the transformation of a territorial "population" into digital citizens. This was a step before they could then be trained to become smart citizens, articulating and performing their access to digital space in various ways to serve the demands of smart cities. Enumeration promoted a reductionist version of citizenship where participation as a "category of governance" (Chatterjee, 2004, p. 69) began to stand for "democratic" urban transformation. Online citizens became what Pollio (2016, p. 528) has called a "thinkable, knowable issue of urban governance" that were the digital version of earlier analogue systems of documentation, form filling, bureaucracy and "red tape" (Gupta, 2012), which enabled the state to make its power and practices illegible (Das & Poole, 2004). This online population was the material of a digital society – fast-tracked digital subjects with a digital signature, who formed the prelude to "smart citizenry".

5.2 Articulation: Performing smart citizenship

The citizen consultations highlighted three key processes in the making of smart citizenship. First that the state's desire to govern its territory through its digital population set the pedagogic terrain of smart citizenship which believed that citizens "need to be educated by city leaders as to the benefits IT can bring" (Hollands, 2015, p. 70). This was evident in the smart city guidelines, which underlined key duties of "smart people".

> The Smart Cities Mission requires smart people who actively participate in governance and reforms. Citizen involvement is much more than a ceremonial participation in governance. Smart people involve themselves in the definition of the Smart City, decisions on deploying Smart Solutions, implementing reforms, doing more with less and oversight during implementing and designing post-project structures in order to make the Smart City developments sustainable. (GoI, 2015, p. 18) [|9]

The introduction of "smart people" in policy reflects a shift from the ordinary citizen to the tech-savvy, entrepreneurial and judicious citizen working for and on behalf of state enterprise, innovation and growth. This production of "smart people" was central to the state's alignment with digital governance. It reflects what Vanolo notes as the desire of the state to

> speak about the citizens of smart cities, and speak in the name of them, but very little is known about citizens' real desires and aspirations. (2016, p. 12)

In this role, smart people were collaborators and endorsers of the smart city, rather than critical and active citizens. Thus, instead of testing smart cities as the site of democratic participation, smart citizens were constructed as allies of state–private sector experiments in urban governance.

Second, the smart city proposals used ICT as the filter to sieve through citizens' concerns with everyday urban problems. A large part of citizen feedback in the online consultations involved suggestions on improving urban basic services, provision of physical and social infrastructure, supply of urban basic services and so on. Yet, for example, while parts of the city (such as informal settlements) were still disconnected from urban basic services, smart city proposals often aimed at 24/7 water supply, public wifi, e-rickshaws and so on. Local planners and consultants in our workshops accepted that although urban basic services were priority themes in citizen consultations, their smart city proposals focused on those that fit within the ICT-driven prescriptive guidelines of the Smart City Mission document.

Finally, online citizen consultations extended historic social inequalities from the urban to the digital realm. It is here that new categories of exclusion emerged from the digital realm where the right to be a smart citizen was premised on the removal of those not deemed to belong to the future smart city. For example, a large part of the online citizen suggestions for Navi Mumbai focused on the importance of law enforcement to ensure a clean and "sanitised city". These comments of varied length and detail called for the elimination of open defecation, removal of hawkers and street markets, provision of green spaces, prohibiting spitting, beautification of neighbourhoods, planting trees, jogging and cycling tracks, enforcing parking provisions, traffic rules and so on. These popular middle-class demands for a "sanitised" city constructed the elimination of physical and moral "dirt" and the control of "disorder" as a route to smart urbanism. If technology was clean, it was believed that the use of technology-driven urbanism in smart cities would also clean the city symbolically and materially. These views were supported by most urban municipalities such as Chandigarh, which regularly demolished "illegal encroachments" of slums, hawkers and pavement dwellers since the announcement of the 100 smart cities challenge in order to become "slum free" (Nayaki, 2016; Sharma, 2016). This exclusionary discourse and practice drew on the logics of "propertied citizenship" (Roy, 2003, p. 463) onto the digital realm in paradoxical ways. While there were no clear guidelines on who owned the data on citizen consultations, and how that data was to be used, ownership of physical property in the city determined how inclusion and exclusion from the future city were discoursed and practised. This was strikingly evident in the implementation of a "smart cleaning" system in Chandigarh.

"Smart cleaning" in Chandigarh is one of the key smart applications (among several others) to be adopted in Chandigarh's smart city projects. The "problem statement" identified as the inefficiency in garbage collection, was attributed to the unauthorised absence of sweepers. In order to make this service more efficient and lead to better satisfaction of municipal services, Chandigarh municipality was advised by smart city consultants to look into the "smart cleaning" model piloted in a nearby satellite city, Mohali. The Mohali model used GIS and mobile apps connected to a citizens' reporting and management tool that tracked attendance, uniform, equipment and performance of sweepers (who were mostly from the lower castes, unorganised and without permanent contracts). One of the most controversial aspects of this tool was the ability of "citizens" to take pictures of sweepers at work, upload it on cloud-based software and hence assist the municipality in monitoring their performance in real time.

The smart cleaning model was highlighted by Chandigarh municipality as the potential of smart applications to deliver efficient and cost-effective urban services to citizens. Smart city consultants noted that this would produce the future smart citizen who used their digital presence to support and transform urban governance. Since Mohali's smart cleaning was identified as "best practice", civil servants and local councillors from Chandigarh had visited its municipality offices in early 2016 (Rohtaki, 2016) to subsequently adopt these in Chandigarh Municipality. The Chandigarh smart cleaning idea was thus sold to the social and political elite alike in imagining a sanitised city, thereby deepening the gap between the "smart citizen" and subaltern citizen. The former claimed their space in a smart urban future using their "digital acts" of vigilante governance to discipline the latter. These were the middle-classes who became citizen-consumers, demanding service efficiency and thus transposing their existing historical material-social privileges from the public sphere onto the digital sphere. It reflects what Vanolo calls "smartmentality" – a hidden ideology of the state which uses the smart city as a "powerful tool for the production of docile subjects and mechanisms of political legitimisation" (2013, p. 883). The smart cleaning system went further in [|10] creating what Lerman calls a "new kind of voicelessness" (2013, p. 59), where subaltern actors had little or no control over their own labour, which was monitored by those who already had greater access to public goods and services.

Chandigarh's smart cleaning system blurs the distinction between participation as a "category of governance" and participation as a "practice of democracy" (Chatterjee, 2004, p. 69) in deeply problematic ways. It indicates that postcolonial citizenship in the new smart city emerges from overlapping layers of urban and digital publics around longstanding postcolonial elite anxieties over "dirt" and "disorder" (Gooptu, 2001; Kaviraj, 1997) in the city. Presented through the objective technocratic rationality of smart apps, it legitimises the surveillance of those on the margins of social power networks by the "practices of democracy" by an elite civil society further empowered by digital citizenship. The "smart citizen" thus becomes a euphemism for an elite citizenship built on class, religious and caste privilege. The subaltern citizen can now no longer make straightforward moral rights claims through political society. Rather they must now find new ways to breach the boundaries between digital and urban publics that define their exclusion from the future city.

5.3 Breaches: The "chatur" citizen

In the face of enumeration and articulation of smart citizenship by the state, what emerges is a new form of citizenship that breaches the norms of the smart city, in order to access and belong to it under a new social contract between the state and citizen. "Breach" here refers to both a verb and a noun – the former "an act of breaking or failing to observe a law, agreement, or code of conduct" and the latter, "a gap in a wall, barrier, or defence" (Oxford English Dictionary). This breach resonates with what Isin and Ruppert (2015, p. 131) suggest as the "witnessing, hacking and commoning" of particular forms of authority and power in digital spaces, which opens up "opportunities, possibilities and beginnings" to enact new forms of citizenship across digital and material spaces. A significant aspect of breaching of smart citizenship in India drew on the ability of the subaltern citizen to "speak" in their own voice and of their own experiences in order to challenge the norms of conduct embodied by "smart people" in the policy. These "speech acts" distinguished them as postcolonial "citizens" (with their moral and legal attributes of community) from a digital "population" (their empirical form in census data, survey, policies and the numerous enumeration logics of the state) (Chatterjee, 2004). In the stories that are emerging from a diverse range of future smart cities across India, it is clear that these breaches are multifarious, multiscalar and commonplace. I present here a few examples from our workshops.

The first breach occurred at the scale and in the spaces of urban local governments. In our Nashik workshop, the Deputy Mayor (a locally elected official) said to the Commissioner (a civil servant), "those who want two rotis [bread] are given half and those don't want rotis are also given half ... You are giving us sweets, but we have diabetes". Through metaphors of hunger, indulgence and disease, he was referring to the rolling out of universal financial packages to all the 100 smart cities despite huge variations in their demographic, social, political and economic contexts. He continued,

> Smart city, smart city, everyone is talking about the smart city. But did anyone try to analyse the reality behind it? The Central Government will give 500 million (Rs), and that is their maximum limit. The State Government will provide 250 million (Rs), so a total of 750 million. Now revenues of [municipal] Corporation are about 3,000 million (Rs), which suggests 2,250 million (Rs) will be citizens' money. But the Central and the State Government will keep control on this 3,000 million.

This assertion for autonomy by the Deputy Mayor illustrates the "voice" of the city within the policy and discursive landscape of Indian smart cities. In this case, the "urban" as a political unit of governance breaches the smart city visions of the Indian state through speech acts that assert its autonomy and identity. This is in contrast with recent claims of smart city investments being a "manna from heaven" (Mukhopadhyay, 2015, p. 79) for local political elites who seemingly clamoured in their requests to the federal state for securing these projects in their cities. The concern raised by the Deputy Mayor suggests that as the geographies of smart cities are complex and uneven, so are the ways that these are received and challenged by the urban authorities. By highlighting the top–down nature of smart city policies and the resultant shrinking of democratic power of the city, the Deputy Mayor's presentation resonated with the observation made by Mukhopadhyay (2015, p. 77) of the loss of autonomy written into the smart city guidelines, which is "silent on a democratic representative local government. Indeed, the word 'mayor' does not once appear". While the citizen pays taxes to the municipality, their elected representatives had no effective decision-making role in disbursing the smart city funds.

Nashik and Navi Mumbai urban authorities particularly sought to challenge and breach the different ways that smart city transformations were imposed as top-down financial transactions which privileged the power of the federal state. Local councillors there argued that acceptance of the smart city challenge would lock them into long-term path dependencies with the corporate sector. [|11] The Deputy Mayor of Nashik's challenge to the federal smart city policies resulted in a few concessions, giving them more representative voice in smart city projects. However, despite being included within the 100 smart cities nominations, Navi Mumbai has opted out of the national programme because of irreconcilable differences with the federal state over the use and control over urban revenues in smart city projects.

The second breach occurred at the scale of everyday life, which sought to redefine the meaning and impact of being "smart" in improving subaltern livelihoods and capacity. In our workshop in Nashik, a slum dwellers' representative said,

> In general, the government is trying to create a separate city for the rich and a separate city for the poor. The prevalence of this ideology is going to be fatal for the future of Indian cities. The Smart City project should adopt a holistic approach to development. We also want Nashik to become smart. There should be an increase in a number of employment opportunities, and every child should get education.

Here the future smart city and the already existing Indian city were made to stand respectively for the city of the rich and city of the poor. In a reflection on the future of slums in smart city visions, this representative drew on an alternative characterisation of the smart city as economically and socially just. They highlighted that they were not claiming "free entitlements"; rather as citizens, they wanted a share in the economic success of the smart city. This was extended in the Navi Mumbai workshop where an undocumented worker said to the Municipal planner,

> What do you mean by smart? Does it mean, if you have money then only you can be called smart? No. We are smart because we do hard work for our living, we do not steal from anyone, and every day we struggling. Suppose today, if we get ten rupees, we think if we spend all of it today then how will our children get to eat tomorrow? That is why we are smart. Because we plan our every move. … Smart is when you plan your children's future and make do with the little money you have. I do that every day and that is why I am smart.

This definition of the smart citizen as a pragmatic and prudent financial planner was a "speech act" which sought to claim right to the future city. Significantly, the narrative above was spoken in Hindi, while "smart" was left untranslated in English. This appropriation of the definition of "smart people" prescribed in the guidelines as "doing more with less" (GoI, 2015, p. 18) into an alternative definition rooted in everyday economic struggles sought to breach the definition of smart from within. This smart citizen was very different from the active "civic-cyber" citizen of Sadoway and Shekhar (2014), who worked within the constraints of digital infrastructures. The smart citizen in our workshops often emerged in the absence of digital participation and "in the absence of a society that guarantees formal and informal security and welfare once provided by community and state policies" (Szeman, 2015, p. 474). Smartness was, as Mukhopadhyay notes, "a set of community attributes that can only be experienced by living in it [city]" (2015, p. 79). As the state sought to reorganise democratic power through its smart cities challenge, this new idiom of citizenship offers a more "knowledge intensive smart city" (Soderstrom & McFarlane, 2016, p. 312) than a city filled with the digital acts of its residents. It offers the potential to breach the technocratic wall of the future smart city to articulate alternative discursive and material imaginations of smart citizenship.

These varied understandings and definitions of the smart citizen created disagreements among stakeholders about the identity of this elusive subject as a digitally empowered or knowledge-empowered active citizen. Discussion centred around the absence of linguistic tools to contextualise and culturally embody smart citizenship in India. The Deputy Mayor of Nashik noted that while the smart citizen was presented by the state as a digitally connected, politically neutral subject, its closest cultural translation *chatur* was in fact a citizen of increased political subjectivity. In Hindi and several other regional languages, "chatur" refers to those who are clever, astute, resourceful and quick-witted. *Chatur* people are understood to be street-smart – they fulfil their tasks with efficiency, skill and insight, use circumstances to their advantage and in general display a capacity to talk their way out of difficult situations. Significantly, a *chatur* person is a politically engaged citizen rooted in multiple publics and spatialities – far removed from the consensual citizens in the service of urban governance. *Chatur* citizens make do with less by seeking out opportunities in their favour. The Deputy Mayor suggested that in India "smart" citizens should be relabelled as "intelligent", citizens since the latter does not embody the shrewd political manoeuvrability of *chatur* citizens.

I suggest that the translation of smart citizenship into the *chatur* citizen presents us with a challenge. The *chatur* citizen is the third potential breach, a discursive and material formation who provides an alternative imagination of citizenship in India, which directs us towards a far more "disordered democracy" (Sundaram, 2004, p. 64) than that envisioned in the "smart citizenry" policy speak. The *chatur* citizen offers a "breach" in the state imaginary of smart citizen precisely because [|12] of its political ambiguity – they blur the lines between subject and citizen, digital and material, governance and governability in the future smart city. And thus, they are also unacceptable to dissenting urban political elite (who prefer the label "intelligent"). In this context, the *chatur* citizen is the "strange language found in the predominant language of urbanization" (Boucher et al., 2008, p. 989) and smart citizenship in India.

The *chatur* citizen, however, is not a completely new entity in India's political landscape. Its qualities have been observed in the postcolonial embodiment of the "men in the middle" (Sud, 2014, p. 593), who are often attributed to thirdworld corruption (Oldenburg, 1987). In India, these men are largely absent from policy discourse, but they play a crucial role in implementation. Variously called aggregator, aratdar, tout, broker, dalaal (Tarlo, 2003) or "pyraveekar" (Ram Reddy & Haragopal, 1985), they are intermediaries who know the art of acquiring political favours to achieve their goals. The middleman can also be seen in the "Marabout" in Senegal (Simone, 2003), a "fixer" who is equally disparaged and respected for his ability to get things done. The middleman indicates that in the context of the smart city, the *chatur* citizen is more than just fiction or conjecture. Despite its disparagement by

stakeholders in the workshops, the figure of the *chatur* citizen can be seen as simultaneously material, digital and discursive – analogous to an urban hacker, or a "clever
programmer" who gains access to forms of participation and demonstration within
the virtual and material spaces of the future smart city, and thus ruptures the normative subjecthood of smart citizenship presented by the state. This *chatur* citizen is
one who finds possibilities of "social collaboration" (Simone, 2003) across digital
and material spaces to disrupt how things are done, who by, where and under what
conditions in the future smart city. They open up new and entangled performances of
compliance and connivance, authority and insecurity, visibility and indiscernibility
across political and digital society. The *chatur* citizen is yet to come in Indian smart
cities, but the discursive landscape has been cast for its emergence.

6 CONCLUSIONS: TOWARDS A *CHATUR* CITIZEN?

The digital turn towards smart cities in postcolonial contexts has initiated a radical
transformation in the understanding of what constitutes citizenship in a new urban
age. In India, this is largely a process of capturing those outside of data structures
within systematic processes of performative citizenship and "push-button democracy" (Hill, 2012, p. 744). The performative demands of smart citizenship suggest
that it entails more than just digital access to a territorial population; rather it involves a coordinated state strategy in transforming how citizens react and respond to
a state's smart city initiatives. The emergence of a *chatur* citizen in this context raises
two key questions: Can there be a smart citizen outside of the structures of power
embodied by digital technologies? Can there be a postcolonial citizen beyond and
outside of historical structures of power and inequality? These two questions are
connected in that they direct us to the very nature of what constitutes postcolonial
urbanism, the digital turn and the practices of citizenship in the postcolony.

India's smart cities programme can be seen as the test-bed of a global smart
citizenship. In other words, if citizens embedded in informalised structures of pa-
tronage and economy can be turned into digitally savvy smart citizens, it can be
possible anywhere. Yet this transformation is a top-down vision that aims at service
efficiency and compliance rather than transgressive practice. Thus its translation
and vernacularisation offers possible openings to articulate another version of smart
citizenship that speaks from a different space and positionality of the subaltern. It
offers us a lens to examine the emergence of a new kind of postcolonial citizenship
whose shapelessness and ambiguity is its enduring feature. I have argued that this
is the *chatur* citizen – presumably a new postcolonial smart citizen who speaks both
from within and beyond the structures of digital governance and smartmentality.
The *chatur* citizen is inherently bound within the historical, social and economic
power structures and it is also distinct from it. It speaks from a position of subal-
ternity to challenge power and yet attempts to replicate the very forms of power
that it challenges through compliance and governance of the self. In so doing, the
chatur citizen is both a complaint subject and an active citizen moving fluidly and
unproblematically between these two positionalities. Their quotidian acts of

governing themselves in everyday life suggests how it is possible to garner support for a large-scale top-down urbanisation project such as the 100 smart cities. Their transgressive speech acts and opportunist politics also underline the faultlines of this large-scale vision of smart urban futures.

The "digital turn" we see emerging in Indian urbanism and its future *chatur* citizen suggest a wider move towards what scholars have recently identified as de-colonisation (Radcliffe, 2017). Indeed, in the context that geographical scholarship itself is now recognised as entrenched in modes of colonial power and knowledge, the decolonial imperative in urbanism is already upon us. Indian smart cities speak to the decolonial moment – articulating and delinking the knowledge in the making of smart cities and citizens from their western origins, even as they are reliant on the global knowledge networks and technologies of smart city making. They claim to delink and decentralise the production of smart cities by locating [|13] knowledge at the scale of the city and citizen, and yet do not build in possibilities of transgres-sion or subversion within it. The *chatur* citizen indicates a process of decolonisation from the hegemonic power of the national programme through the rhetorical sub-version of smart citizenship, but it also ironically reminds us that colonial and his-torical power inequalities cannot be simply erased in contemporary urban transfor-mations. Thus the emergence of a vernacular smart citizen shows that citizenship and subjecthood are not binary categories, rather they are entangled and compli-cated through the digital turn in postcolonial urbanism.

While decolonisation is a claim undertaken by the state and the *chatur* citizen to particular ends and with their own contradictions and faultlines, postcolonial links to historical structures of power embodied by class, caste and social privilege continue to resonate in new urban futures. Both the state and citizen claim to speak from different positions – the state as a postcolony and the *chatur* citizen as the subaltern. The key question both are engaged with – What makes a smart citizen? – is also the key question of 21st-century urbanism and imaginations of global urban futures. However, to understand how a diversity of answers to this question is emerging in the postcolony, we need to move beyond the mere delinking of colonial and western knowledge from local translations. As Jazeel reminds us

> decolonising geographical knowledge requires more. It requires us to think carefully about how to de-link the production of geographical knowledge from the hegemony of our disciplinary infrastructure. (2017, p. 336)

This understanding of decolonisation underlines an examination of the histories of structural violence over marginal and subaltern subjects which produce the land-scapes of smart cities and citizenship in India today. Rethinking smart citizenship from the positionality of the subaltern urban actor puts smart cities and postcolonial cities in dialogue with each other and leads to their mutual transformation, as we see in this paper. Opening up knowledge of smart citizenship beyond policy impo-sitions means, as Chatterjee has argued,

> using the full panoply of modern technologies of communication, switching and mixing lan-guages and media, and making sense of as well as enriching the diverse worlds they inhabit. (2012, p. 49)

This is not merely a provincialising of the smart city as a scholarly project, rather as Radcliffe notes a "rethinking of the world *from* Indigenous places and *from* the marginalised academia in the global South" (2017, p. 329).

So far, "smart" has been the alien language of postcolonial urban theory. For some time now, the "digital turn" has also been seen as existing and operating outside postcolonial urbanism. But smart and digital technologies are now the key policy and political tools of postcolonial states to embrace and imagine new urban futures. Understanding their role in transforming state–citizen relations is key to gaining insights into the workings of postcolonial urban transformations and emergent citizenships. If the Indian state reimagines its urban future in smart cities, I have argued that we need new modes of critical inquiry, methodology and analysis of the configurations of emergent citizenships in the digital urban age to understand how the subaltern citizen is finding ways to become "chatur" in the future city. Rather than delinking geographical knowledge, this will offer us opportunities to open up the postcolonial city to scrutinise the networks of power and knowledge shaping its present histories and futures. Opening up a postcolonial citizenship engaged with both the material and digital worlds (whether materially or rhetorically) as well as with the past and possible futures, will enable more critical scholarship that engages with the key problematic of the postcolonial urban age. This paper has initiated this process of imagining the future *chatur* citizen in a digital urban age, who speak of and to the smart city from the subaltern positionality of a disconnected or analogue citizen. More critical research and scholarship is needed to fully understand the diversity of emergent citizenships in this new postcolonial urban future from critical feminist, subaltern and digital lens in Indian cities.

Acknowledgements

This paper has been facilitated by funding from the Arts and Humanities Research Council (PI ref: AH/N007395/1). I am grateful to stakeholders in our city workshops in Varanasi, Chandigarh, Navi Mumbai and Nashik for the insights and debates that went into the making of this paper. I am particularly indebted to Councillors in Navi Mumbai Municipal Authority for sharing the details of their citizen consultations, which have formed the basis of this paper. I am also grateful to colleagues William Gould, Anu Sabhlok and Rebecca Madgin for co-organising and supporting the workshop activities. I am also grateful to the anonymous reviewers for their incisive and constructive comments that have helped improve this paper. All other mistakes are mine. [|14]

REFERENCES

Appadurai, A. (2001). Deep democracy: Urban governmentality and the horizon of politics. *Environment and Urbanization, 13*, 23–44.

Ash, J., Kitchin, R., & Leszczynski, A. (2016). Digital turn, digital geographies?. *Progress in Human Geography*, https://doi.org/10.1177/0309132516664800

Bhan, G. (2009). 'This is no longer the city I once knew' Evictions, the urban poor and the right to the city in millennial Delhi. *Environment and Urbanization, 21*, 127–142.

Boucher, N., Cavalcanti, M., Kipfer, S., Pieterse, E., Rao, V., & Smith, N. (2008). Writing the lines of connection: Unveiling the strange language of urbanization. *International Journal for Urban and Regional Research, 32*, 989–1027.

Chatterjee, P. (2004). *Politics of the governed: Reflections on popular politics in most of the world*. New York, NY: Columbia University Press.

Chatterjee, P. (2012). After subaltern studies. *Economic & Political Weekly, xlvii*, 44–49.

Das, V., & Poole, D. (2004). *Anthropology in the Margins of the State*. Santa Fe: School of American Research Press.

Datta, A. (2012). *The illegal city: Space, law and gender in a Delhi squatter settlement*. Farnham: Ashgate.

Datta, A. (2015). New urban utopias of postcolonial India: Entrepreneurial urbanization in Dholera smart city, Gujarat. *Dialogues in Human Geography, 5*, 3–22.

Dupont, V. (2008). Slum demolitions in Delhi since the 1990s: An appraisal. *Economic & Political Weekly, 43*, 79–87.

Ghertner, A. (2011a). Rule by aesthetics: World-class city making in Delhi. In A. Roy, & A. Ong (Eds.), *Worlding cities: Asian experiments and the art of being global* (pp. 279–306). Oxford: Wiley-Blackwell.

Ghertner, A. (2014). India's urban revolution: Geographies of displacement beyond gentrification. *Environment and Planning A, 46*, 1554–1571.

Glasmeier, A., & Christophers, S. (2015). Thinking about smart cities. *Cambridge Journal of Regions, Economy and Society, 8*, 3–12.

Gooptu, N. (2001). *The politics of the urban poor in early twentieth-century India*. Cambridge: Cambridge University Press.

Graham, S. (2000). Constructing premium networked spaces: Reflections on infrastructure network and contemporary urban development. *International Journal for Urban and Regional Research, 24*, 183–200.

Graham, S., & Marvin, S. (2001). *Splintering urbanism: Networked infrastructure, technological mobilities and the urban condition*. London: Routledge.

Hardt, M., & Negri, A. (2000). *Empire*. Cambridge, MA: Harvard University Press.

Hill, D. (2012). *Dark matter and trojan horses: A strategic design vocabulary*. Moscow: Strelka Press.

Hoelscher, K. (2016). The evolution of the smart cities agenda in India. *International Area Studies Review, 19*, 28–44.

Hollands, R. (2015). Critical interventions into the corporate smart city. *Cambridge Journal of Regions, Economy and Society, 8*, 61–77.

Isin, E., & Ruppert, E. (2015). *Being digital citizens*. London: Rowman and Littlefield.

Jayal, N.G. (2013). *Citizenship and its discontents: An Indian history*. Cambridge, MA: Harvard University Press.

Jazeel, T. (2014). Subaltern geographies: Geographical knowledge and postcolonial strategy, *Singapore Journal of Tropical Geography, 35*, 88–103.

Jazeel, T. (2017). Mainstreaming geography's decolonial imperative. *Transactions of the Institute of British Geographers, 42*, 334–337.

Kaviraj, S. (1997). Filth and the public sphere. *Public Culture, 10*, 83–113.

Kitchin, R. (2015). Making sense of smart cities: Addressing present shortcomings. *Cambridge Journal of Regions, Economy and Society, 8*, 131–136.

Legg, S. (2007). *Spaces of colonialism: Delhi's urban governmentalities.* Oxford: Wiley-Blackwell.

Lerman, J. (2013). Big data and it exclusions. *Stanford Law Review, 66*, 55–63.

Leszczynski, A. (2016). Speculative futures: Cities, data, and governance beyond smart urbanism. *Environment and Planning A, 48*, 1691–1708.

Levien (2013). Regimes of Dispossession.

Luque-Ayala, A., & Marvin, S. (2015). Developing a critical understanding of smart urbanism?. *Urban Studies, 52*, 2105–2116.

McFarlane, C., & Ola, S. (2017). On alternative smart cities. *City, 21*, 312–328.

McKinsey. (2011). *Big data: The next frontier for innovation, competition, and productivity.* New Delhi: McKinsey Global Institute.

McKinsey Global Institute. (2010). India's Urban Awakening: Building Inclusive Cities, Sustaining Economic Growth (McKinsey Global Institute, 2010).

MoUD. (2014). Draft Concept Note on Smart City Scheme. updated September 2014. Ministry of Urban Development, Delhi.

Mukhopadhyay, P. (2015). The un-smart city. *Seminar, 665*, 75–79.

Naidu, V. (2016). Smart city challenge has kick-started a revolution in urban landscape. *The Economic Times 2.* February (http://blogs.economictimes.indiatimes.com/et-commentary/smart-city-challenge-has-kick-started-a-revolution-in-urban-landscape/) Accessed 4 November 2016.

Nayaki, B.B. (2016). NMMC, Cidco clear slums in Belapur, Nerul. *The Times of India*, 27 May (http://timesofindia.indiatimes.com/city/navimumbai/NMMC-Cidco-clear-slums-in-Belapur-Nerul/articleshow/52455384.cms) Accessed 4 November 2016.

Odendaal, N. (2011). Splintering urbanism or split agendas? Examining the spatial distribution of technology access in relation to ICT policy in Durban, South Africa. *Urban Studies, 48*, 2375–2397.

Oldenburg, P. (1987). Middlemen in third-world corruption: Implications of an Indian case. *World Politics, 39*, 508–535.

Pollio, A. (2016). Technologies of austerity urbanism: The 'smart city' agenda in Italy (2011–2013). *Urban Geography, 37*, 514–534.

Radcliffe, S.A. (2017). Decolonising geographical knowledges. *Transactions of the Institute of British Geographers, 42*, 329–333.

Ram Reddy, G., & Haragopal, G. (1985). The Pyraveekar: 'The fixer' in rural India. *Asian Survey, 25*, 1148–1162.

Rapoport, E. (2015). Globalising sustainable urbanism: The role of international masterplanners. *Area, 47*, 110–115. [|15]

Robinson, J. (2003). Postcolonialising Geography: Tactics and Pitfalls. *Singapore Journal of Tropical Geography, 24*, 273–289.

Robinson, J. (2005). Urban geography: World cities, or a world of cities. *Progress in Human Geography, 29*, 757–765.

Rohtaki, H. (2016). Chandigarh: UT MC keen on replicating Mohali's cleaning system. (http://indianexpress.com/article/cities/chandigarh-ut-mckeen-on-replicating-mohalis-cleaning-system-2853739/) Accessed 4 November 2016.

Roy, A. (2003). Paradigms of propertied citizenship: Transnational techniques of analysis. *Urban Affairs Review, 38*, 463–491.

Roy, A. (2009). The 21st-century metropolis: New geographies of theory. *Regional Studies, 43*, 819–830.

Roy, A. (2011a). Slumdog cities: Rethinking subaltern urbanism. *International Journal of Urban and Regional Research, 35*, 223–238.

Roy, A. (2011b). The blockade of the world-class city: Dialectical images of Indian urbanism. In A. Roy, & A. Ong (Eds.), *Worlding cities: Asian experiments and the art of being global* (pp. 259–278). Oxford: Wiley-Blackwell.

Roy, A., & Ong, A. (2011). *Worlding cities: Asian experiments and the art of being global*. Oxford: Wiley-Blackwell.

Sadoway, D., & Shekhar, S. (2014). (Re)Prioritizing citizens in smart cities governance: Examples of smart citizenship from urban India. *The Journal of Community Informatics, 10*(3).

Sharma, M. (2016). Admn plans to make Chandigarh slum-free by March 2017. *Hindustan Times*, 26 April (http://www.hindustantimes.com/punjab/admn-plans-to-make-chandigarh-slum-free-by-march-2017/story-ro2jt7OoU1CqivNErIPpzL.html) Accessed 4 November 2016.

Shin, H. (2017). Envisioned by the State: The Paradox of Private Urbanism and Making of Songdo City, South Korea. In A. Datta & A. Shaban (Eds.), *Mega-urbanization in the Global South: Fast cities and new urban utopias of the postcolonial state*. London: Routledge.

Simone, A. (2003). *For the city yet to come: Changing African life in four cities*. Durham, NC: Duke University Press.

Sud, N. (2014). The men in the middle: A missing dimension in global land deals. *Journal of Peasant Studies, 41*, 593–612.

Sundaram, R. (2004). Uncanny networks: Pirate, urban and new globalisation. *Economic and Political Weekly, 39*, 64–71.

Sundaram, R. (2009). *Pirate Modernity: Delhi's Media Urbanism*. London: Routledge.

Szeman, I. (2015). Entrepreneurship as the New Common Sense. *South Atlantic Quarterly 114*, 471–490.

Tarlo, E. (2003). *Unsettling memories: Narratives of emergency in Delhi*. London: Hurst and Co.

Vanolo, A. (2013). Smartmentality: The smart city as disciplinary strategy. *Urban Studies, 51*, 883–898.

Vanolo, A. (2016). Is there anybody out there? The place and role of citizens in tomorrow's smart cities. *Futures, 82*, 26–36.

Watson, V. (2014). African urban fantasies: Dreams or nightmares?. *Environment and Urbanization, 26*, 215–231.

VOLUNTEER GEOGRAPHIC INFORMATION IN THE GLOBAL SOUTH: BARRIERS TO LOCAL IMPLEMENTATION OF MAPPING PROJECTS ACROSS AFRICA

Jason C. Young / Renee Lynch / Stanley Boakye-Achampong /
Chris Jowaisas / Joel Sam / Bree Norlander

Abstract: The world is awash in data – by 2020 it is expected that there will be approximately 40 trillion gigabytes of data in existence, with that number doubling every 2 to 3 years. However, data production is not equal in all places – the global data landscape remains heavily concentrated on English-speaking, urban, and relatively affluent locations within the Global North. This inequality can contribute to new forms of digital and data colonialism. One partial solution to these issues may come in the form of crowdsourcing and volunteer geographic information (VGI), which allow Global South populations to produce their own data. Despite initial optimism about these approaches, many challenges and research gaps remain in understanding the opportunities and barriers that organizations endemic to the Global South face in carrying out their own sustainable crowdsourcing projects. What opportunities and barriers do these endemic organizations face when trying to carry out mapping projects driven by their own goals and desires? This paper contributes answers to this question by examining a VGI project that is currently mapping public libraries across the African continent. Our findings highlight how dramatically digital divides can bias crowdsourcing results; the importance of local cultural views in influencing participation in crowdsourcing; and the continued importance of traditional, authoritative organizations for crowdsourcing. These findings offer important lessons for researchers and organizations attempting to develop their own VGI projects in the Global South.
Keywords: VGI, Global South, Crowdsourcing, Public libraries, Mapping

INTRODUCTION

The world is awash in data. By 2020 it is expected that there will be approximately 40 trillion gigabytes (40 zettabytes) of data in existence, with that number doubling every 2 to 3 years (Petrov 2019). For context, Internet users currently share more than 500,000 [|2] photos on Snapchat, watch over 4,000,000 videos on YouTube, and send more than 450,000 tweets *every minute* (Marr 2018). Amongst researchers there is broad consensus that this explosion of digital data, along with increased accessibility of digital information and communication technologies (ICTs), has had a profound effect on political, economic, and social processes across the globe (Benkler 2006; Bimber 2007; Castells 2004). However, neither data production nor the presence of the digital technologies and networks that support them is equal in all places around the world (Castells 2004). The global data landscape remains heavily concentrated on Englishspeaking, urban, and relatively affluent locations within the Global North (Burns 2014; Caquard 2014; Graham and Zook 2013; Young 2019a, b).

Nor is it simply that the datasets most richly and accurately represent locations in the Global North – residents in the North also tend to have the highest ability to *control* how data are produced, owned, analyzed, and shared. This issue of control is perhaps more problematic than the current lack of data, given that data are increasingly emerging from the Global South with attendant interest in leveraging them for economic change, development, and governance (Kshetri 2014; Mann 2017; Taylor 2017; Taylor and Schroeder 2014). Scholars are now calling attention to the emergence of new forms of digital and data colonialism. This research describes both how denizens of the Global South are exploited as data producers to reproduce neoliberal political economies (Ettlinger 2016; Thatcher et al. 2016) and also how existing datasets are used to extend Western cultural hegemonies and development visions to new locations (Burns 2014; Taylor and Broeders 2015; Young 2019a). Unfortunately, there remains widespread agreement that this body of work remains underdeveloped, and much of the current scholarship has focused on critique rather than solutions (Dé et al. 2018).

One partial solution to these issues may come in the form of crowdsourcing. In an ideal world crowdsourcing projects would allow individuals and organizations in the Global South to produce their own data and establish control over what happens to the data that they produce. At times organizations and researchers are too optimistic in their descriptions of crowdsourcing as a 'panacea' for problems afflicting the Global South (e. g., Bott and Young 2012), but even those that take a more critical approach see great promise. Lievano (2017), for example, argues that crowdsourcing makes *more* sense in the Global South given current data gaps, and Ingwe (2017) argues for the increased adoption of the method by civil society organizations across sub-Saharan Africa. Already many map-based crowdsourcing, also known as volunteer geographic information (VGI), projects have been successfully carried out across the continent (Yilma 2019).

Despite this work, many challenges and research gaps remain. There is a paucity of data available within many African countries, and VGI is largely driven by international non-governmental organizations (NGOs) instead of African organizations (Omanga and Mainye 2019). As a result, there is a risk that these projects feed back into the data inequalities and digital colonialism that VGI would ideally resist. Needed are more discussions of how to build the capacity of African organizations and governments to institute and sustain their own comprehensive and broadly implemented VGI programs. What opportunities and barriers do these endemic organizations face when trying to carry out mapping projects driven by their own goals and desires? This paper contributes answers to this question by examining a VGI project that is currently mapping public libraries across the African continent. This project is a collaboration between researchers at the University of Washington Information School and practitioners at the African Library & Information Associations and Institutions (AfLIA), and a primary goal of the work is to build the capacity of AfLIA to sustainably collect VGI over the long run. Recently the project team held a stakeholders' meeting with representatives from 22 different countries, each of whom is coordinating the mapping effort within their respective country. This paper analyzes their feedback on the challenges and opportunities they have faced in implementing the project in their countries, as well as their thoughts on the long-term sustainability of data collection. Their lessons can help to inform other large-scale data collection efforts in the Global South.

VOLUNTEER GEOGRAPHIC INFORMATION IN THE GLOBAL SOUTH

VGI projects like the one described in this paper have been made possible by fundamental shifts in processes [|3] and understandings of geospatial knowledge production, enabled by mobile and Web 2.0 technologies. In this section we describe the emergence of VGI, as well as the related concepts of crowdsourcing, user generated content (UGC), and neogeography. We also review current efforts to extend VGI to countries and communities across the Global South. We argue that there is insufficient literature describing the challenges and opportunities that African organizations face in implementing and sustaining their own VGI projects.

Crowdsourcing, and the related processes described in this section, have been enabled by the development and extension of digital networks which greatly augment the human capacity for information storage, analysis, and communication (Castells 2004). The digital technologies that access these networks are now cheaper than ever, are relatively ubiquitous, and relatively easy to use – meaning that more individuals now have more opportunities to produce, communicate, and use digital media than ever before (Benkler 2006; Bimber 2007; Castells 2004). Furthermore, the digital reach of these networks allow individuals to access global networks of other users, giving them the ability to connect to other users across very large distances. This produces a small-world effect that allows information to travel quickly and widely within those networks to reach broad audiences (Barabasi and Bonabeau 2003; Bennett and Segerberg 2013; Buchanan 2002; Lotan et al. 2011). These technologies thus allow individuals across vast scales to interact and collaborate with one another to co-produce knowledge, in ways that only governments have traditionally been able to do. This new ability to co-produce knowledge is foundational to all the processes described here.

The term crowdsourcing was coined in 2006 in *Wired* magazine to describe an extension of outsourcing (Howe 2006). It was viewed as a novel way in which corporations could use the Internet to access large pools of (often untrained) international labor to complete menial tasks cheaply (Ettlinger 2016). The process was quickly adopted by proponents of open content and collaboration, who have argued that crowdsourcing has the potential to empower regular citizens to produce knowledge and products that disrupt proprietary business models (e. g., Benkler 2006, Lievrouw 2011). Crowdsourcing is largely thought to function via Linus's Law, which says that all problems can be fixed quickly given enough participants in a process. Wikipedia is perhaps the most widely-cited example of open knowledge production through crowdsourcing, but other examples abound (Elwood et al. 2013). The data resulting from crowdsourcing are sometimes referred to as usergenerated content (UGC), which is generally placed in contrast to professionally-generated content (Cooper et al. 2017). Importantly, though, some of the neoliberal elements of crowdsourcing's inception remain, and scholars have criticized how even well-intentioned crowdsourcing efforts can responsibilize citizens to collect their own data and then exploit that data for profit (e.g., Leszczynski 2013). This underscores the colonial potential of crowdsourcing, as highlighted in the introduction.

VGI is regularly framed as a subset of crowdsourcing, although the terms do not perfectly overlap (Cooper et al. 2017). This term was first introduced by geographer Michael Goodchild (2008) to refer to the ways in which citizens can now use GPS units (often embedded in mobile devices) to act as 'voluntary sensors' to collectively produce geospatial intelligence. Goodchild's piece set off a series of discussions within the discipline about how the term should be defined, including debates of whether information is actually being produced voluntarily (e.g., Harvey 2013; van Exel et al. 2011), whether there is a strict binary between VGI and professional geospatial data (e.g., Cinnamon 2015), the relationship between VGI and hacking (e.g., McConchie 2015), and others. Despite these debates, there is widespread agreement that VGI is both a product of and has contributed to broader shifts in how society thinks about and engages with geospatial data. Within the discipline of geography, these shifts have been encapsulated by the term 'neogeography', which broadly refers to the opening up of geospatial knowledge production to new individuals and methods (Capineri 2016; Graham 2010). While VGI and other neogeographic practices are not intrinsically digital in nature, they most often do result in digital data production thereby expanding the socalled geospatial web, or geoweb (Elwood 2008; Elwood et al. 2013). Taken together, these processes represent a fundamental shift in who is actively participating in the negotiation of geospatial knowledge, data, and representations, and therefore shifts in how geospatial data are productive of power relations. For some these shifts hold great potential for the radical democratization of the epistemological [|4] foundations of cartography (e.g., Young and Gilmore 2017), while for others VGI signals the production of new data inequalities and tyrannies (e.g., Elwood et al. 2013).

Debates around the democratizing potential of VGI are perhaps nowhere more relevant than in the Global South, given the historical use of crowdsourcing to exploit international divisions of labor as well as existing digital and data divide. Researchers, international organizations, and corporations have all explored various approaches to bringing the benefits of VGI to the Global South. Perhaps the earliest and largest set of academic research has focused on how VGI can be used for crisis or disaster management within international settings (Goodchild and Glennon 2010). Zook et al. (2010), for example, described how VGI was used to support responses to the 2010 earthquake in Haiti. Geoweb applications including OpenStreetMaps (OSM) and GeoCommons were set up to allow users from around the world to produce maps that could be used to direct the emergency management practices of international organizations that were in the field. OSM users, for instance, used satellite imagery to trace out information about how streets, buildings, and other infrastructure were impacted by the earthquake. Other applications, like

Ushahidi, allowed survivors to use SMS, MMS, or online interfaces to send messages directly to emergency responders (Zook et al. 2010). This allowed survivors of the earthquake to text for help, and for emergency response units to quickly and efficiently locate and respond to incidents. These platforms, and others, have also been used to respond to a range of natural and humanitarian crises, including earthquakes, election tampering, refugee crises, the spread of epidemics, and more (Weyer et al. 2019; Zambrano 2014). Even within this relatively well-developed area of research, though, many gaps remain. Porto de Albuquerque et al. (2019) argues that there still aren't sufficient methodological guidelines for how to implement VGI within humanitarian relief, and that more research is particularly necessary in the area of validation.

Although it has been most popularized within disaster management, the method has also spread to other areas of sustainable development. VGI has proven successful for Global South projects in the areas of mapping land cover and agriculture (Fritz et al. 2009; Lesiv et al. 2018; See et al. 2013), citizen science for conservation (Genovese and Roche 2010; Pocock et al. 2018), urban planning (Diaz 2016; Ruiz-Correa et al. 2017), and more. VGI projects have been viewed as having particular potential across Africa, and as a result the UN Economic Commission for Africa (ECA) produced a 2017 guideline document to help organizations implement their own crowdsourced mapping projects. Numerous VGI projects have been carried out across the continent, including Ushahidi-based election monitoring and Map Kibera in Kenya, iCitizens in South Africa, agricultural support in Uganda, Ebola mapping in West Africa, the mapping of refugee shelters in Somalia, natural disaster response in Tanzania and Malawi, and much more (UNECA 2017; Yilma 2019; Zambrano 2014).

While these projects have demonstrated the potential for VGI across Africa, many research gaps and challenges remain. First, there is still a paucity of data available within and across many African countries. Many of the projects described above are confined to a few specific topic areas (e. g., disaster management, urban governance, agriculture), relatively small scales (e. g., single communities or countries), and projects of relatively short duration (e. g., a single election cycle). This ensures the continuation of data gaps and inequalities across the continent – especially with regard to data that are difficult to glean from satellite imagery and that come from more rural or remote areas. There also tends to be more experimentation with crowdsourcing in countries with more developed ICT infrastructure and higher GDP, such as Kenya and South Africa. Second, and perhaps more importantly, many of these VGI projects are strongly driven by international NGOs that hold Western views of development and data. Omanga and Mainye (2019), for example, describe their negative experience, as African scholars, participating in a research project that sought to evaluate the effectiveness of Ushahidi at monitoring electoral-related violence in Kenya. They found that the actual users of Ushahidi tended to be networks of international NGO employees, rather than Kenyans – and that this greatly biased the knowledge produced. In their words, they found that the relationship between digital innovation, NGOs, and funding agencies "reproduced a hierarchical, top-down 'developmental' logic, whose main inspiration was an uncritical techno-determinist rationality." (Omanga and Mainye 2019) Their experiences are consistent with broader critiques of how supply-side aid (Fechter [|5] and Schwittay 2019) and digital humanitarianism (Burns2014) use development to extend Western and neoliberal rationalities. There is a risk that these projects feed back into the processes of colonialism and neoliberalism that VGI would ideally resist.

More work is therefore needed to understand how African organizations agencies might implement and sustain their own comprehensive VGI projects. As Graham et al. (2014) point out, this research needs not only to identify the technical constraints that African organizations face in implementing technology-based projects, but also the social, political, and even regulatory barriers to the success of VGI in particular African contexts. Other important areas of research include what types of digital platforms can be developed to overcome the resource and ICT constraints faced in many African countries (Chaula 2019); how methods can be developed to minimize bias and uncertainty within African VGI datasets (Basiri et al. 2019; Bordogna et al. 2016; Brown 2017); how VGI projects might be made scalable and sustainable (Arora 2016); and what relationship African-driven VGI projects can and should have to broader patterns of neoliberalization and colonialism (Arora 2016). While this paper cannot answer all of these large questions, it attempts to begin the conversation by examining a VGI project designed to help AfLIA, an international NGO based in Ghana, map all public libraries across the continent. Our hope is that, by describing the challenges and benefits that AfLIA and their partners experienced throughout the project, we will provide a road map for other African organizations interested in implementing their own VGI projects. This case study is described in the next section.

CASE STUDY: MAPPING PUBLIC LIBRARIES ACROSS AFRICA

This paper focuses on a VGI project that has emerged out of a research collaboration between the University of Washington and AfLIA. This project, called Advancing Library Visibility in Africa (ALVA), broadly examines the relationship between public libraries and sustainable development across sub-Saharan Africa. Public libraries and development organizations share many common goals that make them strong potential partners. Both groups seek to build strong community partnerships as they work toward sustainable development goals by increasing access to information and communication technologies (ICTs; Abdulla 1998; Akintunde 2004; Bamgbose and Etim 2015), promoting literacy and lifelong learning (Alabi et al. 2018), providing health and social services (Albright 2007), and much more. Agbo and Ongekweodiri (2014) go so far as to describe libraries as 'engines of development' to underscore the powerful and active role that these institutions might play. In spite of these commonalities, libraries are often overlooked as development partners (Fellows et al. 2012). Many librarians argue that this is often the result of a perception problem – libraries are not framing their own work in terms of development, and development organizations therefore do not see the potential value in partnerships. As a result, libraries are not getting the support that they need in order to effectively implement services that will advance local development (Ashraf 2018; Bradley 2016; Moahi 2019). This makes the role of libraries within development more of an unrealized potential than a reality (Moahi 2019).

ALVA responds to this challenge by asking how public libraries can overcome perception issues and fully demonstrate their value as development partners. Long term the project hopes to build strong data culture and data collection expertise within libraries across Africa, so that they can collect, analyze, and present data that documents the impact they have on their local communities. In the short term, however, the team recognized that a much more basic data need was more pressing – in most countries across sub-Saharan Africa, there is a lack of data on even the number of libraries and their locations. The project team therefore determined that collection of geospatial and organizational information (including contact information) for libraries across the continent was the highest priority. This could then serve as a base layer for presenting additional information (e. g., development impact) about libraries in the future.

Because library location data was not officially collected by government agencies in most countries, the project team chose to explore a VGI approach. However, the project team was concerned that a traditional crowdsourcing approach (targeting the general public) would fail given that many libraries are in remote, rural locations with little ICT infrastructure and that the mapping of public libraries seemed unlikely to garner broad public interest. This [|6] feeling was confirmed by the low number of public libraries that we found on other crowdsourcing platforms, such as OSM, relative to rough estimates published by the International Federation of Library Associations and Institutions (IFLA; IFLA 2018). We therefore chose to adopt a facilitated VGI approach (Cinnamon and Schuurman 2013), in which we solicited the help of targeted library professionals and library networks to crowdsource library locations. Our plan was to develop a public-facing mapping platform to which anyone could contribute data, but then to train in-country 'Champions' within every country across the continent to direct data collection activities within their professional networks.

We first developed a test platform using the Ushahidi platform, which we piloted with a group of 120 participants from one of AfLIA's library training programs. The platform asked users to identify the location of their own library and then to provide attribute information including its name, what type of library it is, contact details, and what types of services it provides to patrons. The goal of the pilot was to determine the general usability of the platform for users, as well as to understand challenges they faced in contributing accurate geolocation information about their library. By the end of the test 28 unique library sites had been contributed across 11 countries. Users were invited to participate in a follow-up survey which asked them about their experience with the platform. We found that multiple users had difficulty with the auto-location feature of the platform and that some participants became confused by some of the terminology and concepts related to the attribute data we were attempting to collect. Based on feedback we chose to develop a new platform using the ArcGIS Online Crowdsource Reporter application, with a more user-friendly interface and a much shorter survey. Users are now only asked to provide the location of the library, its name, its type (public, academic, etc.), and contact information for both the library and user. They also have the option of submitting a photograph of the library. Eventually we would like to transition the platform to open source software, so that it can be sustainably hosted by AfLIA instead of the University of Washington.

We then began the process of selecting our incountry Champions. The goal was to select individuals that are well-connected within the public library sector of their respective country, so that they would be able to advertise the platform to a wide range of librarians around the country. We began the process with Champions from three countries and then slowly added additional countries to the project. After a Champion was selected they were trained by AfLIA on how to use the crowdsourcing application. At that point they were then in charge of organizing crowdsourcing efforts within their country. They chose to do this in varied ways – some organized training sessions for librarians; some distributed training videos or other material to contacts, but didn't formally train them within an interactive setting; some simply distributed the link to the site to their contacts; and some chose not to involve other librarians at all, but instead to travel around the country doing the mapping themselves.

We also developed a separate platform, using ArcGIS Online Crowdsource Manager, that allowed the research team to implement a quality control process for library submissions. Once a library is submitted, researchers at AfLIA take several steps to ensure that it seemed like a reasonable location. First, they check the timestamp of the submission to verify that it was submitted during working hours. Throughout the project we found that if a location was submitted outside of these work hours, then the submitter was often not actually at the library location. Since they would often use the application's autolocation feature, these submissions were often erroneous. Second, using satellite imagery the researchers verify that the submission was on or near a specific building. Third, the researchers compare photographs of the library (either obtained from the submission or found through social media searches) to both satellite imagery of the location and also, when available, Google Street View images. If any of these steps raised questions for the reviewer, then a researcher would reach out to the relevant Champion to get further clarification. The researcher would often share screenshots of satellite imagery or Google Street View images of the submission and ask whether the location seemed to be correct. This would lead either to verification or revision of the location. At the end of this process the library point would be approved. Prior to the stakeholders' meeting described below, Champions from twenty-three countries were participating in the project. [|7]

METHODOLOGY

In October 2019 the research team invited the twenty-three Champions actively involved in the mapping work to participate in a 3-day Champions' Meeting in Accra, Ghana. The purpose of that meeting was to examine the challenges that the Champions faced in implementing the crowdsourcing process within their country, to discuss opportunities produced by either the mapping process or resulting data, and to brainstorm ways to make the project sustainable in the longterm within their country. Research related to this meeting was divided into two portions. First, Champions were asked to participate in a survey about their mapping work ahead of the meeting. Second, Champions both presented on their mapping progress and also engaged in group work during the meeting itself. The individual and group work presentations were recorded for analysis.

The survey was distributed via email to all twenty-three Champions ahead of the meeting. The email provided them with a link that took them to an online survey portal, where they could choose to take the survey in English, French, Portuguese, or Arabic. Once the Champion chose their language and gave their informed consent to participate, they were then taken to a series of questions that solicited either openended or categorical responses. Questions covered the following topics: background on the participant's job within the library field; their prior experience with performing data collection in their jobs; the bureaucratic, technology, financial, or other challenges that they encountered while participating in our crowdsourcing project; the benefits that they have experienced from participation in the project; and the broader impact they believe that the crowdsourced data could have on their country's library sector. All twenty-three of the invited Champions successfully completed the survey.

The meeting itself took place over three days and was attended by twenty-two of the invited Champions. Days 1 and 2 of the meeting were a combination of individual presentations to the whole group, question and answer periods with the whole group, and group work in smaller, 5–6 person groups. In most cases the group work resulted in a presentation back to the rest of the Champions, in order to summarize the major conclusions of the group activity or discussion. On Day 1 the participants discussed the following topics: their general experiences with the data collection process; the challenges they encountered; the personal benefits they experienced through participation in the project; a summary of the results from the survey that they took; and an update on the overall progress of the project. On Day 2 they discussed how they think libraries could utilize the project data in their country and how they believe data collection projects could be made sustainable within the country. On the final day the meeting took a slightly different format. In the morning the Champions visited two libraries in Accra, as a bonding experience and form of professional development. In the afternoon the research team then shared preliminary results from Days 1 and 2 with National Librarians from across Africa, who were visiting Accra for the 3rd Meeting of African Library Ministers. This was an opportunity for the team to get greater buy-in from the government agencies that control library activities. As discussed below, greater buy-in from government was a key suggestion from Champions to ensure long-term sustainability for the project.

Analysis for this paper focused on a subset of data from the meeting. First, the meeting began with presentations by the Champions about their experiences with the data collection process. They were asked to design a presentation using a template that focused on the strategies they used to organize crowdsourcing within their country; the challenges they faced in implementing the project, and how they attempted to overcome those challenges; the benefits they've experienced through their participation; and any thoughts they have on long-term sustainability of the data collection. These presentations were included in analysis. Second, at the end of Day 1 Champions were asked to perform small group work to identify what they considered to be the top three financial problems, technological problems, bureaucratic problems, and benefits related to the project. Prior to this group work, Champions had given their individual presentations about these topics and the research team had also presented a summary of the survey (which also covered these topics). This group work was used as an opportunity to encourage the Champions to reflect on all of those prior discussions, and to try to come to a consensus around the most important challenges and opportunities related to crowdsourcing. Each group created a presentation summarizing the challenges and benefits that they selected, and [|8] these presentations were analyzed for this paper. Third, on Day 2 the small groups were asked to discuss possible strategies for making our data collection process sustainable within their respective countries. The group presentations resulting from those discussions were analyzed for this paper. These three data sets were analyzed alongside the qualitative (openended) results of the survey. The researchers performed an inductive analysis of these data using a grounded theoretical approach (Clarke 2003; Glaser 1978; Kitchin and Tate 2000). They analyzed all four data sources together, looking for common trends that shed light on the key challenges, opportunities, and approaches to sustainability related to implementing our crowdsourcing project within participating African countries. This process was iterative and the researchers triangulated codes across the different data sources to ensure consistency (Baxter and Eyles 2010). Results of this analysis are discussed below. Although analysis reflected discussions of both the challenges and benefits of VGI, this paper focuses primarily on challenges in order to highlight the unique barriers that Global South researchers and practitioners need to plan around when designing their own projects. Nevertheless, our Champions saw large benefits to the work, and these are summarized in the conclusion.

DISCUSSION

By performing the VGI project across twenty-three different countries and giving Champions wide latitude in how they chose to implement it, we allowed for a wide range of experimentation in order to see what techniques most effectively produce crowdsourced data. Despite this flexible approach, our research revealed more similarities than differences across the Champions' experiences. Although they were coming from countries with different linguistic and cultural histories, library governance systems, economic and development levels, and more, the Champions tended to use very similar methods for contacting librarians and they also faced very similar challenges and opportunities. In fact, toward the end of the meeting one Champion commented that the presentations made them feel much better about their slow progress, precisely because they saw that their struggles were common ones across the continent. Our analysis revealed three take-aways that resonated most strongly across both survey results and meeting discussions – poor Internet connectivity was one of the largest issues faced by Champions; local librarians often resisted participation within the project; and the relationship between the project and existing library (government or non-governmental) organizations presented both challenges and opportunities. In many instances these lessons contradict common assumptions about the advantages and methods of crowdsourcing, and therefore highlight the need to develop unique VGI approaches tailored to the unique context of countries across Africa. Each of these take-aways is discussed below.

Internet connectivity

One common assumption about crowdsourcing is that it is able to successfully leverage the ubiquity of digital networks to democratize knowledge production. However, our project highlights the deep limits that digital divides continue to impose on crowdsourcing efforts across Africa. Champions overwhelmingly identified Internet connectivity as a challenge that they faced during the project. For some Champions this was a defining feature of all aspects of the project – they live in countries where the Internet is slow, expensive, and regularly disrupted by power outages. These problems made it more difficult for those Champions to contact any librarians, thereby hindering their general progress. In other countries connectivity tends to be more stable but is still affected by unequal infrastructural geographies. Librarians working in remote and rural locations tend to have little access to Internet connectivity, whether through Wi-Fi or mobile data. During the meeting Groups 1, 2, and 3 also shared that some of these rural libraries suffered from lack of access to computers or mobile phones, making Internet access entirely impossible even if the area had some form of connectivity. To some extent this challenge should not be surprising – research on digital divides has long emphasized how uneven access to the material infrastructure of the Web produces asymmetries in what geographies experience digital empowerment (Crampton 2003, 2009; Crutcher and Zook 2009; Elwood et al. 2013; Gilbert 2010; Graham and Zook 2013). These inequalities are also often experienced more dramatically in Global South contexts (Young 2019a, b). Nevertheless, VGI [|9] literature still tends to be optimistic about how crowdsourcing opens knowledge production up to even the most marginalized populations. Our research indicates that, without appropriate interventions, crowdsourcing approaches alone will produce deeply incomplete and uneven representations of sub-Saharan Africa. Digital divides produce both inequalities *between* countries and between rural and urban areas *within* countries.

Connectivity challenges were exacerbated by lack of funding and a lack of technical literacy amongst librarians. The project provided Champions with a small stipend to compensate them for the time they spent implementing the project in their countries. However, following a traditional crowdsourcing approach, the project did not provide librarian users with financial support. Our assumption was that submission of a single library location would be a quick and relatively simple process, thereby not justifying financial support. Unfortunately, Champions found that this assumption was not borne out in reality. As one argued in their survey, "Calling and communicating with librarians needed data and airtime and this was very costly especially in my country data is very expensive." Rural librarians, in particular, often did not have enough money to make phone calls, much less to use data to submit their library location on our site. As one Champion said in their survey, "At times some Librarians are having no credit to access WhatsApp, to complete the data, but willing to call back once they have credit." In this instance funding only produced delays, since librarians would eventually have data that they could use. In other cases Champions chose to use their own funding to purchase data packages for librarians: "I mostly had to call, which required airtime and in some instances I had to buy data for the participants to connect their phones online." Notably, these costs weren't only (or even mostly) associated with the submission of data – they were largely incurred because Champions had to spend a large amount of time on the phone with librarians, walking them through the location submission process. This was because many librarians had very low levels of technical and cartographic literacy. Champions found that many librarians could not use their phones for anything other than phone calls; that some librarians did not know how to use social network applications; and that they had difficulty navigating the platform itself, even when provided with a full explanation of the process. Other librarians also had a great deal of trouble zooming and panning the map to locate their library. Whenever possible we recommended that they use a smartphone so that they could take advantage of its GPS unit. In these instances the librarian could use the autolocate feature on the map, assuming that they were doing the submission while at their library. However, some librarians only had access to computers or older smartphones, and therefore could not use this feature. Throughout the quality control process we regularly found that manual placement was highly inaccurate. We believe that librarians would often not zoom the map sufficiently, and would simply place their library in the general space of their country or city. This would produce very inaccurate locations when a user zoomed into the submission. While our quality control process was designed to reduce some of the inaccuracies, it is certainly imperfect given the lack of ground truthing. Some risk therefore remains that these rural areas are less accurately represented. Given that these literacy issues also tend to be correlated with rural areas, this means that rural areas are not only more likely to be unrepresented within maps but also more likely to be inaccurately represented.

In order to overcome these issues Champions were forced to use a wide variety of methods to support local librarians. This represents a need for a much higher level of support and intervention than is used by most crowdsourcing projects. First, Champions tended to use many different communication methods to ensure that the librarians understood the process – it was not as simple as advertising a URL within their social media networks. These methods included databased modes of communication (email, social media, WhatsApp, sharing of downloadable videos), phone calls, and in-person meetings. Nearly half of the Champions indicated, in the survey, that they needed to use at least two of these three forms of communication. Five of the twenty-three respondents indicated that they regularly used all three. One Champion explained their communication workflow as the following: "Email sent to libraries explaining the project and its objectives and seeking their collabo-ration; This is followed by phone calls and further clarifications/explanations on the project; in some cases personal meeting is required on site and assistance provided to submit the data." Another Champion followed a similar process, but began with WhatsApp [|10] and phone calls: "Communication through whatsapp [sic] or phone calls; training for data input on the platform if consensus is reached; trave-ling to the place of libraries in case of technical difficulties." We were particularly surprised to see how often the Champions were physically traveling to libraries, since this was not an expectation of the project. Several of the Champions were physically traveling to all library locations and submitting locations themselves. This was facilitated if traveling between the libraries was a regular part of their job. As one Champion said, "Sometimes I take advantage of my routine monitoring trips to collect data and brief staff on the significance of the project." In these in-stances the method more resembled an authoritative data collection effort than it did a crowdsourcing project, since an officially trained project representative was going from site to site to collect the data. However, the Champions emphasized that this was often necessary to overcome the ICT access and technical literacy issues faced at some libraries, particularly in rural areas. The take-away here is that a more tra-ditional VGI approach would have been highly biased toward more affluent African countries and urban areas, as compared to our methodology. This underscores the importance of augmenting VGI projects with higher levels of support, in order to overcome the biases produced by digital divides across Africa.

Librarian motivation

Crowdsourcing projects also require a large and motivated public in order to suc-ceed. Unfortunately, Champions found that there was not a sufficient volunteer or open data culture within this library community to easily sustain crowdsourcing – a second important challenge was motivating public librarians to participate in the project. This challenge is concerning because librarians should, hypothetically, be ideal participants in a project intended to benefit the public library sector. Brown (2017) argues that a focus on the technological components of VGI projects is often misplaced – the most important aspect of these projects lies in understanding how to isolate and motivate a particular public to produce high quality data. This is eas-iest when the mapping work is closely aligned to the livelihoods or everyday needs of those being asked to engage in mapping (Brown 2017; Chuene and Mtsweni 2015). In the context of this project, the best possible users should be librarians and library users. However, our Champions found that many librarians approached the project with a great deal of skepticism and resistance. In some cases, it was unclear why the librarians didn't wish to participate – they were often willing to learn about the project and even undergo training, but they wouldn't follow through with data submission. One Champion said, "One main issue is that of getting Librarians to participate in the project. They are willing to go through the training on filling out the form but for some reasons they don't end up filling it." Champions were often quite persistent in following up in these cases, and would sometimes finally get a librarian to submit data. However, if that data appeared inaccurate during our quality control process, they then wouldn't get any responses from the librarian for revision. As another Champion said, "Reluctance of some libraries [is a problem]; some libraries promise to fill in the form, but they don't even after several follow up calls are made. Others take long to make the correction requested for, some don't even bother." In some cases Champions felt they would need to travel to these libraries and do the mapping work themselves, due to total lack of response or engagement from the librarians.

In other instances, the Champions had a clearer idea of why librarians chose not to engage in the project. Some librarians did not believe that the project would directly benefit them at a personal level, and therefore did not see a point to their participation. Champions were trained to describe how the data could be used for advocacy purposes, to increase the visibility of libraries in the eyes of potential funders. However, this purpose was often not concrete enough. As one Champion said, the librarians often did not want to participate in data collection, "especially because they did not understand the immediate benefit to the library." Instead, this same Champion said, these librarians "requested to know if AfLIA had an intervention plan to support them. When asked to give an indication of their requirements, they stressed donation of books, and computers as well as training in use of ICTs, etc." Others wanted direct, personal benefits instead of benefits to the library sector. One Champion stated that the "main challenge was getting some librarians to submit data. They were eager to know how the project would benefit them individually not as an organization." Others described how [|11] "librarians and library sector leaders seem to cooperate for projects that pay them for volunteering" and "there is always high expectation for financial rewards the moment you try to engage other people". In preparing their summary of the bureaucratic changes they faced, Group 2 tied this dynamic back to the practices of NGOs in their countries. They argued that "NGOs are paying for research data and librarians are expecting Champions to also pay them." Group 1 listed 'motivation costs' as a financial challenge faced by the project. It seemed clear in each of these cases that librarians did not want the abstract benefits often offered as motivation by crowdsourcing benefits, but instead expected direct payment for their labor.

Other librarians feared using the Internet or sharing data. Fear of the Internet, or what several Champions referred to as 'technophobia', often seemed to be related to low levels of technical literacy. As one Champion argued, "some librarians are having fear on Internet, hence they assume it is complicated to do online input, especially the GPS, to locate the Library." Others seem to have a nebulous fear of sharing information about their libraries, perhaps because the information might be used against them (e. g., if data collection strategies are used by governments to restrict funding based on performance). One Champion said that library authorities would not "grant permission to library staff to provide the data on the grounds that such data are confidential and are meant for internal use only". Another Champion even indicated that they faced "lack of data accessibility in certain areas linked to insecurity in these areas due to terrorism." In most instances these reasons were not well explained or developed, since they are only based on the perceptions and assumptions of Champions (who often had low levels of access to the librarians themselves). More research is therefore needed to really understand the motivations of potential crowdsourcing participants in these contexts. The key takeaway, though, is that VGI projects in these countries face large hurdles in getting widespread buy-in from the public. Given that VGI projects tend to be *most* successful with those individuals that most directly interact with the locations being crowdsourced on an everyday basis (Brown 2017), it is worrisome that librarians are so highly resistant to participation. This is particularly the case given that they are receiving high levels of support and encouragement from Champions, which doesn't usually happen in more passive forms of crowdsourcing.

This isn't to say that the Champions were unsuccessful at eventually motivating many of the participants – they used a variety of tactics to overcome this barrier. As discussed in the next section, many Champions found that they could motivate librarians by leveraging their relationships with AfLIA or with national library authorities. For example, Champions requested a letter from AfLIA detailing the organization's support for the project and the benefits of participating. Champions believed that this letter greatly improved their success in soliciting participation. Other Champions obtained similar letters from their own National Librarian, to indicate that incountry authorities were supportive of the project. In other instances Champions were able to successfully motivate librarians by sharing some of the more abstract or indirect benefits of participation, including increasing the visibility of their library, expanding connections to others in their country's library field, and increasing their technological capacity through training to use the platform. Successful approaches for motivation varied by country and even local library, and at times no approach was successful at all. More research is therefore needed to understand the cultural dimensions of VGI participation. VGI projects need to consider how local cultural understandings of payment/volunteering, data, and privacy intersect with motivations to engage in crowdsourcing, and either adapt their project (e. g., provide payment) or work to change that particular culture (e. g., develop nonpayment based motivation strategies grounded in the local culture). Other aspects of our project have begun to explore how cultural understandings of knowledge production impact local data culture (Lynch et al. 2020), but much more research is needed in this area.

Relationships to existing library authorities

Finally, crowdsourcing projects are often viewed as a challenge to or shift away from authoritative forms of data collection. However, our Champions largely viewed it as the opposite – as a method for increasing the involvement of authorities in data collection. Negotiating the relationship between this project and existing library authorities represented both a challenge and an opportunity for all of the Champions. Like the other challenges, this shouldn't be a huge [|12] surprise – all forms of neogeography represent a unique challenge to authoritative forms of data collection, and therefore must negotiate their relationship to authorities. What was surprising was the strength of the desire by both Champions and librarians alike to ultimately fold the project into traditional library hierarchies. A common theme amongst Champions was the need to get authorization for the project from government authorities (e. g., a secretary of general within the government ministry that has authority over public libraries). Approximately half of the Champions needed this authorization so that they could engage in project activities themselves. More importantly, though, the librarians themselves would expect to see that permission was expressly granted before they were willing to participate in the project themselves, in the form of the letters discussed in the last section. One Champion told us, for example, that they encountered "reluctance of library staff to provide data unless permission is granted by employer in a very formal way". At the meeting, Group 4 presented that they felt that these hierarchies were sometimes used as an excuse to justify non-participation. In these cases lack of authorization wasn't actually an issue being faced by the participant, but was a fabricated excuse used to cover up whatever the librarian's real motivation was for not participating. They could then displace blame for their lack of participation onto someone higher up in the hierarchy. Interestingly, the members of Group 3 found that they did not encounter many of these bureaucratic hurdles. They attributed this to the fact that many of them already hold very high positions within their own library sector, and therefore already have the authority to take on any project that they might like. They argued that this provides a strong justification for more formally integrating the project into regular government operations, rather than continuing to frame it as VGI.

Interestingly, while inclusion of the project within existing organizational structures was extremely important for political and symbolic reasons, it did not provide Champions with many additional resources. In particular, most Champions reported that their governments did not have the resources to connect the Champions with librarians that might want to participate in the project. Champions reported that there was no existing database of libraries nor "existing formal and informal communication networks" for networking with librarians. However, Champions saw this as a key opportunity for the project – it gave them the justification to create these networks. Several Champions reported that they had created their own WhatsApp chat group for public libraries for this project, and they were now using it to talk about other opportunities such as "Library event happenings, studying opportunities, [and] funding opportunities." Others reported opportunities for building mentorship opportunities with younger librarians. In the long run they believed that these networks, alongside the data they were collecting, would also be valuable for other training and advocacy activities.

In the end every single group at the meeting recommended increased integration between the project and their respective national governments. The official recommendation that they put forward was that the project's data collection activities should be integrated into the mandate of each country's National Library system, with implementation delegated to an appropriate subdivision. They viewed this as a far more sustainable and effective method of maintaining up-to-date information about library locations in their countries. The take-away here is that the momentum of the project is toward more organizational integration, not less. In this case crowdsourcing has been viewed not as an attempt to privatize data collection practices that have historically been carried out by the government, but instead as a method for government agencies to start collecting particular types of data. Thus far this approach has seemed effective – when recommendations from the Champions were later presented to national library directors and ministers, these government actors seemed (1) to support the activities and (2) to be eager to ensure that their country was keeping up with data collection in relation to other countries. As the project continues to negotiate the relationship between VGI and government, it will draw on the experiences of other governments that have leveraged crowdsourcing methods (see, e. g., Haklay et al. 2014).

CONCLUSION

VGI projects like OSM, crisis mapping in Haiti, and Map Kibera have effectively captured the academic imagination, highlighting the potential power of [|13] crowdsourcing as a tool for knowledge production in the Global South. However, their success has also normalized a particular vision of what a Global South crowdsourcing project might look like – focused on topics of high interest to the global development community; on features that are easily verified by remote sensing; on locations with higher ICT penetration, affluence, or NGO presence; etc. As Omanga and Mainye (2019) point out, these common models of VGI can undermine the method's empowering potential by allowing organizations from the Global North to largely drive what types of knowledges are created through crowdsourcing. In this sense, maps produced through VGI risk, drawing on Spivak (1999), acting as tools through which the Global North continues to ventriloquize the Global South. This, in turn, can turn VGI into a tool for data colonialism rather than democratization (Dé et al. 2018; Fraser 2019; Thatcher et al. 2016; Young 2019a).

This paper asks what a VGI project might look like if it is co-created and co-implemented with African organizations, with the goal of sustainably handing the project over to those organizations over the long run. The focus of the mapping, libraries, is quite different from many of the features commonly mapped through VGI – libraries are largely ignored by development organizations (Fellows et al. 2012), are located in geographies that have little access to ICT infrastructure and little contact with outside organizations, and are difficult to identify through the use of satellite imagery. We have found that crowdsourcing has been a relatively effective method for mapping libraries and for getting library organizations (and government actors) excited about developing more of a data culture around libraries. However, we also found that crowdsourcing in these contexts has faced different challenges and realized different benefits from other similar projects. The project highlighted how dramatically digital divides can bias crowdsourcing results, as well as the degree to which local cultural views influence public motivations to participate in crowdsourcing. Perhaps most importantly, the project showed how crowdsourcing is viewed by some as a way to influence, and even increase the involvement of, government authorities in mapping instead of as a way to privatize existing government data collection efforts.

All of these findings offer lessons for researchers attempting to implement crowdsourcing projects in the Global South. These projects must be carefully designed so that they account for digital divides, local cultural views of volunteerism and open data, and orientations toward government or organizational hierarchy. We would argue that this is best done through consultation and partnership with partners in the Global South, to ensure that the end result is something that reflects their needs and can be sustained. Naturally, more research is also needed in many of these areas, to ensure greater success. What devices, platforms, and training approaches are most successful at reaching remote and rural communities, to ensure their representation within crowdsourcing projects? Low-data usage platforms like WhatsApp are extremely popular in many of these areas, for example, but have not been fully explored in the context of crowdsourcing. What are local understandings of (open) data across different areas of the Global South, and how do they intersect with public motivations to participate in crowdsourcing? How are ICTs creating or transforming data culture, and how does this produce new democratizing or colonial geographies? What is the relationship between crowdsourcing methods and governments that have not historically collected large amounts of authoritative data? These questions, and many others, must be answered before we really understand the implications of these digital data-production methods across many geographies of the Global South.

While this paper focused on the challenges of implementing VGI projects in the Global South, it is worth noting that the stakeholder meeting also highlighted a range of benefits. Benefits discussed by the Champions included increasing the visibility of libraries, expanding connections between participants and others in the library field, and increasing the capacity of librarians. The first benefit, increasing library visibility, is a direct result of having location data about libraries, and is a common benefit of all crowdsourcing projects. The other two benefits are more indirect forms of empowerment that are not directly related to the collected data, but instead to participation in data collection itself (see, e. g., Elwood 2002; Young and Gilmore 2013). Champions argued that the project forced them to create new communication channels (e. g., via WhatsApp) with their colleagues across the country, led them to learn more about the services of other libraries, and more. They believe that this networking will spur future [|14] cooperative efforts across the field. They also found that participation in the project expanded the data and technology skills of themselves and participating librarians. One Champion found that the participating "librarians improved their skills and some were excited to be part of a global research [project]." They argue that these skills would broadly advance their own professional lives. In the end the Champions unanimously agreed that the project's benefits outweighed the challenges, and that it was vital to ensure the project's long-term sustainability. This reflects a strong belief that it is worth navigating difficult challenges to expand data collection and culture in the Global South.

ALVA will continue to explore many of these questions as it continues to expand to additional countries. Since the completion of the Champions' meeting data collection has expanded to an additional seven countries. Just over 700 libraries have been submitted to the site and approved, and an additional 200 locations are proceeding through the quality control process. In the long run the project will explore what additional types of information it can collect about the libraries that have been mapped, with the ultimate goal of building a powerful platform that libraries can use to collect, analyze, and present data that documents the impact they have on their local communities. Our hope is that this will make libraries more visible as local partners and champions of community-driven development. Along the way we expect to uncover many other lessons about data culture and knowledge production across the Global South.

Acknowledgements

We would like to thank all our ALVA Champions, who have made this research possible. The project was funded by a gift from the Bill & Melinda Gates Foundation.

Funding

Funding was provided by the Bill & Melinda Gates Foundation.

Availability of data and material

This dataset includes sensitive human subjects information, which is currently not available for public consumption.

Compliance with ethical standards

Conflicts of interest
The authors declare that they have no conflict of interest.

Informed consent
All participants provided informed consent ahead of their participation on the research.

Human and animal rights
The project used human participants and received prior approval from the University of Washington Human Subjects Division.

REFERENCES

Abdulla, A. D. (1998). The role of libraries in Somalia's reformation. *Libri, 48*(1), 58–66.

Agbo, A. D., & Onyekweodiri, N. E. (2014). Libraries are dynamic tools for national development. *Chinese Librarianship, 38*, 29–35.

Akintunde, S. A. (2004). Libraries as tools for ICT development. In *National conference / annual general meeting of the Nigerian Library Association (NLA), Akure, Nigeria.*

Alabi, A. O., Oyelude, A. A., & Sokoya, A. A. (2018). 'It takes two to tango': Libraries achieving sustainable development goals through preservation of Indigenous knowledge on textile craft making (adire) among women. In *XXIII SCECSAL conference, Entebbe, Uganda.*

Albright, K. (2007). Libraries in the time of AIDS: African perspectives and recommendations for a revised model of LIS education. *The International information & library review, 39*(2), 109–120.

Arora, P. (2016). The bottom of the data pyramid: Big data and the global south. *International Journal of Communication, 10*, 1681–1699.

Ashraf, T. (2018). Transforming libraries into centers of community engagement: Towards inclusion, equality & empowerment. In *IFLA WLIC 2018, Kuala Lumpur.*

Bamgbose, O. J., & Etim, I. A. (2015). Accessing government information in Africa through the Right to Know: The role of the library. *IFLA WLIC 2015: Cape Town.*

Barabasi, A. L., & Bonabeau, E. (2003). Scale-free networks. *Scientific American, 28*(8), 50–59.

Basiri, A., Haklay, M., Foody, G., & Mooney, P. (2019). Crowdsourced geospatial data quality: Challenges and future directions. *International Journal of Geographical Information Science, 33*(8), 1588–1593.

Baxter, J., & Eyles, J. (2010). *Qualitative research methods in human geography.* North York: Oxford University Press.

Benkler, Y. (2006). *The wealth of networks: How social production transforms markets and freedom.* New Haven, CT: Yale University Press. [|15]

Bennett, W. L., & Segerberg, A. (2013). *The logic of connective action: Digital media and the personalization of contentious politics.* Cambridge: Cambridge University Press.

Bimber, B. (2007). How information shapes political institutions. In D. A. Graber (Ed.), *Media power in politics.* Washington, DC: CQ Press.

Bordogna, G., Frigerio, L., Kliment, T., Brivio, P. A., Hossard, L., Manfron, G., et al. (2016). Contextualized VGI creation and management to cope with uncertainty and imprecision. *International Journal of Geo-Information, 5*(234), 1–19.

Bott, M., & Young, G. (2012). The role of crowdsourcing for better governance in international development. *The Fletcher Journal of Human Security, 27*(1), 47–70.

Bradley, F. (2016). A world with universal literacy: The role of libraries and access to information in the UN 2030 Agenda. *International Federation of Library Associations and Institutions, 42*(2), 118–125.

Brown, G. (2017). A review of sampling effects and response bias in internet participatory mapping (PPGIS/PGIS/VGI). *Transactions in GIS, 21*(1), 39–56.

Buchanan, M. (2002). *Small worlds and the groundbreaking theory of networks.* New York: Norton.

Burns, R. (2014). Moments of closure in the knowledge politics of digital humanitarianism. *Geoforum, 53*(2014), 51–62.

Capineri, C. (2016). The nature of volunteered geographic information. In C. Capineri, M. Haklay, H. Huang, V. Antoniou, J. Kettunen, F. Ostermann, & R. Purves (Eds.), *European handbook of crowdsourced geographic information* (pp. 15–33). London: Ubiquity Press.

Caquard, S. (2014). Cartography II: Collective cartographies in the social media era. *Progress in Human Geography, 38*(1), 141–150.

Castells, M. (2004). Informationalism, networks, and the network society: A theoretical blueprint. In M. Castells's (Ed.), *The network society: A cross-cultural perspective.* Northampton, MA: Edward Elgar.

Chaula, J. (2019). Opportunities and challenges for integrating statistical and spatial data in East African Countries. *American Scientific Research Journal for Engineering, Technology, and Sciences, 59*(1), 42–48.

Chuene, D., & Mtsweni, J. (2015). The adoption of crowdsourcing platforms in South Africa. In *IST-Africa 2015 conference proceedings.*

Cinnamon, J. (2015). Deconstructing the binaries of spatial data production: Towards hybridity. *The Canadian Geographer, 59*(1), 35–51.

Cinnamon, J., & Schuurman, N. (2013). Confronting the datadivide in a time of spatial turns and volunteered geographic information. *GeoJournal, 78*(4), 657–674.

Clarke, A. E. (2003). Situational analyses: Grounded theory mapping after the postmodern turn. *Symbolic Interaction, 26*(4), 553–576.

Cooper, A. K., Coetzee, S., & Kourie, D. G. (2017). Volunteered geographical information, crowdsourcing, citizen science and neogeography are not the same. In *Proceedings of the international cartographic association* (pp. 657–674).

Crampton, J. (2009). Cartography: performative, participatory, political. *Progress in Human Geography, 33*(6), 840–848.

Crampton, J. (2003). *The political mapping of cyberspace.* Chicago: University of Chicago Press.

Crutcher, M., & Zook, M. (2009). Placemarks and waterlines: Racialized cyberscapes in post-Katrina Google Earth. *Geoforum, 40*(4), 523–534.

Dé, R., Pal, A., Sethi, R., Reddy, S. K., & Chitre, C. (2018). ICT4D research: A call for a strong critical approach. *Information Technology for Development, 24*(1), 63–94.

Diaz, A. G. P. (2016). *Tweet-sourcing Caracas: Using E-participation for Urban Planning in Global South Cities.* Thesis, Department of Urban Planning, Columbia University.

Elwood, S. (2002). GIS use in community planning: A multidimensional analysis of empowerment. *Environment and Planning A, 34*(5), 905–922.

Elwood, S. (2008). Volunteered geographic information: Key questions, concepts and methods to guide emerging research and practice. *GeoJournal, 72*(3–4), 133–135.

Elwood, S., Goodchild, M., & Sui, D. (2013). Researching volunteered geographic information (VGI). *Annals of the Association of American Geographers, 102*(3), 571–590.

Ettlinger, N. (2016). The governance of crowdsourcing: Rationalities of the new exploitation. *Environment and Planning A, 48*(11), 2162–2180.

Fechter, A.-M., & Schwittay, A. (2019). Citizen aid: Grassroots interventions in development and humanitarianism. *Third World Quarterly, 40*(10), 1769–1780.

Fellows, M., Coward, C., & Sears, R. (2012). Beyond access: *Perceptions of libraries as development partners.* Seattle: Technology and Social Change Group, University of Washington Information School.

Fraser, A. (2019). Curating digital geographies in an era of data colonialism. *Geoforum, 104*, 193–200.

Fritz, S., McCallum, I., Schill, C., Perger, C., Grillmayer, R., Achard, F., et al. (2009). Geo-Wiki.org: The use of crowdsourcing to improve global land cover. *Remote Sensing, 1*(3), 345–354.

Genovese, E., & Roche, S. (2010). Potential of VGI as resource for SDIs in the North/South Context. *Geomatica, 64*(4), 439–450.

Gilbert, M. (2010). Theorizing digital and urban inequalities: Critical geographies of 'race', gender and technological capital. *Information, Communication & Society, 13*(7), 1000–1018.

Glaser, B. G. (1978). *Advances in the methodology of grounded theory: Theoretical sensitivity*. Mill Valley: The Sociology Press.

Goodchild, M. F. (2008). Citizens as sensors: The world of volunteered geography. *GeoJournal, 69*(4), 211–221.

Goodchild, M. F., & Glennon, J. A. (2010). Crowdsourcing geographic information for disaster response: A research frontier. *International Journal of Digital Earth, 3*(3), 231–241.

Graham, M. (2010). Neogeography and the palimpsests of place: Web 2.0 and the construction of a virtual earth. *Tidjchrift voor Economische en Sociale Geografie, 101*(4), 422–436.

Graham, M., & Zook, M. (2013). Augmented realities and uneven geographies: Exploring the geo-linguistic contours of the web. *Environment and Planning A, 45*(1), 77–99.

Graham, M., Hogan, B., Straumann, R. K., & Medhat, A. (2014). Uneven geographies of user-generated information: Patterns of increasing informational poverty. *Annals* [|16] *of the Association of American Geographers, 104*(4), 746–764.

Haklay, M, Antoniou, V., Basiouka, S., et al. (2014). *Crowdsourced geographic information use in government*. Report to GFDRR. London: World Bank.

Harvey, F. (2013). To volunteer or to contribute locational information? Towards truth in labeling for crowdsourced geographic information. In D. Z. Sui, S. Elwood, & M. F. Goodchild (Eds.), *Crowdsourcing geographic knowledge* (pp. 31–42). Berlin: Springer.

Howe, J. (2006). The rise of crowdsourcing. *Wired Magazine, 14*(6), 1–4.

IFLA. (2018). *Libraries and the sustainable development goals: A storytelling manual*. Retrieved January 7, 2020, from https://www.ifla.org/files/assets/hq/topics/libraries-development/documents/sdg-storytelling-manual.pdf.

Ingwe, R. (2017). Crowdsourcing-based geoinformation, disadvantaged urbanization challenges, sub-Saharan Africa: Theoretical perspectives and notes. *Quaestiones Geographicae, 36*(1), 5–14.

Kitchin, R., & Tate, N. J. (2000). *Conducting research in human geography*. Upper Saddle River: Prentice Hall.

Kshetri, N. (2014). The emerging role of Big Data in key development issues: Opportunities, challenges, and concerns. *Big Data & Society, 1*(2), 1–20.

Lesiv, M., Bayas, J. C. L., See, L., et al. (2018). Estimating the global distribution of field size using crowdsourcing. *Global Change Biology, 25*(1), 174–186.

Leszczynski, A. (2013). Situating the geoweb in political economy. *Progress in Human Geography, 36*(1), 72–89.

Lievano, K. (2017). Is Volunteered Geographic Information (VGI) a better option for developing countries like Panama than for developed countries like the United States? *Latitude: Multidisciplinary Research Journal, 9*, 8–12.

Lievrouw, L. (2011). *Alternative and activist new media*. Malden, MA: Polity Press.

Lotan, G., Graeff, E., Ananny, M., Gaffney, D., Pearce, I., & Boyd, D. (2011). The revolutions were tweeted: Information flows during the 2011 Tunisian and Egyptian revolutions. *International Journal of Communication, 5*, 1375–1405.

Lynch, R., Young, J., Jowaisas, C., et al. (2020). Data challenges for public libraries: African perspectives and the social context of knowledge. *Information Development*. https://doi.org/10.1177/0266666920907118.

Mann, S. (2017). Big Data is a big lie without little data: Humanistic intelligence as a human right. *Big Data & Society, 4*(1), 1–10.

Marr, B. (2018). How much data do we create every day? The mind-blowing stats everyone should read. *Forbes*. Retrieved January 7, 2020, from https://www.forbes.com/sites/bernardmarr/ 2018/05/21/how-much-data-do-we-create-every-day-the-mind-blowing-stats-everyone-should-read/#3affce0960ba.

McConchie, A. (2015). Hacker cartography: Crowdsourced geography, OpenStreetMap, and the hacker political imaginary. ACME: *An International EJournal for Critical Geographies, 14*(3), 874–898.

Moahi, K. H. (2019). A framework for advocacy, outreach and public programming in public libraries in Africa. In P. Ngulubes (Ed.), *Handbook of research on advocacy, promotion, and public programming for memory institutions.* Pennsylvania: IGI Global.

Omanga, D., & Mainye, P. C. (2019). North-South collaborations as a way of 'not knowing Africa': Researching digital technologies in Kenya. *Journal of African Cultural Studies, 31*(3), 273–275.

Petrov C. (2019). Big Statistics (2019). *Techjury.* Retrieved January 7, 2020, from https://techjury. net/stats-about/big-data-statistics/.

Pocock, M. J. O., Roy, H. E., August, T., et al. (2018). Developing the global potential of citizen science: Assessing opportunities that benefit people, society and the environment in East Africa. *Journal of Applied Ecology, 56*(2), 274–281.

Porto de Albuquerque, J., Yeboah, G., Pitidis, V., & Ulbrich, P. (2019). Towards a participatory methodology for community-generated data generation to analyse urban health inequalities: A multi-country case study. In *Proceedings of the 52nd Hawaii international conference on system sciences.*

Ruiz-Correa, S., et al. (2017). SenseCityVity: Mobile crowdsourcing, urban awareness, and collective action in Mexico. *Pervasive Computing, 16*(2), 44–53.

See, L., McCallum, I., Fritz, S., et al. (2013). Mapping cropland in Ethiopia using crowdsourcing. *International Journal of Geosciences, 4*(6), 6–13.

Spivak, G. (1999). *A critique of postcolonial reason.* Cambridge: Harvard University Press.

Taylor, L. (2017). What is data justice? The case for connecting digital rights and freedoms globally. *Big Data & Society, 4*(2), 1–14.

Taylor, L., & Broeders, D. (2015). In the name of development: Power, profit and the datafication of the global South. *Geoforum, 64*, 229–237.

Taylor, L., & Schroeder, R. (2014). Is bigger better? The emergence of big data as a tool for international development policy. *GeoJournal, 80*(4), 1–16.

Thatcher, J., O'Sullivan, D., & Mahmoudi, D. (2016). Data colonialism through accumulation by dispossession. *Environment and Planning D: Society and Space, 34*(6), 990–1006.

UNECA. (2017). Volunteered Geographic Information in Africa, United Nations Economic Commission for Africa (UNECA) (p. 2017). Ethiopia: Addis Ababa.

van Exel, M., Dias, E. & Fruijtier, S. (2011). Proposing a redefinition of the social geographic information domain – Why perpetuating the use of 'VGI' will lead to misconceptions and information clutter. In A. Çöltekin, & K. C. Clarke (Eds.), Position papers on virtual globes or virtual geographical reality: How much detail does a digital earth require? Proceedings of the ASPRS/ CaGIS 2010 Workshop (pp. 29–36), Orlando, Florida, USA.

Weyer, D., Bezerra, J. C., & De Vos, A. (2019). Participatory mapping in a developing country context: Lessons from South Africa. *Land, 8*(134), 1–16.

Yilma, A. D. (2019). Volunteer geographic information in Africa. International Archives of the Photogrammetry, Remote Sensing and Spatial Information Sciences, XLII-2/W13, 1615–1620. [|17]

Young, J. C. (2019a). The new knowledge politics of digital colonialism. *Environment and Planning A: Economy and Space, 51*(7), 1424–1441.

Young, J. C. (2019b). Rural digital geographies and new landscapes of social resilience. *The Journal of Rural Studies, 70*, 66–74.

Young, J. C., & Gilmore, M. P. (2013). The spatial politics of affect and emotion in participatory GIS. *Annals of the Association of American Geographers, 103*(4), 808–823.

Young, J.C., & Gilmore, M.P. (2017). Participatory uses of geospatial technologies to leverage multiple knowledge systems within development contexts. *World Development, 93*, 389–401.

Zambrano, R. (2014). Crowdsourcing and human development: The role of governments. In *ICE-GOV2014* (pp. 170–177).

Zook, M., Graham, M., Shelton, T., & Gorman, S. (2010). Volunteered geographic information and crowdsourcing disaster relief: A case study of the Haitian earthquake. *World Medical & Health Policy, 2*(2), 7–33.

Publisher's Note

Springer Nature remains neutral with regard to jurisdictional claims in published maps and institutional affiliations.

MÜNDIGKEIT IN EINER KULTUR DER DIGITALITÄT – GEOGRAPHISCHE BILDUNG UND „SPATIAL CITIZENSHIP"

Christian Dorsch / Detlef Kanwischer

1. KULTUR DER DIGITALITÄT

„Like air and drinking water, being digital will be noticed only by its absence, not its presence" (Negroponte 1998, o. S.). Mit diesem Zitat hat Nicholas Negroponte (1998) eine Entwicklung prognostiziert, die den gegenwärtigen gesellschaftlichen Wandlungsprozess illustriert. Im Kern geht es darum, dass der Leitmedienwechsel von der Gutenberg- zur Internetgalaxis abgeschlossen ist, wie es das Zitat von Petar Jandrić u. a. (2016) verdeutlicht: „We are increasingly no longer in a world where digital technology and media is separate, virtual, ‚other' to ‚natural' human and social life" (893). Mit anderen Worten: Das Digitale bildet heutzutage keine virtuelle Parallelwelt mehr, sondern ist in unseren Alltag integriert. Es kommt somit zu einer Neukonfiguration zwischen dem gesellschaftlichen und individuellen Handeln und der Digitalität. Das Digitale ist hierbei keine isolierbare Entität mehr, sondern konstitutiv. In Anbetracht dieser weitreichenden Veränderungen rücken Fragen nach neuen Formen des Verhältnisses zwischen Individuum, Technik und Gesellschaft in den Mittelpunkt des Bildungsinteresses. Dies bezieht sich insbesondere auf die Eruierung von mündigkeitsorientierten Bildungskonzepten im Umgang mit der Digitalisierung, wie sie u. a. in der politischen Bildung in den letzten Jahren diskutiert wurden (Achour/Massing 2018; Jantschek/Waldmann 2017). Hiermit wäre der erkenntnisleitende [|24] Ausgangspunkt unseres Beitrags markiert, der sich insbesondere auf die Perspektive der geographischen Bildung bezieht.

Felix Stalder (2017) beschreibt die Gegenwart bereits als „Kultur der Digitalität". In ihr sind Referentialität, Gemeinschaftlichkeit und Algorithmizität die dominierenden Prinzipien: Schon immer beziehen sich kulturprägende Akteur/-innen, wie Künstler/-innen oder Musiker/-innen, in ihren Werken aufeinander, werten diese auf und verändern sie (Referentialität). Die allumfassende digitale Verfügbarkeit kulturellen Materials ermöglicht es nun auch Laien, am Computer oder Smartphone durch das Auswählen und Neuarrangieren von Bestehendem neue Bedeutungen zu generieren. Ein kollektiv getragener Referenzrahmen dient dabei als Verstärker und Bühne für die neuen Produkte. Nur in der Gemeinschaft, beispielsweise innerhalb der Community von Wikipedia, stehen den Kulturproduzent/-innen die Ressourcen und Handlungsoptionen zur Verfügung und nur durch sie können die geschaffenen Bedeutungen dauerhaft werden (Gemeinschaftlichkeit). Gleichzeitig wirken in der Gemeinschaft „Dynamiken der Netzwerkmacht, die Freiwilligkeit und Zwang, Autonomie und Fremdbestimmung in neuer Weise konfigurieren" (Stalder 2017, 13). Automatisierte Entscheidungsverfahren in Form von Algorithmen reduzieren den Informationsfluss und machen ihn für die menschliche Wahrnehmung sichtbar. Algorithmen entscheiden letztendlich, was Grundlage des menschlichen Handelns wird, also auch, was künstlerisch aufgegriffen und reproduziert wird (Algorithmizität). Stalder (2017) beschreibt hiermit die drei gegenwärtig dominierenden Praktiken, die Kultur konstituieren und die auch für die Untersuchung von Bildungsprozessen in der digitalisierten Gegenwart relevant sind.

Diese Entwicklung lässt sich anhand eines kleinen Fallbeispiels illustrieren. Am Vorabend einer Demonstration der Bürger/-innenbewegung „Patriotische Europäer gegen die Islamisierung des Abendlandes" (Pegida) am 7. Mai 2018 konnte man auf die Frauenkirche in Dresden projiziert lesen: „Durchhalten, freundliches Dresden, ihr seid nicht alleine!" Am selben Tag wurde in den sozialen Medien unter dem Hashtag #DurchhaltenDresden dazu aufgerufen, den rechten Raumkonstruktionen eine eigene Sicht entgegenzusetzen: „Auf Facebook, auf Twitter und natürlich in Dresden selbst. Verbindet Dich eine schöne Geschichte mit Dresden? […] Ein Geschäft oder etwas anderes ganz Besonderes, was jeder gesehen haben sollte? Teile es mit uns! Wir bewerten diesen Ort und Dresden in den sozialen Netzwerken positiv […] oder teilen ihn mit unseren Freunden" (Reconquista Internet 2018a). Die für diese Aktion verantwortliche Gruppe nennt sich „Reconquista Internet" und versteht sich als überparteiliches Gegenmodell zu der rechtsradikalen Plattform „Reconquista Germanica", deren Mitglieder [|25] schon seit Längerem durch gezielte Aktionen die Diskurse in sozialen Medien beeinflussen. Ziel ist es, die Instrumente des Internets zu nutzen, um den rechtskonservativen bzw. rechtsradikalen Raumkonstruktionen im digitalen, aber zunehmend auch im physischen Raum etwas entgegenzusetzen. Mithilfe gezielter Aktionen im „Realraum", der Verbreitung sogenannter Memes und GIFs (Bilder und Videos) und durch das massenhafte Weiterverbreiten von Nachrichten (tweets) auf der Plattform Twitter soll „Hass und Ignoranz mit Vernunft und Liebe" (Reconquista Internet 2018b) begegnet und der Diskurs in den sozialen Medien verändert werden. Die Nutzer/-innen von „Reconquista Internet" erstellen für ihre Aktionen Memes, in denen sie sich auf Kunstwerke, Werbeanzeigen und anderes kulturelles Material beziehen, um durch deren Manipulation eine neue Botschaft zu transportieren. Dabei wird also im Sinne der Referentialität „bereits mit Bedeutung versehenes Material […] verwendet, um neue Bedeutung zu schaffen" (Stalder 2017, 97). Gleichzeitig profitieren die Aktivist/-innen von der gemeinschaftlichen Organisation der Plattform: Ressourcen, die ihnen dabei helfen, „die geteilte Weltsicht in Handlungen umsetzen" (Stalder 2017, 148) zu können, werden erst durch die Gesamtzahl der Mitglieder innerhalb der Gemeinschaft geschaffen. Schließlich steht das Projekt „Reconquista Internet" in Wechselwirkung zu der von Stalder so bezeichneten Algorithmizität: Algorithmen, die wiederum von Individuen mit bestimmten Intentionen programmiert wurden, bestimmen, welche Informationen Internetnutzern/-innen angezeigt werden. Sie treffen also eine scheinbar auf den/die Nutzer/-in zugeschnittene Vorauswahl in der Datenflut des Internets. In ihrem kreativen Umgang mit Algorithmen, indem sie z. B. die sogenannten „Trends" bei Twitter durch gezieltes „Retweeten" von Nachrichten manipulieren, nehmen die Aktivist/-innen auf der einen Seite Einfluss auf diese Algorithmizität und nutzen sie in ihrem Sinne. Auf der anderen Seite sind auch sie in ihren Handlungsräumen durch Algorithmen geprägt, da sie nur auf diejenigen Nachrichten mit Aktionen reagieren können, die ihnen in den algorithmengesteuerten Nachrichtenportalen, beispielsweise auf der Social-Media-Plattform Facebook, angezeigt werden.

Wie aus dem Fallbeispiel deutlich wurde, erlaubt die Digitalisierung einerseits neue Wege der demokratischen Auseinandersetzung, lässt neue gemeinschaftliche Formationen entstehen und erleichtert Partizipation an der gesellschaftlichen Entwicklung, andererseits stellen sich z. B. Fragen nach der Beeinflussung demokratischer Entscheidungen durch Algorithmen, der Machtposition dahinterstehender Unternehmen bzw. Individuen und der Gewährleistung informationeller Selbstbestimmung. Die Verantwortlichen in Schulen und Hochschulen müssen die dabei auftretenden Praktiken aufgreifen und für Bildungsprozesse fruchtbar machen. Gleichzeitig müssen sie dafür [|26] sorgen, dass Schüler/-innen und Studierende Fähigkeiten erlernen, die ihnen dabei helfen, die Entwicklungen zu hinterfragen und ihre Interessen zu kommunizieren. Ziel muss es sein, in dieser Kultur mündig agieren und die digitale Gesellschaft gestalten zu können. Es stellt sich jedoch die Frage, welche Bildungskonzepte sich hierfür anbieten. Um diese Frage für die geographische Bildung zu beantworten, werden wir eingangs Dimensionen einer mündigkeitsorientierten Bildung herausarbeiten, diese nachfolgend in Bezug zur digitalen Gesellschaft setzen, um darauf aufbauend Bildungskonzepte für eine Förderung der mündigen Auseinandersetzung in einer Kultur der Digitalität zu diskutieren. Abschließend werden wir ein kurzes Fazit ziehen.

2. DIMENSIONEN MÜNDIGKEITSORIENTIERTER BILDUNG

Mündigkeit ist nach dem Zweiten Weltkrieg zu einem Leitziel der deutschen Schulbildung geworden. Die normative Leitidee einer „Erziehung zur Mündigkeit" (Adorno 1970) soll insbesondere in der Schule umgesetzt werden. In diesem Begriffspaar wird die Widersprüchlichkeit des Begriffs im Bildungskontext deutlich. Andreas Gruschka (1994) verweist z. B. auf die Diskrepanz zwischen den individuellen Bildungsbemühungen, die autonome und kritisch denkende Individuen hervorbringen sollen, und der Institution Schule, die den Lernenden ihre Lerngegenstände als indiskutabel vorsetzt. Trotz des aufgezeigten Widerspruchs hat das Konstrukt Mündigkeit in den letzten Jahrzehnten nichts von seiner Anziehungskraft eingebüßt. Gleichwohl wird der Begriff je nach ideengeschichtlichem pädagogischem Hintergrund mit unterschiedlichen Bedeutungen aufgeladen. Angesichts der Vieldeutigkeit des Begriffs werden wir im Folgenden unterschiedliche Dimensionen vorstellen, die in den zahlreichen Definitionen immer wieder mit Mündigkeit verknüpft werden. Ohne den Anspruch zu erheben, Mündigkeit in all ihren Facetten abzubilden, sind die drei Dimensionen Struktur- und Selbstreflexivität (Dewey 1951; Adorno 1970; Eis 2015), Sich-seiner-selbst-bewusst-Sein (Roth 1971; Tugendhat 1979; Klafki 2007) und Autonomie (Adorno 1970; Deci/Ryan 1993) die Kernelemente eines Konstrukts, das wir als mündigkeitsorientierte Bildung bezeichnen. Die einzelnen Dimensionen werden im Folgenden charakterisiert.

2.1 Struktur- und Selbstreflexivität

Für Dewey besteht Reflexion in einem „regen, andauernden, sorgfältigen Prüfen von etwas, das für wahr gehalten wird" (Dewey 1951, 6). Am Anfang der Reflexion steht ein „Zustand der Beunruhigung" (ebd.). Jede mögliche Handlungsoption [|27] werde in der Reflexion dahingehend geprüft, ob sie geeignet sei, das Problem zu lösen. Für diesen Beurteilungsprozess sei Geduld erforderlich und die Fähigkeit, den Zustand der Unsicherheit so lange zu ertragen, bis eine geeignete Lösung gefunden sei (Dewey 1951, 12–14). Scott Lash (1996) unterscheidet zwischen der „Strukturreflexivität" – sie bezieht sich als Reflexion auf die „sozialen Existenzbedingungen der Handelnden" und die „Regeln und Ressourcen" der gesellschaftlichen Struktur – und der „Selbstreflexivität", die sich auf das eigene Handeln und Denken bezieht. An die Stelle der vormodernen heteronormen Bestimmung ist in der Moderne die „Eigenbestimmung" getreten, ohne die Selbstreflexivität nicht möglich sei (Lash 1996, 203). Heinrich Roth (1971) sieht die Selbstreflexivität als Voraussetzung für die Entwicklung einer mündigen Handlungsfähigkeit, die dem Individuum gestatte, sein Selbst bewusst wahrzunehmen. Sie befähige den Menschen zudem, produktive und kreative Lösungen für Konflikte zu entwickeln (Roth 1971, 382). Für Theodor W. Adorno (1970) stellt die kritische Selbstreflexion gar den einzig sinnvollen Zweck der Erziehung dar. Ähnlich sehen auch Jörissen/Marotzki (2009) in ihr das einzig sinnstiftende Moment von Bildung: In der durch Krisen geprägten Moderne, in der tradierte Sinnbezüge erodiert werden, können Orientierungssysteme nur zeitlich begrenzt wirken – die eine „richtige Weltsicht" existiert nicht mehr. Vielmehr müsse das Individuum die verschiedenen Perspektiven seiner Umwelt wahrnehmen und dadurch seinen eigenen Standpunkt relativieren (Jörissen/Marotzki 2009, 16–18). Bezüglich des Zeitpunkts des Reflektierens lassen sich darüber hinaus nach Donald A. Schön (1983) zwei Unterscheidungen machen: „Reflection in action" meint das unmittelbare Reflektieren während der Handlung. Es geht also darum, eine Situation spontan auf Basis von Erfahrung meistern zu können. „Reflection on action" dagegen findet nach der Aktion statt und bezeichnet das nachträgliche Dokumentieren und Reflektieren von Erfolgs- und Misserfolgskriterien. Generell ist Reflexivität notwendig, um sich seiner selbst bewusst zu sein, womit wir zur nächsten Dimension kommen.

2.2 Sich seiner selbst bewusst sein

Bei Kant ist es erst das „Bewusstsein seiner selbst" (Kant 1900 ff., 79), das es ermöglicht, anderes von sich zu unterscheiden und zu ordnen. Auch Roth (1971) hebt die besondere Rolle des Selbst hervor. Das mündige Individuum verfüge neben einem Orientierungs- und Wertungssystem über ein „entwickeltes und ausgebautes

Steuerungs- und Kontrollsystem" (Roth 1971, 220). Das Ich, dessen der mündige Mensch sich bewusst sei und das ihn vom Säugling unterscheide, sei darin die zentrale Instanz. Ebenso thematisiert Ernst Tugendhat (1979) die Verknüpfung zwischen dem Selbst [|28] und der Selbstreflexivität. Er unterscheidet ein epistemisches Selbstbewusstsein und ein „Sichzusichverhalten". Ersteres umfasst das Wissen, das eine Person über sich selbst hat, z. B. darüber, wo sie geboren ist, aber auch, ob sie mutig oder feige ist (Tugendhat 1979, 27). Kern des freiheitlichen Lebens sei es darüber hinaus, sich die sogenannten „praktischen Fragen" stellen zu können: Wie will ich leben, welche Person will ich sein, wie strebe ich am ehesten auf meine eigene Weise mein eigenes Wohl an? Diese Fragen gründeten auf der Möglichkeit, zu sich Stellung nehmen zu können, wofür wiederum Reflexivität nötig sei (Tugendhat 1979, 30). Dies bezeichnet er als „Sichzusichverhalten" bzw. als ein „reflektiertes Selbstverhältnis". Klafki (2007) nennt die „vernünftige Selbstbestimmung" das zentrale Ziel von Bildung. Letztlich könne der Mensch diese nur aus eigenem Antrieb heraus erreichen: Bildung sei „zugleich Weg und Ausdruck solcher Selbstbestimmungsfähigkeit" (Klafki 2007, 20). Sie umfasst zum einen die Bestimmung der eigenen persönlichen Lebensbeziehungen und zum anderen Mitbestimmungsfähigkeit in Gesellschaft und Politik (Klafki 2007, 52).

2.3 Autonomie

Um als Individuum autonom handeln zu können, ist für Taylor Lorenz (1988) entscheidend, zu ergründen, was der „authentische Wunsch oder Zweck" ist – anders gesagt: Der Mensch muss sich seiner selbst bewusst sein. Wer einfach triebgesteuert dem stärksten Bedürfnis nachgeht, handele unfrei und nicht autonom (Taylor Lorenz 1988, 134). Deci und Ryan (1993) sehen ähnliche Voraussetzungen für autonomes Handeln. In ihrer self-determination-theory gilt Autonomie neben Kompetenzerleben und sozialer Eingebundenheit als eines von drei Grundbedürfnissen des Menschen. Je mehr man eine Handlung als frei erlebe, d. h. den individuellen Bedürfnissen und Wünschen entsprechend, desto autonomer bzw. selbstbestimmter sei sie (Deci/Ryan 1993, 225). Eine Lernmotivation könne nur dann effektiv sein, wenn sie den „Prinzipien des individuellen Selbst" entspreche, also nicht aufoktroyiert sei (Deci/Ryan 1993, 235). Für Adorno ist im Zusammenhang mit Mündigkeit das Kernelement von Autonomie die Fähigkeit, sich bestehenden Routinen und Herrschaftsformen zu widersetzen: Widerstand ist für ihn sogar gleichbedeutend mit Autonomie als „die einzig wahrhafte Kraft gegen das Prinzip von Auschwitz […], die Kraft zur Reflexion, zur Selbstbestimmung, zum Nicht-Mitmachen" (Adorno 1970, 93). Gleichwohl besteht zwischen Erziehung und Autonomie ein naheliegender Widerspruch, der sich u. a. darin zeigt, dass sich Lernende nicht von Lehrkräften durch Erziehung vereinnahmen lassen, da ihr Streben nach Autonomie stärker ist (Bernfeld 1973, 143). Trotz dieses Widerspruchs gab und gibt es immer wieder Ansätze in der Pädagogik, mit [|29] denen die Autonomie der Lernenden gefördert werden soll. Werner Helsper (2018) schlägt z. B. vor, das Selbst- und Weltbild der Lernenden immer wieder zu irritieren und herauszufordern, um damit

„Krisen des Wissens" herauszufordern. Aufgabe der Lehrkraft sei es dann, unterstützend Krisenlösungen anzubieten.

So schillernd und abstrakt der Begriff der Mündigkeit auch ist, die Analyse der drei Dimensionen einer mündigkeitsorientierten Bildung verdeutlicht, dass diese durch Bildungsprozesse gefördert werden können. Bevor wir entsprechende Bildungsansätze diskutieren, möchten wir jedoch die Rolle der einzelnen Dimensionen in der digitalen Gesellschaft beleuchten.

3. MÜNDIGKEIT IN DER DIGITALEN GESELLSCHAFT

Die Autor/-innen der im vorigen Punkt vorgestellten theoretischen Ausführungen zum Begriff der Mündigkeit und den drei Dimensionen mündigkeitsorientierter Bildung beziehen sich zunächst auf den gesellschaftlichen Kontext und die Herrschaftsstrukturen ihrer jeweiligen Zeit, aus der sie stammen. So schrieb beispielsweise Kant vor dem Hintergrund der Ständegesellschaft in Westeuropa, während Adorno maßgeblich durch die Erfahrung der nationalsozialistischen Diktatur und des Zweiten Weltkriegs beeinflusst wurde. Natürlich stellt sich somit die Frage, inwiefern ihre Theorien für die digitale Gesellschaft weiter Relevanz haben, inwiefern also Mündigkeit heute noch „gebraucht" wird. Betrachtet man die einzelnen Aspekte mündigkeitsorientierter Bildung, also Struktur- und Selbstreflexivität, Sich-seiner-selbst-bewusst-Sein und Autonomie, fällt die Antwort scheinbar leicht, da sich zahlreiche Arbeiten aus der Philosophie, Pädagogik, Psychologie und anderen Fachrichtungen mit ihnen beschäftigen. Aber auch der Begriff der Mündigkeit selbst wird heute wieder häufiger in den Mund genommen, wenn vom „mündige[n] Bürger in der digitalen Welt" (Beer 2016) oder „digitale[r] Mündigkeit" (Körber-Stiftung o. J.) die Rede ist. Inwiefern die kulturellen Praktiken des Digitalen mit den Dimensionen einer mündigkeitsorientierten Bildung verknüpft sind und welche möglichen Bewältigungsstrategien diskutiert werden, wird nachfolgend aufgezeigt.

3.1 Struktur- und Selbstreflexivität in der digitalen Gesellschaft

Für alle drei Praktiken, die Stalders Konzept der Kultur der Digitalität beinhaltet, ist Struktur- und Selbstreflexivität notwendig. So verlangt z. B. die Auswahl und Inwertsetzung kultureller Artefakte wie Kunstobjekte, Texte, Videos und Musik im Sinne der Referentialität die Fähigkeit der Reflexivität, wenn bestehende Artefakte auf ihr virales Potenzial hin beurteilt werden müssen, bevor sie verändert und verbreitet [|30] werden können. Algorithmen müssen, sobald sie den Entscheidungsraum des Individuums einschränken, ebenso im Fokus der Reflexion sein. Ihre Funktionsweise gilt es dabei zu verstehen, um mögliche Formen des Widerstands gegen sie zu erproben (Allert/Richter 2017, 23).

Auch die gestiegenen Partizipationsmöglichkeiten in der digitalen Gesellschaft verlangen reflexive Fähigkeiten: So müssen Nutzer/-innen in der Lage sein, ihre eigenen Sichtweisen auf Umwelt und Raum mittels digitaler Medien wie z. B. Karten,

Blogs, Meinungsforen u. Ä. zu produzieren und zu kommunizieren und damit den bisher meist durch Expert/-innen vorgenommenen (Raum-)Deutungen eine eigene Perspektive entgegenzustellen, um beispielsweise einen Entscheidungsprozess zu beeinflussen. Dies lässt sich auf einfachster Ebene über digitale (Karten-)Plattformen realisieren, auf denen Nutzer/-innen Informationen eintragen: Dabei kann es sich um Mängel in der städtischen Infrastruktur handeln, wie beispielsweise defekte Straßenlaternen, Schlaglöcher in Fahrradwegen etc. Solche „Mängelmelder" existieren bereits in vielen Städten. Ebenso gibt es komplexere Plattformen, die dazu auffordern, Vorschläge für die Stadtentwicklung zu machen, über die online abgestimmt wird und die bei ausreichender Unterstützung in den Magistrat eingebracht werden. Die Menschen reflektieren und entscheiden also darüber, wie sich die Gesellschaft, in der sie leben, entwickeln soll. Dabei gilt es, die Fülle dieser Entscheidungen und der zugehörigen Informationen nach subjektiver Relevanz zu hierarchisieren und in einen biografischen Zusammenhang zu stellen, um sie überhaupt erfassen zu können. Die Vertreter der reflexiven Moderne (Giddens 1995, 151–155) haben schon verdeutlicht, dass sich der Grad der eigenen Reflexivität erhöhen muss, je weniger der Mensch an Traditionen und soziale Einbettungen gebunden ist: Der/Die Einzelne müsse sich „unter den von abstrakten Systemen gebotenen Strategien und Alternativen umsehen, um die eigene Identität auszumachen" (Giddens 1995, 155). Seine/Ihre Biografie wird dann – anstelle sozialer Kontexte und Gemeinschaften – zum Referenzrahmen (Jörissen/Marotzki 2009, 36). Gleichsam sollten die Auswirkungen der Digitalisierung – wie der Umgang mit persönlichen Daten durch Unternehmen, aber auch durch Kommunen – Ziel der Strukturreflexivität sein: Die informationelle Selbstbestimmung ist ein Gut, das in Zukunft essentiell für ein mündiges Handeln in der Gesellschaft sein wird.

3.2 Sich seiner selbst bewusst sein in der digitalen Gesellschaft

Das Digitale wurde in der individualisierten Gesellschaft immer wieder als Heilsbringer gefeiert: Es sollte die Kluft schließen zwischen der Gemeinschaft und dem ermächtigten Selbst, das befreit ist „von den Hindernissen der Hierarchie und den durch Raum [|31] und Zeit auferlegten Grenzen" (Zuboff 2017, 169). Shoshana Zuboff (2017) warnt jedoch davor, dass die geschaffenen digitalen „Verbindungen" sich immer stärker zu einem Element des Überwachungskapitalismus entwickeln, der von wenigen privaten Unternehmen dominiert werde. Statt die Freiheitsräume des autonomen Selbst zu vergrößern, stehen diese Verbindungen ihm nun fundamental entgegen. Einerseits eröffnen sich folglich durch die Digitalisierung neue Räume der Selbstbestimmung und Selbstgestaltung, andererseits werden neue Möglichkeiten der Überwachung und Fremdsteuerung geschaffen (Hoffmann-Riem 2017, 122): Wenn beispielsweise Gemeinschaften in digitalen Räumen in der Regel auf Freiwilligkeit beruhen, existieren dennoch formelle und informelle Regeln, die autonomes Handeln des Einzelnen erschweren können. Diese werden nicht immer gemeinschaftlich festgelegt, sondern oftmals von Einzelnen (z. B. Moderator/-innen) ausgelegt und dementsprechend durchgesetzt. Wer die Protokolle übernimmt, kann interagieren. Wer dies nicht tut, dessen Beiträge finden keine Beachtung. Diese Bedingungen stellen insbesondere das Selbst des Individuums vor Herausforderungen. Nur wer seine Interessen kennt, kann diese in der Gemeinschaft auch vertreten. Hierzu müssen zuvor unterschiedliche Perspektiven eingenommen und der eigene Standpunkt – das Selbst – in Relation zu anderen, aber auch zur eigenen Biografie erkannt werden.

Inwiefern Algorithmen daran beteiligt sind, wie Menschen ihr Selbst wahrnehmen, zeigt Lucas D. Introna (2017) anhand von Onlinewerbung: Ziel der Werbeunternehmen ist es, „beeindruckbare Subjekte" zu generieren. Sogenannte Adserver erlauben es, mittels in den Browsern platzierter Cookies nachzuvollziehen, welche Art von Hyperlink der/die Nutzer/-in bevorzugt anklickt und welche Internetseiten er/sie besucht. Werbeanzeigen können somit mithilfe von Algorithmen immer weiter personalisiert werden, um das Subjekt zu „beeindrucken". Introna (2017) betont dabei, dass das „beeindruckbare Subjekt" auch immer eine aktive Beteiligung voraussetzt: Die Nutzer/-innen spielen das Spiel mit, indem sie die angezeigten Inhalte durch Anklicken oder Ignorieren „kuratieren" (Introna 2017, 64). Sie entwickeln somit ihre eigene Subjektivität aktiv mit: „So verstanden sind die auf dem Bildschirm erscheinenden Werbungen nicht ‚nur' Werbungen, sie sind gleichzeitig Anregungen zu dem Subjekt, das ich werden will" (Introna 2017, 69). Somit existiert eine ambivalente Freiwilligkeit nicht nur in den digitalen Gemeinschaften, sondern auch im Umgang mit Algorithmen. Auch auf das Selbstverständnis von Gemeinschaften nehmen Algorithmen Einfluss, teilweise produzieren sie diese auch, wie Tarleton Gillespie (2014, 188) zeigt. Wenn Amazon dem/der Nutzer/-in z. B. anzeigt „Menschen wie du kauften auch …", dann konstruiert der dahinterstehende Algorithmus eine Gemeinschaft, der wir uns zuge- [|32] hörig fühlen sollten und die nur für diesen Moment existiert. Das Zustandekommen dieser Gemeinschaft ist genauso intransparent wie das der sogenannten Twitter-Trends. Hier identifiziert ein Algorithmus im Nachrichtendienst Twitter beliebte Hashtags, über die eine imaginäre Öffentlichkeit gerade kommuniziert. Wer tatsächlich zu dieser Öffentlichkeit gehört und wer davon ausgeschlossen wird, bleibt im Dunkeln. Gleichzeitig stellt Gillespie (2014) fest: „Some algorithms go further, making claims about the public they purport to know, and the users' place amid them" (Gillespie 2014, 189). Algorithmen stellen den/die Nutzer/-in also in den Kontext einer Gemeinschaft aus scheinbar ähnlichen Individuen mit ähnlichen Interessen, einem ähnlichen Selbst. Dadurch verändert sich gleichzeitig das Selbst dieser Person.

3.3 Autonomie in der digitalen Gesellschaft

Astrid Messerschmidt (2017) führt die oben genannten „Dynamiken der Netzwerkmacht" in den digitalen Gemeinschaften weiter aus: Die Ubiquität des Digitalen bedinge eine gesteigerte „Verfügbarkeit des Einzelnen", die das Individuum immer wieder zu einer „souveränen Selbstdarstellung" (beispielsweise in den sozialen Medien) veranlasse (Messerschmidt 2017, 131). Sie knüpft dabei an Adornos Kritik der Kulturindustrie an: Diese Industrie übe einen enormen Druck durch die

„Steuerung auch der gesamten Innensphäre" auf die Menschen aus (Adorno 1970, 144). Die damit einhergehende heteronome Ausrichtung der Gesellschaft stellt die Möglichkeit autonomen Handelns generell infrage.

Auch die zunehmende Kommerzialisierung des Internets führt nach Colin Crouch (2008) dazu, dass Partizipationsmöglichkeiten nur vorgetäuscht werden. Die genannten Prozesse der Kultur der Digitalität finden in zunehmend „postdemo-kratischen" Räumen statt, also in einem von privilegierten Eliten bestimmten Gemeinwesen, in der bürgerliche Mitbestimmung nichts weiter als eine von PR-Experten gesteuerte Inszenierung darstelle (Crouch 2008, 10). In den sozialen Massenmedien wie Facebook, Twitter oder WhatsApp erscheint der „Output", also die Möglichkeiten, welche die Dienste für ihre Nutzer/-innen schaffen, zwar attraktiv, auf der Input-Seite hingegen, also beispielsweise bei Entscheidungen bezüglich der Programmierung der Algorithmen oder des Datenschutzes, existiert keinerlei Mitspracherecht (Stalder 2017, 214).

Gleichzeitig ermöglichen immer leistungsfähigere Computer die Analyse von immer größer werdenden Datenmengen (Big Data): Die zuvor deskriptiv erfassten Daten – die z. B. Auskunft darüber geben, wie sich ein Individuum, eine Gruppe oder eine Gesellschaft in einer Situation x verhalten hat – werden prädiktiv ausgewertet, um die Frage zu beantworten, wie die genannten Akteure aufgrund ihrer bisher erfassten [|33] Aktionen zukünftig in einer Situation y wahrscheinlich handeln werden. Anschließend ist es mittels präskriptiver Analytik von Big Data beispielsweise möglich, Handlungsempfehlungen darüber abzuleiten, wie z. B. durch gezielte Aktionen Wahlen oder persönliche Meinungen beeinflusst werden können (Hoffmann-Riem 2017, 125). Die Algorithmen der sozialen Medien schlagen ihren Nutzer/-innen Lesenswertes vor oder produzieren sogar eigene News. Teilweise kreieren sogenannte Bots passgenau für entsprechende Nutzer/-innengruppen eigene Nachrichten. Durch Geolokationsdaten ändern sich – je nachdem, wo sich die Person gerade aufhält – Inhalt und Angebot der Nachrichten zusätzlich. Die Welt wird von den Algorithmen für die Nutzer/-innen nicht mehr nur repräsentiert, sondern immer eigens neu generiert (Stalder 2017, 189). Die Parameter dieser Selektion sind meist unbekannt. Somit schwindet der menschliche Ermessensspielraum für Entscheidungen, und Autonomie wird beschränkt (Kurz/Rieger 2017, 91).

Als mögliche Form des Widerstands gegen die Einschränkung von Autonomie und gegen Verstöße gegen das Recht auf informationelle Selbstbestimmung sieht Stalder (2016) einzig das Preisgeben von Informationen durch Mitarbeiter/-innen in den Unternehmen bzw. Behörden, indem sie beispielsweise interne Dokumente, die z. B. Überwachung oder den Missbrauch von Daten abbilden, nach außen geben und durch Medien publik machen (Stalder 2017, 242). Aus unternehmerischer und politischer Sicht sind erfahrungsgemäß nur wenige Beschränkungen und gesetzliche Flankierungen der Digitalisierung zu erwarten. Die Fähigkeit Widerstand zu leisten als Form des autonomen Handelns erhält dadurch eine zusätzliche Relevanz. Wie welche Fähigkeiten im Kontext von Bildungsprozessen gefördert werden können, wird im folgenden Punkt vorgestellt.

4. KONZEPTE FÜR DIE BILDUNG IN DER DIGITALEN GESELLSCHAFT

Die Strategie der Kultusministerkonferenz „Bildung in der digitalen Welt" (Kultus-ministerkonferenz 2016) wird immer wieder zitiert, wenn es um zukunftsweisende Konzepte für das Bildungssystem geht, obwohl das Papier mit einem eher einseiti-gen Bildungsbegriff operiert. Zwar heißt es auch hier zum Bildungsauftrag der all-gemeinbildenden Schulen in fast kantischer Wortwahl: Die Schüler/-innen sollten zu einem „selbstständigen und mündigen Leben in einer digitalen Welt befähigt werden" (Kultusministerkonferenz 2016). Doch zeigt Felicitas Macgilchrist (2017), dass das Papier den Individuen in der digitalen Welt nur einen beschränkten Hand-lungsspielraum zugesteht. Dies wird aus den von den Verfasser/-innen für notwen-dig erachteten [|34] Fähigkeiten und den Rollen, die sie hierzu den Schüler/-innen zuteilen, sichtbar: So werden Schüler/-innen in den Ausführungen als Nutzer/-in-nen, Produzent/-innen (bzw. „Maker") und nur am Rande als Kritiker/-innen der digitalen Systeme bezeichnet. Insbesondere die ersten beiden Rollen zeugen von einem funktionalistischen Medienverständnis, das vor allem auf die ökonomische Verwertbarkeit der zu erlernenden Kompetenzen zielt (Macgilchrist 2017, 147).

Ein Ansatz, der sich überfachlich mit der Frage auseinandersetzt, wie Bildungs-prozesse in der digitalen Gesellschaft fruchtbar sein können, stammt von Heidrun Allert und Christoph Richter (2017). Sie setzen sich konkret mit den Bedingungen der Kultur der Digitalität auseinander und analysieren, welche Anforderungen durch sie an die schulische Bildung gestellt werden. Sie sehen die gestaltende und produktive Auseinandersetzung mit der Unbestimmtheit digitaler Kulturen im Zen-trum der Medienpädagogik. Digitale Lern- und Portfolioplattformen wie Moodle oder Mahara versuchen, mit ihren Sharing-, Feedback- und Forenfunktionen zwar Praktiken der Gemeinschaftlichkeit abzubilden, um damit Lernprozesse zu initiie-ren, doch bezweifeln Allert/Richter (2017) die diesbezügliche Wirksamkeit von Lernplattformen. Bildungsprozesse, die in digitalen Medien stattfinden (z. B. das Schreiben eines Blogs, die Erstellung eines Videobeitrags oder das Twittern über einen Sachverhalt), tragen in immer anderer Konstellation zur Identitätsbildung und Selbstformung bei. Sie könnten jedoch in der Schule oder Universität mit ihren institutionellen Rahmenbedingungen nicht reproduziert werden (Allert/Richter 2017, 23). Das Lernen finde während der genannten Aktivitäten in „sozio-techno-ökonomischen Systemen" statt, z. B. im Austausch mit den eigenen Followern, im Versuch, einen Algorithmus auszutricksen oder in der Gruppe eine Online-Initiative zum Erfolg zu führen. Hierbei komme es weniger auf technisch-operative Fähigkei-ten an als vielmehr auf Kreativität. Kreative Praktiken bezeichnen in diesem Sinne „kollektiv reproduzierte Handlungs- und Deutungsmuster zum produktiven Um-gang mit Situationen, die unbestimmt, ambivalent, handlungs- und deutungsoffen sind" (Allert/Richter 2017, 28). Diese können in der Schule vermittelt werden. Kre-ativität und Innovation, d. h. das Hinterfragen und Neugestalten bestehender Routi-nen, sind zudem eine Voraussetzung für mündiges bzw. autonomes Handeln (Gryl 2013, 19), um die eigenen Interessen wirkungsvoll vertreten zu können.

Neben den überfachlichen Konzepten der Medienpädagogik wurden in den letzten Jahren auch in der Geographiedidaktik Ansätze entwickelt, die zu einer mündigkeitsorientierten Bildung in einer Kultur der Digitalität beitragen können. Detlef Kanwischer und Antja Schlottmann (2017) zeigen am Beispiel von ortsbezogenen Posts, wie in [|35] den sozialen Medien Räume konstruiert werden. Ortsbezogene Posts in den sozialen Medien generieren, synthetisieren und interpretieren somit lokale Informationen (Kanwischer/Schlottmann 2017, 63). Die Erkenntnis, dass soziale Medien Räume konstruieren, lässt sich für Bildungsprozesse fruchtbar machen, indem die Schüler/-innen mündigkeitsorientierte räumliche Erfahrungen machen (Kanwischer/Schlottmann 2017, 76). Hierfür ist eine Entsubjektivierung und Abstraktion notwendig, was über die Strukturelle Medienbildung (Jörissen/ Marotzki 2009) mit ihren vier Strukturierungsdimensionen möglich ist: 1) in Bezug auf die „Grenzen des Wissens", wenn darüber reflektiert wird, wie Hashtags verschiedene Medien miteinander kombinieren und somit durch ihre unterschiedliche Lesart eine Vielzahl von Deutungen möglich werden; 2) in Reflexion auf den Handlungsbezug, wenn z. B. die Handlungsoptionen, die sich aus den sozialen Medien ergeben, im Fokus stehen; 3) in Bezug auf Grenzziehungen, wenn das Verhältnis von Subjekt und Raum durch neue digitale Medien neu konfiguriert wird, und 4) zuletzt in Bezug auf Biografisierungsprozesse, wenn die Frage nach der eigenen Identität und ihren biografischen Bedingungen in sozialen Medien virulent wird (Kanwischer/Schlottmann 2017, 72–75).

Eine Antwort auf die Frage, welche konkreten Fähigkeiten in der Kultur der Digitalität aus geographiedidaktischer Perspektive benötigt werden, liefert der Spatial-Citizenship-Ansatz (Gryl/Jekel 2012; Schulze u. a. 2015): Der Spatial Citizen ist in der Lage, mithilfe digitaler Geomedien gesellschaftliche Diskurse zu initiieren. Bürger/-innen sollen so ermächtigt werden, sich öffentliche Räume mündig anzueignen und an räumlichen Gestaltungsprozessen zu partizipieren (Gryl u. a. 2017, 6). Die häufig zu beobachtende unreflektierte und lediglich technisch-methodische Verwendung digitaler Geomedien im Geographieunterricht veranlasste die Autorengruppe zur Konzeption dieses Ansatzes. Er beruht auf einer konstruktivistischen Sicht auf Karten und andere Geomedien: Einerseits bilden Geomedien stets konstruierte und subjektiv geprägte Perspektiven auf Räume ab und erfordern daher stets eine kritische Reflexion. Andererseits sind sie „machtvolle Kommunikationsinstrumente" (Jekel u. a. 2015, 7), sodass der produktive Umgang mit ihnen eine entscheidende Fähigkeit in gesellschaftlichen Diskursen ist. Die zu fördernden Fähigkeiten gehören dabei im Kompetenzmodell von Uwe Schulze u. a. (2015) zu sechs Kompetenzdimensionen, die relevante Hinweise geben, welche Rolle der Geographieunterricht für die Bildung in der digitalen Gesellschaft spielt. Die erste Dimension, die technisch-methodischen Kompetenzen, umfasst die Fähigkeiten, Karten und digitale Kartendienste zur Navigation einzusetzen, gleichzeitig aber auch Analyseaufgaben damit zu lösen. Zudem soll der Spatial Citizen in der Lage sein, Geomedien zur Prosumption und Produktion [|36] zu nutzen, also z. B. bestehende Karten im Sinne der Referentialität (Stalder 2017) zu verändern oder zu erweitern und eigene räumliche Darstellungen zu erstellen. Schließlich ist er befähigt, diese Produkte z. B. in den sozialen Medien unter Berücksichtigung der dort bestimmenden Algorithmizität einzubinden. Die zweite Kompetenzdimension umfasst zum einen Reflexivität als das „Wissen um die Konstruiertheit von Geomedien" (Jekel u. a. 2015, 7) und die Fähigkeit zur Dekonstruktion der dahinterstehenden Intentionen. Zum anderen betrifft sie den reflektierten Umgang mit Geomedien inklusive des Bewusstseins darüber, wie die eigene Wirklichkeit aus diesen Quellen konstruiert wird. Ergänzt werden sollte diese Dimension allerdings um das Wissen über die Algorithmizität und die aus ihr folgenden Konsequenzen: Auch Geomedien wie Google Maps werden durch Algorithmen geprägt, insbesondere in Bezug darauf, was den Nutzer/-innen innerhalb der Karten angezeigt wird – z. B. in der Frage, welches Café oder Restaurant besonders prominent dargestellt wird. Die Kommunikationskompetenz als dritte Dimension hilft dabei, die eigenen Raumkonstruktionen zielorientiert zu verbreiten, um somit am räumlichen Diskurs teilhaben zu können (Schulze u. a. 2015, 156). Dies kann beispielsweise in Form alternativer Bedeutungszuweisungen an den Raum auf gemeinschaftlichen Plattformen stattfinden. Um mit den dort auftretenden und oben erläuterten „Dynamiken der Netzwerkmacht" (Stalder 2017, 13) umgehen zu können, bedarf es nicht zuletzt auch der sprachlichen, parasprachlichen und nichtsprachlichen Fähigkeiten, „sozial angemessen, d. h. unter Beachtung gesellschaftlicher Konventionen, Normen und Regeln zwischen Vertretern verschiedener (kultureller) Gemeinschaften oder institutioneller Gruppen kommunizieren zu können" (Schulze u. a. 2015, 157). Die vierte und fünfte Dimension sind im Kompetenzmodell fachspezifisch bzw. fachübergreifend: Zunächst ist dies (4.) die räumliche Dimension als genuin geographische Perspektive, die das Wissen über relationale und absolute Raumkonzepte einschließt. Sie besagt, dass erstens Räume durch räumliche Handlungen und Kommunikation konstruiert werden und zweitens räumliches Denken die Voraussetzung dafür sei, (absolute) räumliche Beziehungen nachvollziehen und analysieren zu können. Die fachübergreifende Dimension der politischen Bildung (5.) ist insbesondere im Kontext mündigkeitsorientierter Bildung von Relevanz. Zu den „emanzipatorischen Prinzipien politischer Bildung" (Schulze u. a. 2015, 157) gehören demokratischen Grundsätze wie die Tatsache, dass gesellschaftliche Rahmenbedingungen grundsätzlich verhandelbar sind und somit einer Partizipationsfähigkeit bedürfen. Diese Prozesse müssen unter Anerkennung grundlegender Menschenrechte stattfinden, sodass die Autonomie des/der Einzelnen auch im Spatial-Citizenship-Ansatz beschränkt ist, sobald sie die Freiheit und Unversehrtheit [|37] anderer beeinträchtigt. Zuletzt wird der Spatial Citizen als Lehrende/-r in Schule und Hochschule angesprochen. Er/sie ist fähig, Lernumgebungen so zu gestalten, dass Lehr-Lern-Prozesse den Ansprüchen des Spatial-Citizenship-Ansatzes entsprechen, und bereit, sich dementsprechend fortzubilden. Die mit der Digitalisierung einhergehenden Prozesse machen den Spatial-Citizenship-Ansatz nicht nur für den Geographieunterricht höchst relevant. Es eröffnen sich auch viele Anknüpfungspunkte für die politische Bildung und das historische Lernen.

5. FAZIT

Wie aufgezeigt, gibt es Bildungskonzepte, mit denen eine mündigkeitsorientierte Bildung in einer Kultur der Digitalität gefördert werden kann. Die Bildungspolitik verfolgt jedoch – trotz des verabschiedeten „Digital-Pakts" – andere Zielsetzungen. Ein Blick in den Koalitionsvertrag der aktuellen Regierung veranschaulicht dies: „Wir brauchen eine Digitale Bildungsoffensive, die die gesamte Bildungskette in den Blick nimmt und das gesunde Aufwachsen, die digitale Selbstbestimmung und individuelle aktive Teilhabe, den Umgang mit Daten […] zum Ziel hat" (CDU u. a. 2018, 39). Dieses Zitat aus dem aktuellen Koalitionsvertrag der Bundesregierung verdeutlicht, dass den Herausforderungen, vor denen das Individuum in der Kultur der Digitalität steht, vonseiten der Bundesregierung mit einer „Bildungsoffensive" begegnet werden soll, die auf die Förderung von „Selbstbestimmung" und „aktive Teilhabe" zielt. Das gewählte Vokabular verweist auf ein Bildungsverständnis, das die Mündigkeit der Schüler/-innen, Auszubildenden und Studierenden fördern will. Gleichzeitig stellt sich aber auch die Frage, wie solche Kompetenzen gefördert werden sollen. Hierzu versucht der Koalitionsvertrag auch Antworten zu geben: Ein Lösungsansatz ist „die flächendeckende digitale Ausstattung aller Schulen" (CDU u. a. 2018, 39). Außerdem sollen regionale Kompetenzzentren eingerichtet werden, um „technisches und pädagogisches Know-how zu vermitteln" (CDU u. a. 2018, 40). Ein weiteres Beispiel bezieht sich auf die Hochschulen: „Wir wollen dafür sorgen, dass auch an Hochschulen mehr Online-Lernangebote und digitale Inhalte entstehen. Alle Studierenden brauchen künftig digitale Kompetenzen. Sie sollen digitale Wissens- und Lernangebote selbstständig nutzen und gestalten können sowie Datenanalyse und grundlegende Programmierkenntnisse beherrschen" (CDU u. a. 2018, 40). Es gibt einen auffälligen Kontrast zwischen der Zielsetzung und den daraus abgeleiteten notwendigen Instrumenten: Das Ziel der mündigen Bürger/-innen, die in der digitalen Welt selbstbestimmt handeln, soll dadurch erreicht werden, dass diese ihre technisch- [|38] instrumentellen Fähigkeiten schulen. Ob damit den Herausforderungen und Risiken der Digitalisierung begegnet werden kann, ist zu bezweifeln. Eine solche Sichtweise verkennt, dass es bei der erfolgreichen Initiierung digitaler Bildungsprozesse, die ein mündiges Agieren ermöglich sollen, um mehr geht als um technische Aufrüstung, Informatikkenntnisse oder um die digitale Verbreitung von Lernmaterialien und -inhalten. Eine mündigkeitsorientierte Bildung wird damit konterkariert. Aber auch im Kontext der gesellschaftswissenschaftlichen Bildung müssen wir darauf achten, dass wir den Begriff Mündigkeit mit Bildungskonzepten wie z. B. dem Ansatz des Spatial Citizenship füllen, damit der Mündigkeitsbegriff nicht auf dem Altar der Kompetenzorientierung geopfert wird, oder wie es die Autorengruppe Fachdidaktik (2016) formuliert: „Quantifizierung verdrängt Mündigkeit" (Autorengruppe Fachdidaktik 2016, 18). Mündigkeit hat nach wie vor eine konstitutive Bedeutung für sozialwissenschaftliche Bildungsprozesse. Um diese für die digitale Gesellschaft zu bestimmen, bedarf es jedoch offenbar einer neuen Initiative: Reconquista Mündigkeit.

Förderhinweis

Diese Studie entstand im Kontext des Projekts „The Next Level – Lehrkräftebildung vernetzt entwickeln" der Goethe-Universität Frankfurt/M. Das Projekt wird im Rahmen der gemeinsamen „Qualitätsoffensive Lehrerbildung" von Bund und Ländern aus Mitteln des Bundesministeriums für Bildung und Forschung gefördert.

Dr. Christian Dorsch ist wissenschaftlicher Mitarbeiter in der AG Geographiedidaktik am Institut für Humangeographie der Goethe-Universität Frankfurt.

Prof. Dr. Detlef Kanwischer ist Professor für Geographie und ihre Didaktik am Institut für Humangeographie der Goethe-Universität Frankfurt.

LITERATUR

Achour, Sabine / Massing, Peter 2018: Smart Democracy. Politikum 3/2018. Berlin.

Adorno, Theodor W. 1970: Erziehung zur Mündigkeit. Vorträge und Gespräche mit Hellmut Becker 1959–1969. 1. Aufl. Frankfurt.

Allert, Heidrun / Richter, Christoph 2017: Kultur der Digitalität statt digitaler Bildungsrevolution. In: Pädagogische Rundschau, Heft 1, S. 19–32.

Autorengruppe Fachdidaktik 2016: Was ist gute politische Bildung? Leitfaden für den sozialwissenschaftlichen Unterricht. Schwalbach/Ts.

Beer, Kristina 2016: Heise online Polittalk: Der mündige Bürger in der digitalen Welt. Verfügbar unter: <https://www.heise.de/newsticker/meldung/heise-online-Polittalk-Der-muendige-Buergerin-der-digitalen-Welt-3328667.html> (4.2.2019).

Bernfeld, Siegfried 1973: Sisyphos oder die Grenzen der Erziehung. 10. Aufl. Theorie. Frankfurt/M.

CDU u.a. 2018: Ein neuer Aufbruch für Europa. Eine neue Dynamik für Deutschland. Ein neuer Zusammenhalt für unser Land. Koalitionsvertrag zwischen CDU, CSU und SPD. Verfügbar unter: <https://www.bundesregierung.de/breg-de/themen/koalitionsvertrag-zwischen-cdu-csuund-spd-195906> (16.11.2018).

Crouch, Colin 2008: Postdemokratie. Edition Suhrkamp. Berlin, Frankfurt/M. [|39]

Deci, Edward L. / Ryan, Richard M. 1993: Die Selbstbestimmungstheorie der Motivation und ihre Bedeutung für die Pädagogik. In: Zeitschrift für Pädagogik, Heft 39.

Dewey, John 1951: Wie wir denken. Eine Untersuchung über die Beziehung des reflektiven Denkens zum Prozeß der Erziehung. Sammlung Erkenntnis und Leben, Bd. 5. Zürich.

Eis, Andreas 2015: Mythos Mündigkeit – oder Erziehung zum funktionalen Subjekt? In: Widmaier, Benedikt / Overwien, Bernd (Hg.): Was heißt heute Kritische Politische Bildung? Schwalbach/ Ts., S. 69–77.

Giddens, Anthony 1995: Konsequenzen der Moderne. 1. Aufl. Frankfurt/M.

Gillespie, Tarleton 2014: The Relevance of Algorithms. In: Gillespie, Tarleton u.a. (Hg.): Media technologies. Essays on communication, materiality, and society. Inside technology. Cambridge, Mass., S. 167–193.

Gruschka, Andreas 1994: Bürgerliche Kälte und Pädagogik. Moral in Gesellschaft und Erziehung. Schriftenreihe des Instituts für Pädagogik und Gesellschaft, Münster, Bd. 4. Wetzlar.

Gryl, Inga 2013: Alles neu – innovativ durch Geographie- und GW-Unterricht? In: GW-Unterricht, Heft 131, S. 16–27.

Gryl, Inga u.a. 2017: Limits of Freedom – Defining a Normative Background for Spatial Citizenship. In: GI_Forum, Heft 2, S. 3–12.

Gryl, Inga / Jekel, Thomas 2012: Re-centering GI in Secondary Education: Towards a Spatial Citizenship Approach. In: Cartographica, 47 (1), S. 18–28.

Helsper, Werner 2018: Lehrerhabitus. Lehrer zwischen Herkunft, Milieu und Profession. In: Paseka, Angelika u. a. (Hg.): Ungewissheit als Herausforderung für pädagogisches Handeln. Wiesbaden, S. 105–135.

Hoffmann-Riem, Wolfgang 2017: Re:claim Autonomy. Die Macht digitaler Konzerne. In: Augstein, Jakob (Hg.): Reclaim autonomy. Selbstermächtigung in der digitalen Weltordnung. Edition Suhrkamp, Bd. 2714. Berlin, S. 121–139.

Introna, Lucas D. 2017: Die algorithmische Choreographie des beeindruckbaren Subjekts. In: Seyfert, Robert / Roberge, Jonathan (Hg.): Algorithmuskulturen. Über die rechnerische Konstruktion der Wirklichkeit. Kulturen der Gesellschaft. Bielefeld, S. 41–74.

Jandrić, Petar u. a. 2016: Postdigital science and education. In: Educational Philosophy and Theory, Jg. 50, Heft 10, S. 893–899.

Jantschek, Ole / Waldmann, Klaus (Hg.) 2017: Shape the Future. Digitale Medien in der politischen Jugendbildung. Non-formale politische Bildung, Band 8. Schwalbach/Ts.

Jekel, Thomas u. a. 2015: Education for Spatial Citizenship: Versuch einer Einordnung. In: GW-Unterricht, Heft 137, S. 5–13.

Jörissen, Benjamin / Marotzki, Winfried 2009: Medienbildung – eine Einführung. Theorie – Methoden – Analysen. 1. Aufl. Bad Heilbrunn.

Kant, Immanuel 1900 ff.: Kritik der reinen Vernunft. Kants Gesammelte Schriften. Hg. von der Königlich Preussischen Akademie der Wissenschaften, Bd. 4. Berlin, Leipzig.

Kanwischer, Detlef / Schlottmann, Antje 2017: Virale Raumkonstruktionen. Soziale Medien und Mündigkeit im Kontext gesellschaftswissenschaftlicher Medienbildung. In: Zeitschrift für Didaktik der Gesellschaftswissenschaften zdg, Jg. 8, Heft 2.

Klafki, Wolfgang 2007: Neue Studien zur Bildungstheorie und Didaktik. Zeitgemäße Allgemeinbildung und kritisch-konstruktive Didaktik. 6. Aufl. Weinheim.

Körber-Stiftung o. J.: Digitale Mündigkeit. Verfügbar unter: <https://www.koerber-stiftung.de/themen/digitale-muendigkeit> (4.2.2019). [|40]

Kultusministerkonferenz 2016: Bildung in der digitalen Welt. Strategie der Kultusministerkonferenz.

Kurz, Constanze / Rieger, Frank 2017: Autonomie und Handlungsfähigkeit in der digitalen Welt. Crossing the creepy line? In: Augstein, Jakob (Hg.): Reclaim autonomy. Selbstermächtigung in der digitalen Weltordnung. Edition Suhrkamp, Bd. 2714. Berlin, S. 85–97.

Lash, Scott 1996: Reflexivität und ihre Doppelungen. Struktur, Ästhetik und Gemeinschaft. In: Beck, Ulrich u. a. (Hg.): Reflexive Modernisierung: eine Kontroverse. Frankfurt/M., S. 195–286.

Macgilchrist, Felicitas 2017: Die medialen Subjekte des 21. Jahrhundert. Digitale Kompetenzen und/oder Critical Digital Citizenship. In: Allert, Heidrun u. a. (Hg.): Digitalität und Selbst. Interdisziplinäre Perspektiven auf Subjektivierungs- und Bildungsprozesse. 1. Aufl. Pädagogik. Bielefeld, S. 145–165.

Messerschmidt, Astrid 2017: Widersprüche der Mündigkeit. Anknüpfungen an Adornos und Beckers Gespräch zu einer „Erziehung zur Mündigkeit" unter aktuellen Bedingungen neoliberaler Bildungsreformen. In: Ahlheim, Klaus / Heyl, Matthias (Hg.): Adorno revisited. Erziehung nach Auschwitz und Erziehung zur Mündigkeit heute. Kritische Beiträge zur Bildungswissenschaft, Band 3. Hannover, S. 126–147.

Negroponte, Nicholas 1998: Beyond Digital. In: Wired, Jg. 6, Heft 12.

Reconquista Internet 2018a: faq, zuletzt geprüft: 20.3.2019.

Reconquista Internet 2018b: RI Pressemitteilung.

Roth, Heinrich 1971: Pädagogische Anthropologie. Band 2: Entwicklung und Erziehung. 1. Aufl. Bd. 2. Hannover.

Schieren, Stefan / Pohl, Kerstin 2016: Big Data. Politikum 1/2016. 1. Aufl. Berlin.

Schön, Donald A. 1983: The reflective practitioner. How professionals think in action. New York.

Schulze, Uwe u. a. 2015: Spatial Citizenship – Zur Entwicklung eines Kompetenzstrukturmodells für eine fächerübergreifende Lehrerfortbildung. In: Zeitschrift für Geographiedidaktik zdg, 43 (2), S. 139–164.

Stalder, Felix 2017: Kultur der Digitalität. Berlin.

Taylor, Charles 1988: Negative Freiheit? Zur Kritik des neuzeitlichen Individualismus. Frankfurt/M.

Tugendhat, Ernst 1979: Selbstbewußtsein und Selbstbestimmung. Sprachanalytische Interpretationen. 1. Aufl. Suhrkamp-Taschenbuch Wissenschaft, Bd. 221. Frankfurt/M.

Zuboff, Shoshana 2017: Auf der Suche nach dem autonomen Selbst. In: Augstein, Jakob (Hg.): Reclaim autonomy. Selbstermächtigung in der digitalen Weltordnung. Edition Suhrkamp, Bd. 2714. Berlin, S. 167–172.

SMART EARTH:
A META-REVIEW AND IMPLICATIONS
FOR ENVIRONMENTAL GOVERNANCE

Karen Bakker / Max Ritts

Abstract: Environmental governance has the potential to be significantly transformed by Smart Earth technologies, which deploy enhanced environmental monitoring via combinations of information and communication technologies (ICT), conventional monitoring technologies (e. g. remote sensing), and Internet of Things (IoT) applications (e. g. Environmental Sensor Networks (ESNs)). This paper presents a systematic meta-review of Smart Earth scholarship, focusing our analysis on the potential implications and pitfalls of Smart Earth technologies for environmental governance. We present a meta-review of academic research on Smart Earth, covering 3187 across the full range of academic disciplines from 1997 to 2017, ranging from ecological informatics to the digital humanities. We then offer a critical perspective on potential pathways for evolution in environmental governance frameworks, exploring five key Smart Earth issues relevant to environmental governance: data; real-time regulation; predictive management; open source; and citizen sensing. We conclude by offering suggestions for future research directions and trans-disciplinary conversations about environmental governance in a Smart Earth world.
Keywords: Eco-informatics, Environmental governance, Smart earth, Ecology, ICT, IoT, Information and communications technology, Internet of things, Sensors, Digital

1. INTRODUCTION

Over the past two decades, researchers and practitioners in earth sciences, ecology, and cognate disciplines have been creating innovations in environmental monitoring technologies that combine Information and Communication Technologies (ICT) with conventional monitoring technologies (e. g. remote sensing), and Environmental Sensor Networks (ESNs, which are spatially distributed monitoring networks containing high densities of sensors and actuators). These technologies, which we collectively label "Smart Earth," have proliferated due to the rapid decrease in cost of cloud-based computing and innovations in Machine to Machine (M2M) infrastructure (Hogan et al., 2012; White, 2016), enabling unprecedented environmental management applications. Simply put, Smart Earth is the set of environmental applications of the Internet of Things, and is thus analogous to the widely discussed "Smart City," (Marvin et al., 2015), but articulated across a much wider range of ecosystems and land use types.

Smart Earth technologies enable terabytes of environmental data to be derived from terrestrial, aquatic, and aerial sensors, satellites, and monitoring devices, relying on a rapidly diversifying set of sources – including "wearables" and biotelemetric technologies devised for humans, animals, and even insects. New cloud-based

Web platforms have been created that enable the aggregation, analysis, and real-time display of these unprecedented streams of environmental data. Scientists are also applying innovations in AI, Big Data analytics, machine learning, 3D object-recognition algorithms, and genetic learning to the study and administration of ecological processes (Koomey et al., 2013; Gabrys, 2016; Goodchild, 2007; Kitchin, 2014; Gale et al., 2017; Pettorelli et al., 2014; Schwab, 2017; Zyl et al., 2009). Collectively, these developments have dramatically increased scientists' ability to assess spatio-temporal changes in abiotic conditions as well as biotic communities.

We contend that the volume, integration, accessibility, and timeliness of the data provided by Smart Earth technologies potentially creates the conditions for significant changes in environmental governance. To date, the majority of research on this topic has focused on the potential implications for conservation and waste reduction, pollution mitigation, mapping environmental degradation, geosecurity, and disaster management (Goodchild and Glennon, 2010; Resch et al., 2014; Koomey et al., 2013). However, although a few scholars have engaged with questions of the implications of these technologies for environmental governance (e. g. Gabrys, 2016), this issue remains relatively under-studied from a multi-disciplinary perspective. This paper seeks to address this gap.

Our paper begins from the premise that Smart Earth technologies have the potential to disrupt existing modes of environmental governance. Here, environmental governance is defined from an analytical (rather than normative) perspective as the set of social actors and [|202] institutions (including laws, rules, norms, customs), as well as data-gathering and decision-making processes, engaged in environmental decision-making (Bridge and Perreault, 2009; Ostrom, 1990). Our definition is broadly aligned with social scientists engaged in the study of environmental governance at a global scale (e. g. the Earth System Governance Project), notably those who study the institutional and epistemological realignments of environmental governance globally (e. g. Biermann et al., 2010, 2012). Our analysis of potential pathways for innovation in environmental governance coupled with Smart Earth technologies is related to and inflected by, but distinct from, governance trends such as the partial redistribution of decision-making power from state to non-state actors (e. g. the emergence of non-state market-driven governance systems), and the rescaling of governance above and below the nation-state (Biermann et al., 2012; Cashore, 2002; Cohen and McCarthy, 2015; Reed and Bruyneel, 2010).

The purpose of this meta-review is to provide a synthesis of key issues and critiques that Smart Earth poses for environmental governance. Smart Earth enables a series of shifts: the time-space compression of data availability and decision-making (which in turn enables automated real-time regulation and new prediction capabilities); the multiplication of modalities and agencies of environmental sensing; the proliferation of new environmental governance actors; and, potentially, a much higher degree of transparency in data collection, accessibility, and integration. Taken together, these innovations create the conditions for potentially significant transformations in environmental governance.

Consider, as an example, Sustainability Standards Organisations (SSOs). New forms of access to real-time, continuous information on environmental data from "virtual" monitoring platforms are challenging the "static, limited, and closed "analog" model of auditing conventionally employed by [SSOs]" (Gale et al., 2017). In the past, SSO audits were conducted through brief, intermittent field visits by small teams of auditors and experts. Smart Earth technology creates the potential for continuous monitoring and assessment of the validity of sustainability claims. This in turn enables the emergence of private regulatory bodies and real-time auditing processes which will drive changes in SSOs (Auld and Gulbrandsen, 2010; Carse and Lewis, 2017). The SSO example illustrates the co-evolution of technology and governance occurring across different environmental domains and scientific disciplines, including established fields such as landscape ecology and geography, as well as emergent sub-fields such as environmental digital humanities, animal biotelemetry, and citizen sensing.

Our paper presents a systematic meta-review of this literature. Our intention in conducting this review is to identify the key issues that Smart Earth poses for environmental governance. To conduct this metareview, as detailed in Section 2, we surveyed the scholarly literature (1997–2017) across the full range of academic disciplines to create a database of 3187 articles (discussed in Section 3). In Section 4, we present key issues and critiques relevant to environmental governance debates, including: data (the opportunities and challenges of using big data to provide temporally and spatially comprehensive coverage for monitoring, in contrast to intermittent and low-density monitoring); real-time regulation (including real-time and potentially automated decision-making through the use of mobile platforms to communicate to field-based actors and receptors, such as ship captains, farmers, fishers, and hunters); enhanced predictability, particularly in situations where data was previously unavailable; the technical and ethical implications of open data; and the evolution of citizen engagement through new modalities such as citizen sensing, which incorporate new variables (such as noise and sound) that extend our ability to "sense" the environment (Helmreich, 2015). Section 5 concludes by offering suggestions for future research directions regarding environmental governance in a Smart Earth world.

2. METHODS

Our analysis presents the results of a meta-review of the academic literature on Smart Earth. We conducted a manual search of 17 journals spanning a range of disciplines including computer science, environmental studies, ecology, eco-informatics, and social studies of science. Our manual search included the following journals: Ambio, Annual Review in Environmental Resources, Ecological Informatics, Environmental Humanities, Environment and Planning A, Environment and Planning D, Journal of Applied Ecology, Big Data and Society, Annals of the American Association of Geographers, Global Environmental Change, Global Environmental Politics, International Journal of Digital Earth, PNAS, Nature, Science, Social Studies of Science, Trends in Ecology and Evolution.

Through this review, we identified the keywords most frequently used with respect to Smart Earth, as well as commonly-used terms related to earth processes relevant to Smart Earth topics: remote sensing, eco-informatics (and ecological informatics), Big Data, biomonitoring, citizen sensing, cloud computing, data visualization, fiber optic, Internet of Things, drones, citizen science, fourth industrial revolution, Digital Earth, biomonitoring, and Program Earth. Keywords relevant to earth processes related to Smart Earth-related topics: ecosystem services, environment, ecology, Anthropocene, planet, habitat, species, biodiversity, animal migration, geology, geomorphology, conservation, ecosystem, species distribution, migration, and climate. We then conducted a search using these keywords across the full range of disciplines in the natural sciences, social sciences, and humanities, on Web of Science and Google Scholar. Using paired combinations of keywords, we generated 176 discrete paired search terms. With each paired search, we identified the top 100 most cited papers, inclusive of the period 1997–2017, the period which best captures the onset phase of Smart Earth research. This strategy identified the most highly-cited papers (7892 papers in total). Each paper's abstract was reviewed to determine whether or not the paper focused on Smart Earth issues, resulting in a database of 3187 articles, the citations from which were used to generate a content cloud (Fig. 1). These articles span the natural and social sciences, and humanities, and include such disciplines as ecology, environmental humanities, geography, geomorphology, and marine biology; and such topics as animal migration studies, eco-informatics, pollution monitoring, remote sensing, and science and technology studies (STS).

3. SMART EARTH: OVERVIEW

This section provides an overview of the Smart Earth literature, which is characterized by a focus on Smart Earth techniques and technologies. Twenty years ago, many of the technologies that are now aggregated under Smart Earth were nascent

Fig. 1. Smart Earth – Content Cloud. [|203]

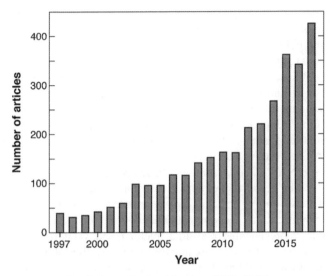

Fig. 2. Smart Earth publications (1997–2017).

or nonexistent. Conversations had yet to form – or at least, become formalized – across many of the scientific disciplines that deploy and debate these technologies. Over the past two decades, new interdisciplinary journals have emerged, including *Ecological Informatics* (founded in 2006), the *International Journal of Digital Earth* (2008), and *Environmental Humanities* (2012). New disciplines have formed (e. g. Digital Environmental Humanities), emblematic research groups have been founded (e. g. MediaLab at Sciences Po, founded by leading social theorist Bruno Latour), and there have been notable transformations in the character of sub-disci-plinary fields: for example, Geo-Information Sciences has grown out of Geographic Information Systems.

As Fig. 2 indicates, publications on Smart Earth have rapidly increased in num-ber. Our bibliometric analysis indicates that use of terms like "Big Data" has in-creased across a range of disciplines. Although we restricted our searches to peer-re-viewed, English-language publications, our meta-review likewise illustrates a di-versity of disciplines and disciplinary approaches in research on Smart Earth: Agri-culture, Anthropology, Applied Physics, Astronomy, Astrophysics, Atmospheric Sciences, Biochemistry, Biology, Biotechnology, Business, Cell Biology, Commu-nication, Computer Science, Conservation, Ecology, Economics, Energy and Fuels, Engineering, Entomology, Environmental Sciences, Environmental Studies, Evolu-tionary Biology, Fisheries, Forestry, Geography, Geology, Limnology, Marine Biol-ogy, Mechanics, Microbiology, Nuclear Chemistry, Oceanography, Ornithology, Plant Physiology, Plant Sciences, Remote Sensing, Robotics, Soil Science, Spec-troscopy, Toxicology, Water Resources, and Zoology.

Entirely new sub-disciplines – such as the Environmental Humanities, a field designed to explore conjunctions across environmental history, philosophy, human geography, and political ecology – are now regularly engaging Big Data questions (e. g. Gabrys et al., 2016). Remote sensing continues to enjoy a great prominence in Smart Earth publications, but is by no means the only technical application being explored in Smart Earth. Companies like Microsoft and Google are now investing significant amounts ($ 50 Million from Microsoft alone) to develop AI-supported "Earth algorithms" (Joppa, 2017). Other initiatives include Nokia's "Sensor Planet," IBM's "A Smarter Planet," HP Labs' "Central Nervous System for the Earth" (CeNSE), NASA's "Earth Observing System Data and Information System" (EOS-DIS), and Cisco/NASA's collaborative "Planetary Skin Institute," together with a rapidly proliferating ecosystem of apps (Jepson and Ladle, 2015).

Many of these initiatives focus on the interface between sensors and internet-based communications technologies which, combined with cloud-based data storage, enable unprecedented real-time tracking and visualization (Gale et al., 2017). Smart Earth initiatives also frequently combine well-established approaches – such as remote sensing and long-term ecological monitoring – with newer technologies (e. g. animal biotelemetry, bacteria-based biosensors) and modalities of data collection (such as drones, Google Earth, and citizen sensing) (Table 1).

Table 1: Examples of Smart Earth technologies

Type	Example	Description
Wearables	Flow	Flow is a smart, wearable air pollution tracker using sensors to monitor the user's real-time exposure to air pollution and an integrated app to help users find cleaner air (Flow, 2017).
Animal Biotelemetry	Save the Elephants: Geo-fencing	Geo-fencing, a virtual fence programmed with GPS positions, has been integrated into elephant tracking collars to get real-time information of their location, allowing for immediate action if an elephant moves out of the reserve into areas more susceptible to poaching (Save the Elephants, 2017).
Plant Biotelemetry (Cyberplants)	PLEASED	The project Plants Employed As Sensing Devices (PLEASED) embeds EEG sensors into plants to measure environmental parameters, which can be used to monitor fires and avalanches (PLEASED, 2017; Manzella et al., 2013).
Insect Biotelemetry	Bees with Backpacks	By attaching small RFID 'backpack' sensors to over 5000 honey bees the data collected on bees' location and movement will provide insight into the decline of Australian regional bee populations (Landau, 2016).

Type	Example	Description
Mobile apps	giveO2	The app giveO2 automatically tracks the user's means of transportation, identifies their carbon footprint using GPS, and gives users the opportunity to purchase carbon credits to offset their carbon output (GiveO2, 2018).
Fixed sensors	Instant Detect	Instant Detect has been successfully deployed in Kenya to tackle poaching thanks to its network of multiple fixed camera traps using a central satellite node to instantly send photographs and data (Zoological Society of London, 2018).
Mobile sensors	Argo	Argo consists of 3800 free-drifting ocean floats with mobile sensors attached that transmit data on salinity and temperature from the upper 2000 meters of the ocean to satellites (Argo, 2017).
Sensor Web	Ocean Observatories Initiative (OOI)	Dubbed the 'Fitbit for the oceans' (Paul, 2015), OOI's cyber-infrastructure is exploring the little-known parts of Earth's oceans with sensors and autonomous drones collecting over 200 types of data (Ocean Observatories Initiative, 2016).
Remote sensing	GasFinder3	GasFinder3 utilizes laser-based remote sensing to continuously monitor gas concentrations in the path of the laser, and has been successfully applied around oil and gas sites in the Arctic region (Boreal Laser Inc, nd; LOOKNorth, 2012).
Virtual reality	Conservation in Virtual Reality	Conservation International's 360-degree virtual reality films immerse viewers in environmental conservation efforts around the globe, including the Amazon rainforest and Indonesian reefs (Conservation International, 2017).
Artificial Intelligence (AI)	Green Horizon	Green Horizon's cognitive computing systems use machine learning and real-time air quality data to analyze and create visual maps displaying the source and dispersion of pollutants across Beijing (IBM, 2018).

The most frequent targets of these applications are: natural resources (such as forests); species prioritized for conservation (such as marine mammals); earth "boundary conditions" and "ecosystem services" (such as freshwater, atmospheric carbon); and environmental security (e. g. natural disasters such as floods and fires) (Fig. 3).

A paradigmatic example of Smart Earth technologies is EarthNC's (2016) SharkNet (www.sharknet.com/), which displays the near realtime locations of great white sharks and marine mammals, gathering data from a network of mobile robots and moored listening stations connected to sensors carried by marine wildlife. Like many marine animal tracking efforts, SharkNet emerged from longstanding collaboration among scientists interested in monitoring animals across large spatial

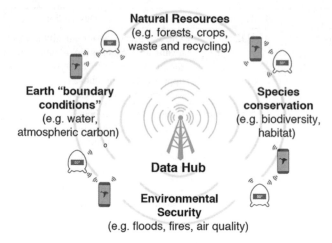

Fig. 3. Smart Earth – Data Hub. [|204]

ranges. Related "citizen sensing" projects have geotagged fish, such as important commercial species like Pacific salmon, whose movements can likewise be detected by networks of underwater cameras (Matabos et al., 2017).

Although the focus of Smart Earth has tended to be on terrestrial processes, ocean environments have also been the subject of innovation (ONC, 2014; IBM, 2014; Favali et al., 2015). Across a range of ecosystems and disciplines, scientists and engineers have created new devices to assess changing oceanographic conditions, including an automated network christened "FitBit for the Oceans" (Ocean Observatory Initiative, 2016). The network incorporates cable and sensor technologies to measure geological, physical, chemical and biological variables in the ocean and seafloor and produces over 200 different kinds of data. These marine developments are emblematic of the ways in which Smart Earth technologies enable comprehensive data acquisition infrastructures to monitor environmental conditions, emergent risks, and geo-hazards (see: ONC, 2014; Helmreich, 2015; Lehman, 2016; Lindenmayer et al., 2017).

Another salient finding of our review is the variability of scales at which Smart Earth is being defined and measured. Scale is critical in assessing Smart Earth because of the spatial variability of the ecological processes involved in environmental governance. For instance, the designers of BeachObserver propose that coastal communities "selfmonitor" locally-occurring marine debris, whereas projects like Argos aim for systemic reviews of marine-ecological change occurring at the planetary scale (Benson, 2015). A single review is probably unable to capture all the scalar variability in Smart Earth, although certain projects (e.g. BirdTracker) are able to straddle multiple scales. However, the challenge of integrating technologies operating at multiple and distinct scales is obvious and will be discussed further below.

4. ENVIRONMENTAL GOVERNANCE IN A SMART EARTH WORLD: ISSUES AND CRITIQUES

This section of the paper offers insights into the key implications of Smart Earth for environmental governance. We explore key issues and critiques stemming from our meta-review, organized by five themes data; "real-time" analysis and regulation; the changing nature of prediction; the meaning and extent of "open data"; and the role of nonscientists, notably "citizen sensors."

4.1 Data

Smart Earth technologies require new multi-scalar data architectures which can provide computing resources and/or services such as data replication and storage (Coleman, 2010; Kitchin and McArdle, 2016; Li and Chen, 2017). The exponential growth in data generated by Smart Earth technologies is illustrated by the collaborative work of IBM and research scientists on Lake George in the north-eastern United States (Coldsnow et al., 2017; McGuinness et al., 2014). The collaboration between biologists, environmental scientists, engineers, physicists, computer scientists, and meteorologists implemented an unprecedented array of sensors to produce 468 million depth measurements (compared with 564 data points in previous models). As this example illustrates, Smart Earth reverses a prior constraint: in the future, it is the abundance (rather than scarcity) of data that will be a defining challenge for effective environmental governance.

In response, many scientists and innovators are focusing on developing automated decision-making systems. The growing embrace of automated management techniques (including water and pesticide application, and crop rotation) in the agricultural sector is a good example of these trends (Zhang and Pierce, 2013). However, multiplying realtime data streams and interoperable technologies does not necessarily engender more efficient or transparent modes of environmental governance. Angwin (2013), for example, cites a coder behind the National Security Agency's (NSA) analytics who speculates that his agencies' data collection efforts may actually be hindering its surveillance efforts; the NSA is simply awash in too much data.

Galaz and Mouazen (2017) point to a related paradox: the most powerful algorithms underlying automated decision-making systems are likely to be of limited accessibility or transparency – raising the risk of replication of biases in decision-making. As many scholars have pointed out, AI and machine-learning driven approaches often contain implicit bias; incomplete datasets and flawed algorithms can prove counterproductive and ecologically damaging (Caliskan et al., 2017; Galaz and Mouazen, 2017). Moreover, algorithms do not reconcile value-laden tensions between competing uses and ecosystem services (e.g. economic versus spiritual values) leaving the question of arbitration incomplete. Perhaps counter-intuitively, this creates a need for new forms of human supervision as the automation of decision-making intensifies (Galaz and Mouazen, 2017, p. 629).

A related issue that threatens to inhibit the uptake of Smart Earth data in environmental governance projects is the lack of data standards across different institutional cultures (Michener, 2015). Our meta-review revealed persistent appeals for cross-disciplinary collaboration on data standards and data sharing (e. g. Frew and Dozier, 2012; Hampton et al., 2013; Michener, 2015). It is thus surprising to note that "silo-ing" persists across academic uptakes of Smart Earth. This can be partially attributed to the long-standing challenges of combining ecological data sets. Pre-ICT approaches were characterized by a diversity of standards; for example, one set of standards applied to the longer-term data gathered by professional scientists (often for the purposes of testing single hypotheses), while another set of standards often applied to shorter-term data collected by professional environmental consultants in response to specific ecological threats (such as pollution incidents), and distinct (or no) standards were applied to data collected by citizen scientists (such as fish and wildlife harvest data) which was housed separately and rarely accessed or used by scientists or governments (Goodchild, 2007). The challenge of combining and communicating data gathered through disparate data collection efforts remains unresolved, although significant efforts have been made to address this issue by a variety of organizations, including NASA's Socioeconomic Data and Applications Center (SEDAC, a Data Center in NASA's Earth [|205] Observing System Data and Information System (EOSDIS)) and the National Ecological Observatory Network.

This data-sharing challenge may nevertheless be largely resolved in the near-term future. "Big ecology" policies, combined with diminished costs for information technologies and new cloud-based data archiving tools and repositories, have proliferated in recent years (Michener, 2015), supported by an increasing number of eco-informatics scientists, and also by the work of organisations such as the Ecological Society of America's Committee on the Future of Long-term Ecological Data (FLED) (Michener et al., 1997; Porter, 2010). Scholars of ecological informatics predict that the new generation of data sharing networks will grow exponentially faster than its predecessors (Michener, 2015; Porter, 2010; Reichman et al., 2011; Zimmerman, 2008). For example: after several decades of operation, the US Long Term Ecological Research Network (LTER) had 6000 shared datasets (Porter, 2010); in contrast, by mid-2017, after less than a decade of operation, DataONE exceeded 1 million data objects. This growth has been facilitated by the expansion of networking capacities among researchers, along with corresponding changes in the cyberinfrastructure landscape. Important data science innovations – including data and metadata standards, persistent identifiers, and search/discovery tools – have enabled more widespread data-sharing than in the past (Michener, 2015). (A detailed exploration of these aspects of ecological informatics is beyond the scope of this paper but for a representationally-focused critique see: Demos, 2017).

In summary, much of the Smart Earth literature focuses on addressing data gaps, the need for more collaborative data sharing, and issues relating to data quality. The underlying assumption is that more comprehensive and higher quality data will lead to more effective environmental governance. As a wealth of literature in science and technology studies shows, however, this assumption is problematic

(c. f. Jasanoff, 2003; Gabrys, 2016). The political commitments associated with measurement-related decisions – in particular, the question of who selects variables to measure and for whom they are selected – are rarely discussed. Because data gaps are likely to become more acute in a Smart Earth world, the problem that "what is measured, matters" or "what is counted, counts" (and by extension, what is not counted does not count) is likely to intensify. These critiques are central to the question of whether Smart Earth will truly enhance and not simply "update" environmental governance.

4.2 Real-time regulation

The concept of "real-time" – e. g. the increasingly instantaneous "actual" time elapsed in the performance of a computation – is central to Smart Earth governance (De Longueville et al., 2010). Interest in real-time regulation can be traced to the managerial discourses shaping urban policy in the late 1990s (for reviews see: Komninos, 2002; Gabrys, 2014), which focused on real-time pricing in "Smart Grids" and "Smart" water distribution systems (Momoh, 2012; Kratz et al., 2006). In the last decade, a rapid decline in the cost of monitoring technologies (driven by innovation in computing and communications) has increased the capacity to conduct real-time assessment of environmental changes (Koomey et al., 2013). Managers are evaluating the success of real-time location information via software applications on location-aware devices, such as cellphones (e. g. Ratti et al., 2006), laptops (Goodchild and Glennon, 2010), acoustic sensors (Farina and Gage, 2017), and apps (Zickuhr, 2013). Formative efforts to explore the implementation of "real-time speed advisory signs" in transportation systems (Haque et al., 2013, p.25), are now common in larger environmental governance contexts, such as marine governance, where ship rerouting efforts proceed in response to wildlife detections in a number of contexts (the Enhancing Cetacean Habitat and Observation, or ECHO, project noted in Ritts, 2017); and pipeline logistics, where flows can be rerouted into different distribution terminals across vast inter-regional networks (e. g. Cowen, 2014).

A paradigmatic example here is the work of Conserve.io, an organisation that assists conservation groups with leveraging mobile, cloudbased and big data technologies through data collection at scale (involving both crowdsourced data from humans and passive environmental sensors). The mobile applications ("apps") produced by Conserve.io (2015) use visual analytics to enhance situational awareness and real-time responses to environmental changes. Whale Alert (2012) (http://www.whalealert.org/) gathers marine mammal observations (both volunteered and professionally-sourced) in marine shipping zones, and uses the data to alert ship captains of vessel proximity to high density whale zones in real-time, with the goal of providing real-time information to reduce whale strikes.

Whereas management and critical policy literatures focus on realtime resource distribution concerns – e. g. coordinating flows of energy, bodies, and commodities – ecologists and biologists have been more engaged with novel possibilities for species "tracking" (Benson, 2012; Gabrys, 2016). Thanks to new advances, tracking efforts can now distill real-time patterns. New technologies, such as conservation drones (Sandbrook, 2015), augmented virtual environments (Jian et al., 2017), and conservation apps (Jepson and Ladle, 2015) are being evaluated as adaptive solutions to monitoring and law enforcement problems. It is increasingly common for large scale observatories – whether located underwater (e. g. VENUS) or in the sky (ARGOS) – to host multiple sensor arrays for multiple research communities (Starosielski, 2015; Benson, 2015).

The challenge (and opportunity) posed by real-time data streams is one of the most salient issues for environmental governance in a Smart Earth world. Real-time data streams and real-time analysis make the idea of "real-time" regulation possible: e. g. the capacity to rapidly shift resources and monitoring capacities in response to unforeseen developments. For example, Little et al. (2015) survey real-time spatial management approaches to reduce fisheries bycatch and discards. Kumar et al. (2015) explore the rise of low-cost sensing for managing urban air pollution. The developments they capture likewise sound a note of caution. Real-time regulation poses significant administrative challenges, insofar as organizing simultaneous temporal attributes (or tracking efforts), including time of acquisition, integration/dwell time, sampling interval, and aggregation time span, can easily overwhelm computing power or lead to insufficient data (Frew and Dozier, 2012; Barnes et al., 2013; Benson, 2015). It is likewise impossible to guarantee that ecological wellbeing will form the basis of real-time responses, which may in fact embolden more tactical political decisionmaking favoring certain actors. This has spurred cross-disciplinary demands for new protocols, so that meaningful results can be obtained and acted upon (e. g. Koomey et al., 2013; Snyder et al., 2013).

A second, related issue is the logistical challenge of sharing insights across spatially and institutionally distributed research communities (Hochachka et al., 2012). Fox and Hendler (2011) observe a growing mismatch between the resource cost of creating scientific visualizations, and the more rapidly decreasing costs of data generation (per unit of data generated). Because digital devices are functional only to the extent of their integration into superannuating real-time networks (Wilson, 2014), the problem of "dark data" – i. e. data rendered invisible (and hence unusable) in Smart Earth practices – looms (Hampton et al., 2013; see also: Roche et al., 2015). Real-time regulation presupposes the constant availability of power-sources for its effective operation: an expectation that more and more coastal communities are routinely disabused of (Parenti, 2012). The sudden absence of power demanded by real-time coordination systems could multiply distributional challenges in the face of a brownout or similar disruption. Cubitt's (2017) claim that insufficient electricity, not oil, is the leading threat to coordinations of global governance takes on added salience in Smart Earth. More research is needed to build on important efforts, such as Graham's (2010) study of infrastructural failure, to investigate the challenge posed by the multiplication effects of small energy stoppages (even at the orders of seconds and nano-seconds) to real-time [|206] environmental governance.

4.3 Prediction

Many papers discussing Smart Earth highlight the benefits of predicting conditions for sustainable development and resource use (Snyder et al., 2013; Schwab, 2017). In distinction from previous approaches to predictability, ecological changes are increasingly conceived as predictable and "programmable" (Frew and Dozier, 2012; Gabrys, 2016; Leszczynski, 2016; Murai, 2010). This innovative emphasis on programmability as an inherent – and in some formulations paramount – characteristic of predictability builds on developments in "adaptive monitoring" and "adaptive management" (e. g. Lindenmayer and Liekns, 2010), as well as "anticipatory governance," which Guston (2011, p.1) defines as "a broad-based capacity extended through society that can act on a variety of inputs to manage emerging knowledgebased technologies while such management is still possible" (see also: Carruth and Marzec, 2014). For example, many publications provide examples of enhancements to predictive capacities enabled by new web platforms (e. g. Oak-Mapper, WhaleAlert; Global Forest Watch), from which managers and scientists alike can source data inputs for environmental niche models, which are able to predict risks, disturbances and terrestrial transformations (Kluza et al., 2007; Clark et al., 2009).

Smart Earth also creates innovative possibilities for predictability for an expanded set of environmental variables. For example, the literature has many examples of novel prediction capabilities across environmental phenomena which were previously characterized by limited predictability – with seasonal whale migrations, and the sonification of advancing "P-Waves" being two prominent examples (e. g. Walsh and Mena, 2013; Pötzsch, 2015; Gale et al., 2017; ONC, 2014). Prediction has also garnered considerable attention in the remote sensing community (Khatami et al., 2016), where new measurement techniques and modeling and visualization capacities are being used to describe and predict local weather patterns in the context of climate change (Mairota et al., 2015), species diversity estimates (Rocchini et al., 2015), lake monitoring (Dörnhöfer and Oppelt, 2016), and management priorities (Pettorelli et al., 2014). Again, innovation is key: Fractal measures, Fourier decomposition, wavelet measures, and spatial autocorrelation (Cushman et al., 2010) represent powerful advances over discrete patch-matrix-corridor models, and serve to generate continuous predictive evaluations of landscape patterns, which can then feed into long-term projections and detailed anticipations of future ecological forms (Mairota et al., 2015). There are a range of studies centered on the prediction, with increasingly high degrees of accuracy, of changing forest structure and aboveground biomass (Zald et al., 2016); food web structure (Woodward et al., 2013); and animal migration (Panzacchi et al., 2016).

Despite their potential, the actual success rate of predictive efforts has been questioned. Hummel et al. (2013) note that future trends in water consumption cannot simply be modeled through population inputs and the provision of known-aquifers. Such determinations will have to contend with "unexpected developments" (p. 122) in price structures and consumption patterns, among other variables. Hochachka et al. (2012, p. 132) regard as a "persistent general challenge" the fact that

"relationships between fine-scale environmental predictors and the observed responses of species tend to vary across large spatial and temporal extents." Predictive capacities are necessarily limited by the present-day assumptions of their programmers, and they are often unable to internalize challenges that arise outside
their framing contexts (see: Amoore, 2013; Leszczynski, 2016). Finally, as Jasanoff
(2003) notes, technologies of predictive analysis have historically tended to pre-
empt political discussion on the basis of scientific claims to objectivity. The challenge of flexibility alongside predictive capacity can be expected to loom as a major
tension in Smart Earth governance moving forward.

   In short, Smart Earth enables new modes of prediction which create new possibilities for environmental governance. Adaptation and anticipation have become
more central to environmental governance in the increasingly predicted (if not necessarily more predictable) "timespace" of a Smart Earth world. These developments
align with current debates over the implications of what Mahony (2014) calls the
"predictive state" (Guston, 2011; Gabrys, 2015, 2016; Pötzsch, 2015) and associated forms of geographical intelligence (Crampton et al., 2013; Wood, 2013;
Thatcher, 2014). Carruth and Marzec (2014) join others in focusing these concerns
around ethical issues of privacy, freedom, and security (e. g. Wood, 2013; DeLoughrey, 2014). Cubitt's (2017, p.159) remark that "databases predict the predictable" is
a reminder that certain ecological forms and processes may be excluded under automated tracking systems.

4.4 Open source

A Smart Earth world abounds in rapidly circulating, often messy and insufficiently
inventoried "open source" data (Gunningham and Holley, 2016; EPA, 2013; Mairota et al., 2015; Rocchini et al., 2017). In the ecological sciences, there is now a
pervasive conviction that biodiversity conservation will be augmented by the provision of open-access data (Morris and White, 2013; Turner et al., 2015). Projects
like eBird tout the value of sharing small, localized citizen-science based observations which, when aggregated, propose broader understandings of ecological phenomena (e. g. Dickinson and Bonney, 2012; EPA, 2013; Snyder et al., 2013). The
"Air Quality Egg" project, an EU-supported effort to collaboratively devise a
"smart" air quality sensor network, is being proclaimed as a best practice example
of bottom-up environmental governance (Zandbergen, 2017). In contrast, "Open
source" is defined by other scholars as a means to facilitate customization in environmental governance (e. g. Bradley and Pullar, 2015; Gale et al., 2017), ensuring
the local determination of environmental decision-making at a time of plane-
tary-scale organization. At its most utopian, open source proposes that "ubiquitously available" data can serve as both a key feature of environmental governance
and an "essential component of democracy" (Mooney and Corcoran, 2014, p.534).

   Efforts to survey the dizzying array of open source archives currently engaged
within governments, research projects and NGOs have resulted in several detailed
reviews (Roche et al., 2015; Gale et al., 2017; Welle Donker and van Loenen, 2017).

Michener (2015) examines how "Big Data" informational policies have historically encouraged the present deluge of open source data – noting in particular the foundational significance of Long-Term Research Networks (LTRNs), Ecological Observatory Networks (EONs), and Coordinated Distributed Experiments and Observations Networks (CDEOs). Bastin et al. (2013) survey the range of open-source software and standards (e. g. PostGIS, OpenLayers, Web Map Services, Web Feature Services and GeoServer) that support new assessments of land-cover change. Despite the fact the scientists have long used decentralized groups of non-professionals to gather ecological information (Connors et al., 2012), many observers note that recent "success stories" remain without truly democratic cultures of "collaboration" (Brondizio et al., 2016), and "sharing" (Faniel and Zimmerman, 2011). Collaboration and data sharing as such remain considerably removed from the actual opportunities enabled by Smart Earth (Ellison, 2010; Reichman et al., 2011; Hampton et al., 2013; Volk et al., 2014). Reviews of marine-based sciences have made similar observations (e. g. Costello, 2009; Starosielski, 2015), as have recent publications within environmental and digital humanities (e. g. Borgman, 2009; Cubitt, 2017). A commonly raised concern is the absence of proper institutional support for collaborative data sharing (e. g. Michener and Jones, 2012; Roche et al., 2015; Specht et al., 2015), which in certain cases might translate into prescriptions for greater incentives to share (Hampton et al., 2013); and the embrace of novel administrative frameworks (e. g. Verburg et al., 2016). Humanities scholars have been particularly keen to critique "crowdsourcing" as a source of ecological knowledge that tends to reinforce the expert [|207] hierarchies it proposes to disable (e. g. Gabrys, 2015; Swanstrom, 2016; Pearson et al., 2016).

For many observers, open source data leads inexorably to the problem of data standards, an issue with far-reaching consequence to Smart Earth environmental governance. Roche et al., (2015) surveyed 100 datasets associated with studies in journals that commonly publish ecological and evolutionary research, finding that 56% of the articles were linked to incompletely archived datasets. Calls for improved metadata have become increasingly common in ecology and biodiversity science (e. g. Frew and Dozier, 2012; Specht et al., 2015; O'Brien et al., 2016). "Good news narratives," showcasing the purported benefits of Smart Earth technologies (Arts et al., 2015, p.661), often obscure questions of quality controls and who will set them. The question of just what constitutes "good enough data" (Gabrys et al., 2016) prefigures a growing politics of "open source" legitimacy in Smart Earth environmental governance. At a time when increasing amounts of data freely circulate, and many scientists advocate for the continued diversification of research models (e. g. Verburg et al., 2016), the integration of data from multiple sensors poses institutional challenges and social tensions (e. g. Snyder et al., 2013). Moreover, what role do different data sources play in the actual conduct of environmental governance? In the Digital Fishers (2018) project (dmas.uvic.ca/DigitalFishers), for example, citizen scientists are solicited for their ability to identify species types on freely-available video data streaming from underwater cabled observatories. But the resulting citizen science is more proper characterized as a kind of volunteered filtering that allows scientists-experts to better identify the videos worth reviewing themselves (Matabos et al., 2017). The actual provision of open source data articulates a fundamental ambiguity of the Smart Earth governance regime: are data flows simply feeding into the predetermined interests of large multinationals (Crampton et al., 2014; Wood, 2013), or are they truly sites of continual "bottom up" modifications and "hacker-led" transformations (Hemmi and Graham, 2014)?

4.5 Citizen science/sensing

Smart Earth engages a broad set of actors. Researchers in disciplines as diverse as ecology, oceanography, geomorphology, political science, and sociology are now applying the Smart Earth concept. Smart Earth also engages non-professional researchers. For example, through "Volunteered Geographic Information," or VGI, community groups and NGOs are developing spatially-distributed, high-quality research outputs (Goodchild and Glennon, 2010). Some scholars have also explored the potential contribution of Artificial Intelligence (AI) to ecology, natural resource management, and wildlife conservation (Rykiel, 1989; Castelli et al., 2015; Millie et al., 2014), thereby invoking parallels with related transitions from "Web 2.0" to "Web 3.0", in which environmental governance is enabled by a connective intelligence that articulates sensors, humans, non-humans, data, and decision-making applications.

This transition will be intensified by Smart Earth's multiplication of sensing practices. Recent innovations create the potential for universal biotic sensors, in which every organism potentially performs as a sensor to be integrated into an array of data hubs and initiatives (such as the Sensor Web, Participatory Geoweb initiatives, and Google Earth) (Goodchild, 2007; Huang and Liang, 2014; Sieber, 2006). Smart Earth research agendas deploy many technologies which make use of the filtering and transducing capacities of bio-sensors (e. g. eyes, ears, skin etc.), as well as providing actionable data to users (e. g. via geospatial visualizations and geovisual analytics) (Helbig et al., 2017; Khan et al., 2013). In short, Smart Earth expand sensing modalities – tactile, auditory, even olfactory – by both humans and non-humans.

The most widespread application of this technology to date is "citizen sensing," which refers to intimate, experiential monitoring of environments by human users. Often, citizen sensing builds on existing citizen science initiatives, including those administered by non governmental groups like the Citizen Science Center (2017) and government agencies like the Environmental Protection Agency (2013). As Gabrys (2016) suggests, "citizen sensing" can be usefully understood as a subset of broader phenomenon of citizen science: "research projects in which the public is enlisted in scientific endeavors" (Hochachka et al., 2012, p.130; cf. Snyder et al., 2013). In Smart Earth, citizen sensing has become a privileged means by which IT-enhanced data collection, learning, decision-making, and participation scales up from discrete local encounters to governance initiatives.

To improve the quality and usefulness of citizen-sensed data, and to validate it against third-party critiques, scientists are now devoting vast resources to training practitioners and elaborating research programs designed for non-experts (e. g. EPA, 2013; ONC, 2014; see also: Kinchy et al., 2014; Lave, 2015). By channeling sensing efforts into area monitoring, animal tracking, or impact assessment, citizen-sensing not only facilitates environmental governance and reduces institutional overhead (citizens are rarely compensated for their efforts in these schemes), it also helps to resolve one of the central challenges for Smart Earth: data is accumulating much faster than computational processing capacity (Woodward et al., 2013; Starosielski, 2015). For many actors, the effective exploitation of such data requires new computational solutions – to which committed citizens appear ideally suited (Goodchild and Glennon, 2010; Gabrys, 2016; Matabos et al., 2017).

Marine-related activity is revealing for the way such "citizen sensing" activities are being solicited in coastline-areas – environments undergoing rapid change and populated by new environmental "risks" (e. g. marine debris, oil spills, whale-vessel strikes). In North America's Pacific Northwest, coastal residents have become central in efforts to construct the "Smartest Coast in the World" (D. Moore, 2015a). Likewise, in China we find growing efforts to utilize citizen capacities to improve "overall marine operational situational awareness" (Heesemann et al., 2014, p.153; See: Guo et al., 2010, 2017). The opportunity to "sense" and not merely "collect" data is a highly relevant enticement within these schemes. Sensing becomes a privileged means to attract and intrigue non-expert researchers about changing littoral regions. For example, the Ocean Observatories Initiative (OOI) entrains viewers to its websites to listen to underwater sounds captured by hydrophones in real time.[1] Hydrophone deployment and (ostensibly) public monitoring activities are a central part of ONC's ECHO partnership (ONC, 2014), which is being touted across the marine research community (Ritts, 2017).

The worldwide growth of Smart Earth technologies potentially converts every citizen into an environmental sensing device (Goodchild, 2007; Elwood and Leszczynski, 2013; Georgiadou et al., 2014). This raises important ethical questions. What does it mean to perform sensory activity according to a pre-determined repertoire of "smart" practices? How are such determinations inflected by the structural inequalities of race and gender, and even species? Some scholars have worried that projects touting citizen sensing might do little to actually democratize decision-making (e. g. Whitman and Pain, 2012). Insofar as citizen sensing efforts like BirdReturn elide formal state regulation (Gabrys, 2014), or cultivate relationships with private entities, questions of arbitration, justice, and collective accountability are salient (Drusch et al., 2012; Georgiadou et al., 2014). In other words: what are the *citizen politics* of citizen science? How are "citizenspokespersons" nominated and legitimated? How would conflictual forms of citizen science be managed via Smart Earth environmental governance?

Privacy issues are also significant. Smart Earth is characterized by new modes of citizen-activated management (e. g. Jepson and Ladle, 2015): for example, in projects like "A Smarter Planet" and CeNSE, individual citizens increasingly operate as essential operational and functional elements of environmental regulation. In

1   See, for example: https://soundcloud.com/oceannetworkscanada.

transforming [|208] citizenship into citizen sensing, the public becomes a constitutive element of an emerging "computational apparatus" (Gabrys, 2016). Citizens interacting with Smart Earth technologies will voluntarily submit to surveillance, providing data whilst having their online actions thoroughly indexed. Such "non-voluntary" systems of locational disclosure are built into many applications that individuals implicitly consent to when participating in Smart Earth activity – something Apple's "Locationgate" scandal made abundantly clear (Cottrill, 2011). Because such disclosures cannot easily be controlled by individuals through settings adjustments to any one device, volunteered geo-data about Earth Processes enable "geosurveillance," including by state security organisations and private contractors (Kitchin, 2014). There is thus potential for exploitative modes of recruitment and usage if proper checks are not established.

5. CONCLUSION

As explored above, Smart Earth technologies create the conditions for potentially significant shifts in environmental governance. Our review has provided insight into some of the ensuing challenges, debates, and critiques. In this concluding section, we briefly summarize key points and areas for future research.

Our analysis has emphasized the point (as reiterated by a significant amount of social science research) that better data does not necessarily lead to better governance. Indeed, we have suggested that algorithms might selectively reduce the sphere of possible intervention and analysis within a particular landscape. Mitigating against this tendency for the purposes of robust citizenship as well as sustainable ecosystem management requires democratizing access to environmental information, the nature of which is constantly evolving in light of technological change. In parallel, a comprehensive analysis of regulatory gaps is required, aligned with an analysis of the growth in integrative architectures currently being positioned as global frameworks for storing, analyzing, and disseminating Smart Earth data. In this context, a more comprehensive understanding of Smart Earth requires analysis of the changing nature of multi-actor and multilevel environmental governance – a key point, but one which was beyond the scope of this metareview.

Another key gap in the literature is a critical analysis of the role of the state, historically a key player in facilitating multi-scalar processes of environmental change (Robertson and Wainwright, 2013). In Canada for instance, it is notable that the Smart Oceans™ initiative enjoys considerable state support; and harmonizes both with state regulatory goals (e. g. *Oceans Protection Plan*, 2016), and with state-led efforts to commercialize Canada's marine technology sector (National Research Council, 2007; Strangway, 2013). Future work needs to critically evaluate the role of the state in enabling Smart Earth processes in different geographical and cultural contexts; for example, China has been extremely active in Digital Earth initiatives (Guo, 2012).

Further research is also required on the political-economic dimensions of Smart Earth governance. For example, Smart Earth creates not only new ways of sensing and administering environments, but also new categories of environmental assets. Some scholars have been concerned by the possibility that Smart Earth technologies may be harnessed to increase the efficiency of resource extraction, rather than serve environmental conservation purposes (J. W. Moore, 2015b; Malm, 2016; Demos, 2017). Further analysis on these issues could usefully draw upon debates of evolving environmental governance frameworks, including debates over Schwab's (2017) "Fourth Industrial Revolution" as well as the politics of adaptive management and resilience (Walker et al., 2004; Rockström et al., 2009), multi-scalar environmental governance (Bulkeley, 2005; Cash et al., 2006), and network fragility (Graham, 2010).

Questions of ethics also merit more scrutiny. Smart Earth governance implies a shift not only from "government to governance", but also from "manual to automated" eco-governance. Emergent regimes of state-sponsored surveillance consolidated around environmental big data – such as the Smart Oceans^{TM} project noted above – are mobilizing in support of security objectives rather than equitable access or efficiency (Amoore, 2013). Smart Earth also raises fundamental issues of socio-environmental justice. Elderly residents and those unable to own a smartphone face diminished opportunities to participate in Smart Earth governance since they "do not [necessarily] register as digital signals" (Crawford et al., 2014, p. 1667). Such social inequalities risk becoming entrenched through iterative forms of Smart Earth governance. As Leszczynski (2016) explains: "Algorithmic governmentality cannot divest itself of actual realities of socio-spatial stratification to which the derivative is theoretically indifferent" (1693).

Last but not least, issues of Smart Earth-generated e-waste (Cubitt, 2017) promise to be major problems in the coming years, and case studies of these issues (and their solutions) are scarce to date. Smart data will be derived from an expanded array of sensors, continuously sampling the physical world; its processing will in turn require real time big-data analytics with greater energy demands. Innovation in batteries, power-saving technologies, and backups will be increasingly essential to the functioning and performance of "actually existing" smart grids, app-based conservation efforts, and the like (Shelton et al., 2015). E-waste will also pose new ecological problems for system managers and government institutions. There are considerable problems inherent in the Smart Earth proliferation of screen-based technologies, owing to their material externalities. A 2015 report by the Natural Resources Defense Council (NRDC) found that the idle-load electricity demands of digital consumer electronics (televisions, computers, printers, game consoles, etc.) accounted for 51 percent of an average American household's energy budget (2015). An earlier report (Natural Resources Defence Council (NRDC), 2012) noted that 85 percent of electronics are now thrown out rather than recycled, leading some to calls for North Americans to adopt the radio in the place of the television, as the former creates substantially lower ecological costs (e. g. Smith, 2015). However, our review did not identify a single academic publication quantifying the e-waste associated with Smart Earth – a significant gap.

Given these concerns, Galaz and Mouazen (2017) are well-justified to call for a code of conduct (which they term a "bio-code") that allows citizens and institutions an opportunity to take stock of the proliferation of new social relationships and ethical challenges created by Smart Earth forms of governance. Data-sharing policies and ecological measurements standards, key mechanisms by which Smart infrastructure attains the obscurity its planners routinely "seek" (Jackson and Bobrow, 2015, 1770), require new forms of visibility in public education and debate. Jasanoff's (2003) demand for "technologies of humility" continues to resonate as a forceful appeal for new kinds of mergers between "the 'can do' orientation of science and engineering" and "the 'should do' questions of ethical and political analysis" (244). In this framing, ethics is not an "after thought" or addition to design but a crucial input across the life cycle of a given system – particularly one as ambitious and far-reaching as Smart Earth.

Acknowledgements

Research assistance from Jessica Hak Hepburn, Donna Liu, Andrea Lucy, and Adele Therias is gratefully acknowledged, as is funding from the Social Sciences and Humanities Research Council of Canada.

REFERENCES

Amoore, L., 2013. The Politics of Possibility. Duke University Press, Durham, NC.
Angwin, J., 2013. NSA struggles to make sense of flood of surveillance data. The Wall Street Journal (December). Available from: https://www.wsj.com/articles/nsastruggles-to-make-sense-of-flood-of-surveillance-data-1388028164. (Accessed 2.21.18).
Argo, 2017. Argo Home [Online]. Available from: http://www.argo.ucsd.edu/ (Accessed 2.21.18). [|209]
Arts, K. A. J., van der Wal, R., Adams, W. M., 2015. Digital technology and the conservation of nature. Ambio 44 (4), 661–673. https://doi.org/10.1007/s13280-015-0705-1.
Auld, G., Gulbrandsen, L. H., 2010. Transparency in nonstate certification: consequences for accountability and legitimacy. Glob. Environ. Politics 10 (3), 97–119. https://doi.org/10.1162/GLEP_a_00016.
Barnes, C. R., Best, M. M. R., Johnson, F. R., Pautet, L., Pirenne, B., 2013. Challenges, benefits, and opportunities in installing and operating cabled ocean observatories: perspectives from NEPTUNE Canada". IEEE J. of Ocean. Eng 38 (1), 144–157. https://10.1109/JOE.2012.2212751.
Bastin, L., Buchanan, G., Beresford, A., Pekel, J.F., Dubois, G., 2013. Open-source mapping and services for Web-based land-cover validation. Ecol. Inform. 14, 9–16. https://doi.org/10.1016/j.ecoinf.2012.11.013.
Benson, E., 2012. One infrastructure, many global visions: the commercialization and diversification of Argos, a satellite-based environmental surveillance system. Soc. Stud. Sci. 42 (6), 843–868. https://doi.org/10.1177/0306312712457851.
Benson, E., 2015. Generating infrastructural invisibility: insulation, interconnection, and avian excrement in the Southern California power grid. Environ. Humanit. 6 (1), 103–130. https://doi.org/10.1215/22011919–3615916.
Biermann, F., Betsill, M. M., Vieira, S. C., Gupta, J., Kanie, N., Lebel, L., Liverman, D., Schroeder, H., Siebenhüner, B., Yanda, P. Z., Zondervan, R., 2010. Navigating the anthropocene: the earth system governance project strategy paper. Curr. Opin. Environ. Sustain. 2 (3), 202–208.

Biermann, F., Abbott, K., Andresen, S., Bäckstrand, K., Bernstein, S., Betsill, M. M., Bulkeley, H., Cashore, B., Clapp, J., Folke, C., Gupta, A., Gupta, J., Haas, P. M., Jordan, A., Kanie, N., Kluvankova-Oravska, T., Lebel, L., Liverman, D., Meadowcroft, J., Mitchell, R. B., Newell, P., Oberthür, S., Olsson, L., Pattberg, P. H., Sanchez-Roiguez, R., Schroeder, H., Underdal, A., Carmago Vieira, S., Vogel, C., Young, O. R., Brock, A., Zondervan, R., 2012. Navigating the anthropocene: improving earth system governance. Sci. 335 (6074), 1306–1307.

Boreal Laser Inc. n. d. Gasfinder3-OP: Portable Open-path TDL Analyzer [Online] Available from: http://www.boreal-laser.com/products/portable-open-path-tdlanalyzer/ (Accessed 2.21.18).

Borgman, C. L., 2009. The digital future is now: a call to action for the humanities [Online]. Digit. Humanit. Q. 3 (4), 1–30. Available from: https://works.bepress.com/borgman/233/ (Accessed 2.21.18).

Bradley, A., Pullar, N., 2015. Technology & Tenure: Sharing the Results of a Communitydriven Initiative in Cambodia. XIV World Forestry Congress 2015., Durban.

Bridge, G., Perreault, T., 2009. Environmental governance. In: Noel, C. (Ed.), A Companion to Environmental Geography. Wiley-Blackwell, West Sussex, Malden, pp. 475–497.

Brondizio, E. S., Ostrom, E., Young, O. R., 2016. Connectivity and the governance of multilevel socio-ecological systems: the role of social capital. In: Christophe, B., Pérez, R. (Eds.), Agroressources et écosystèmes: Enjeux sociétaux et pratiques managériales. Septentrion, Villeneuve d'Aseq, pp. 33–69. Available from: http://books.openedition.org/septentrion/9176 (Accessed 2.21.18).

Bulkeley, H., 2005. Reconfiguring environmental governance: towards a politics of scales and networks. Political Geogr. 24 (8), 875–902. https://doi.org/10.1016/j.polgeo.205.07.002.

Caliskan, A., Bryson, J. J., Narayanan, A., 2017. Semantics derived automatically from language corpora contain human-like biases. Science 356 (6334), 183–186. https://doi.org/10.1126/science.aal4230.

Canada's Oceans Protection Plan [Online]. Transport Canada. Available from: https://www.tc.gc.ca/eng/canada-oceans-protection-plan.html, (Accessed 7.24.2018).

Carruth, A., Marzec, R. P., 2014. Environmental visualization in the anthropocene: technologies, aesthetics, ethics. Public Cult. 26 (2–73), 205–211. https://doi.org/10.1215/08992363–2392030.

Carse, A., Lewis, J. A., 2017. Toward a political ecology of infrastructure standards: or, how to think about ships, waterways, sediment, and communities together. Environ. Plan. A 49 (1), 9–28. https://doi.org/10.1177/0308518X16663015.

Cash, D. W., Adger, W. N., Berkes, F., Garden, P., Lebel, L., Olsson, P., Pritchard, L., Young, O., 2006. Scale and cross-scale dynamics: governance and information in a multilevel world. Ecol. Soc. 11 (2), 8. https://doi.org/10.5751/ES-01759–110208.

Cashore, B., 2002. Legitimacy and the privatization of environmental governance: how non-state market-driven (NSMD) governance systems gain rule-making authority. Governance 15 (4), 503–529.

Castelli, M., Vanneschi, L., Popovič, A., 2015. Predicting burned areas of forest fires: an artificial intelligence approach. Fire Ecol. 11 (1), 106–118. https://doi.org/10.4996/fireecology.1101106.

Citizen Science Center, 2017. Citizen Science Center – Welcome [Online]. Available from: http://www.citizensciencecenter.com/ (Accessed 2.21.18).

Clark, C. W., Ellison, W. T., Southall, B. L., Hatch, L., Van Parijs, Sofie M., Frankel, A., Ponirakis, D., 2009. Acoustic masking in marine ecosystems: intuitions, analysis, and implication. Mar. Ecol. Prog. Ser. 395, 201–222. https://doi.org/10.3354/meps08402.

Cohen, A., McCarthy, J., 2015. Reviewing rescaling: strengthening the case for environmental considerations. Prog. Hum. Geogr. 39 (1), 3–25.

Coldsnow, K. D., Relyea, R. A., Hurley, J. M., 2017. Evolution to environmental contamination ablates the circadian clock of an aquatic sentinel species. Ecol. Evol. 1–11. https://doi.org/10.1002/ece3.3490.

Coleman, D. C., 2010. Big Ecology: the Emergence of Ecosystem Science. University of California Press, Berkeley. https://doi.org/10.1525/j.ctt1ppn52.

Connors, J.P., Lei, S., Kelly, M., 2012. Citizen science in the age of neogeography: utilizing volunteered geographic information for environmental monitoring. Ann. Assoc. Am. Geogr. 102 (6), 1267–1289. https://doi.org/10.1080/00045608.2011.627058.

Conservation International, 2017. Conserv. in Virtual Real [Online]. Available from: https://www. conservation.org/stories/vr/Pages/default.aspx (Accessed 2.21.18).

Conserve.iO, 2015. Conserve.iO: Technology for a Better Planet [Online]. Available from: http:// conserve.io/ (Accessed 2.21.18).

Costello, M.J., 2009. Distinguishing marine habitat classification concepts for ecological data management. Mar. Ecol. Prog. Ser. 397, 253–268. https://doi.org/10.3354/meps08317.

Cottrill, C., 2011. Location Privacy: Who protects? Urban Reg. Inf. Syst. Assoc. 23 (2), 49–59.

Cowen, D., 2014. The Deadly Life of Logistics: Mapping Violence in Global Trade. University of Minnesota Press, Minneapolis. https://doi.org/10.5749/j.ctt7zw6vg.

Crampton, J.W., Graham, M., Poorthuis, A., Shelton, T., Stephens, M., Wilson, M.W., Zook, M., 2013. Beyond the geotag: situating 'big data' and leveraging the potential of the geoweb. Cartogr. Geogr. Inf. Sci. 40 (2), 130–139. https://doi.org/10.1080/15230406.2013.777137.

Crampton, J.W., Roberts, S.M., Poorthuis, A., 2014. The new political economy of geographical intelligence. Ann. Assoc. Am. Geogr. 104 (1), 196–214. https://doi.org/10.1080/00045608.2013.843436.

Crawford, K., Miltner, K., Gray, M.L., 2014. Critiquing big data: politics, ethics, epistemology. Int. J. Commun. 8, 1663–1672.

Cubitt, S., 2017. Finite Media: Environmental Implications of Digital Technologies. Duke University Press, Durham. https://doi.org/10.1080/00207233.2017.1312994.

Cushman, S.A., Gutzweiler, K., Evans, J.S., McGarigal, K., 2010. The gradient paradigm: a conceptual and analytical framework for landscape ecology. In: Cushman, S.A., Huettmann, F. (Eds.), Spatial Complexity, Informatics, and Wildlife Conservation. Springer, Tokyo, pp. 83–108.

De Longueville, B., Annoni, A., Schade, S., Ostlaender, N., Whitmore, C., 2010. Digital Earth's nervous system for crisis events: real-time sensor web enablement of volunteered geographic information. Int. J. Digit. Earth 3 (3), 242–259. https://doi.org/10.1080/17538947.2010.484869.

DeLoughrey, E., 2014. Satellite planetarity and the ends of the earth. Public Cult. 26 (2–73), 257–280. https://doi.org/10.1215/08992363–2392057.

Demos, T.J., 2017. Against the Anthropocene: Visual Culture and Environment Today. Sternberg Press, Berlin, New York.

Dickinson, J.L., Bonney, R., 2012. Comstock pub. Associates, Ithaca. Citizen Science: Public Participation in Environmental Research.

Digital Fishers [Online]. Available from: http://www.oceannetworks.ca/learning/getinvolved/citizen-science/digital-fishers (Accessed 2.21.18).

Dörnhöfer, K., Oppelt, N., 2016. Remote sensing for lake research and monitoring – recent advances. Ecol. Indic. 64, 105–122. https://doi.org/10.1016/j.ecolind.2015.12.009.

Drusch, M., Del Bello, U., Carlier, S., Colin, O., Fernandez, V., Gascon, F., Hoersch, B., Isola, C., Laberinti, P., Martimort, P., Meygret, A., Spoto, F., Sy, O., Marchese, F., Bargellini, P., 2012. Sentinel-2: ESA's optical high-resolution mission for GMES operational services. Remote Sens. Environ. 120, 25–36. https://doi.org/10.1016/j.rse.2011.11.026.

EarthNC, 2016. EarthNC: Next Generation Navigation [Online]. Available from: http://earthnc. com/apps/shark-net-app (Accessed 2.21.18).

Ellison, A.M., 2010. Repeatability and transparency in ecological research. Ecol. 91 (9), 2536–2539. https://doi.org/10.1890/09–0032.1.

Elwood, S., Leszczynski, A., 2013. New spatial media, new knowledge politics. Trans. Inst. Br. Geogr. 38 (4), 544–559. https://doi.org/10.1111/j.1475–5661.2012.00543.x.

Environmental Protection Agency, 2013. Draft Roadmap for Next Generation Air Monitoring [Online]. Available from: https://www.epa.gov/sites/production/files/2014–09/documents/roadmap-20130308.pdf (Accessed 2.21.18).

Faniel, I.M., Zimmerman, A., 2011. Beyond the data deluge: a research agenda for largescale data sharing and reuse. Int. J. of Digit. Curation 6 (1), 58–69. https://doi.org/10.2218/ijdc.v6i1.172.

Farina, A., Gage, S.H., 2017. Ecoacoustics: the Ecological Role of Sounds. John Wiley & Sons, Hoboken.

Favali, P., Beranzoli, L., De Santis, A., 2015. Seafloor Observatories: a New Vision of the Earth From the Abyss. Springer, Berlin, Heidelberg. https://doi.org/10.1007/978-3-642-11374-1.

Fox, P., Hendler, J., 2011. Changing the equation on scientific data visualization. Science 331 (6018), 705–708. https://doi.org/10.1126/science.1197654.

Frew, J.E., Dozier, J., 2012. Environmental informatics. Ann. Rev. Environ. Resour. 37 (1), 449–472. https://doi.org/10.1146/annurev-environ-042711–121244.

Gabrys, J., 2014. Programming environments: environmentality and citizen sensing in the smart city. Environ. Plan. D: Soc. Space 32 (1), 30–48. https://doi.org/10.1068/d16812.

Gabrys, J., 2015. Programming environments: environmentality and citizen sensing in the smart city. In: Marvin, S., Luque-Ayala, A., McFarlane, C. (Eds.), Smart Urbanism: Utopian Vision or False Dawn?. Routledge, New York, pp. 88–107.

Gabrys, J., 2016. Program Earth: Environmental Sensing Technology and the Making of a Computational Planet. University of Minnesota Press, Minneapolis.

Gabrys, J., Pritchard, H., Barratt, B., 2016. Just good enough data: figuring data citizenships through air pollution sensing and data stories. Big Data Soc. 3 (2), 1–14. https://doi.org/10.1177/2053951716679677.

Galaz, V., Mouazen, A.M., 2017. 'New wilderness' requires algorithmic transparency: a response to Cantrell et al. Trends Ecol. Evol. (Amst.) 32 (9), 628–629. https://doi.org/10.1016/j.tree.2017.06.013.

Gale, F., Ascui, F., Lovell, H., 2017. Sensing reality? New monitoring technologies for global sustainability standards. Glob. Environ. Politics 17 (2), 65–83. https://doi.org/10.1162/GLEP_a_00401.

Georgiadou, Y., Lungo, J.H., Richter, C., 2014. Citizen sensors or extreme publics? Transparency and accountability interventions on the mobile geoweb. Int. J. Digit. Earth 7 (7), 516–533. https://doi.org/10.1080/17538947.2013.782073.

GiveO2. n.d. GiveO2 [Online] Available from: http://www.giveo2.com/ (Accessed 2.21.18). [|210]

Goodchild, M.F., 2007. Citizens as sensors: the world of volunteered geography. GeoJournal 69 (4), 211–221. https://doi.org/10.1007/s10708-007-9111-y.

Goodchild, M.F., Glennon, J.A., 2010. Crowdsourcing geographic information for disaster response: a research frontier. Int. J. Digit. Earth 3 (3), 231–241. https://doi.org/10.1080/17538941003759255.

Graham, S., 2010. When infrastructures fail. In: Graham, S. (Ed.), Disrupted Cities: When Infrastructure Fails. Routledge, New York, pp. 1–26.

Gunningham, N., Holley, C., 2016. Next-generation environmental regulation: law, regulation, and governance. Ann. Rev. Law Soc. Sci. 12 (1), 273–293. https://doi.org/10.1146/annurev-lawsocsci-110615–084651.

Guo, H., 2012. China's Earth observing satellites for building a Digital Earth. Int. J. Digit. Earth 5 (3), 185–188. https://doi.org/10.1080/17538947.2012.669960.

Guo, H.D., Liu, Z., Zhu, L.W., 2010. Digital Earth: decadal experiences and some thoughts. Int. J. Digit. Earth 3 (1), 31–46. https://doi.org/10.1080/17538941003622602.

Guo, H., Liu, Z., Jiang, H., Wang, C., Liu, J., Liang, D., 2017. Big Earth Data: a new challenge and opportunity for Digital Earth's development. Int. J. Digit. Earth 10 (1), 1–12. https://doi.org/10.1080/17538947.2016.1264490.

Guston, D., 2011. Anticipatory governance: a strategic vision for building reflexivity into emerging technologies. In: Resilience 2011 Conference. Arizona State University CBIE, Tempe. Available from: https://cbie.asu.edu/resilience-2011/program/files/Panels/AnticipatoryGovernance/Guston.pdf (Accessed 2.21.18.

Hampton, S.E., Strasser, C.A., Tewksbury, J.J., Gram, W.K., Budden, A.E., Batcheller, A.L., Duke, C.S., Porter, J.H., 2013. Big data and the future of ecology. Front. Ecol. Environ. 11 (3), 156–162. https://doi.org/10.1890/120103.

Haque, M.M., Chin, H.C., Debnath, A.K., 2013. Sustainable, safe, smart – three key elements of Singapore's evolving transport policies. Transp. Policy (Oxf) 27 (20), 20–31. https://doi.org/10.1016/j.tranpol.2012.11.017.

Heesemann, M., Insua, T.L., Scherwath, M., Juniper, K.S., Moran, K., 2014. Ocean networks Canada: from geohazards research laboratories to smart ocean systems. Oceanography 27 (2), 151–153. https://doi.org/10.5670/oceanog.2014.50.

Helbig, C., Dransch, D., Böttinger, M., Devey, C., Haas, A., Hlawitschka, M., Kuenzer, C., Rink, K., Schäfer-Neth, C., Scheuermann, G., Kwasnitschka, T., 2017. Challenges and strategies for the visual exploration of complex environmental data. Int. J. Digit. Earth 10 (10), 1–7. https://doi.org/10.1080/17538947.2017.1327618.

Helmreich, S., 2015. Sounding the Limits of Life: Essays in the Anthropology of Biology and Beyond. Princeton University Press, Princeton.

Hemmi, A., Graham, I., 2014. Hacker science versus closed science: building environmental monitoring infrastructure. Inf. Commun. Soc. 17 (7), 830–842. https://doi.org/10.1080/1369118X.2013.848918.

Hochachka, W.M., Fink, D., Hutchinson, R.A., Sheldon, D., Wong, W., Kelling, S., 2012. Data-intensive science applied to broad-scale citizen science. Trends Ecol. Evol. (Amst.) 27 (2), 130–137. https://doi.org/10.1016/j.tree.2011.11.006.

Hogan, T., Bunnell, T., Pow, C.P., Permanasari, E., Morshidi, S., 2012. Asian urbanisms and the privatization of cities. Cities 29 (1), 59–63. https://doi.org/10.1016/j.cities.2011.01.001.

Huang, C.Y., Liang, S., 2014. The Open Geospatial Consortium Sensor Web PivotViewer: an innovative tool for worldwide sensor web resource discovery. Int. J. Digit. Earth 7 (9), 761–769. https://doi.org/10.1080/17538947.2014.887798.

Hummel, D., Adamo, S., de Sherbinin, A., Murphy, L., Aggarwal, R., Zulu, L., Liu, J., Knight, K., 2013. Inter- and transdisciplinary approaches to population-environment research for sustainability aims: a review and appraisal. Popul. Environ. 34 (4), 481–509. https://doi.org/10.1007/s11111-012-0176-2.

IBM Technology Underpins Project to Make British Columbia's the "Smartest Coast on the Planet" [Online]. Available from: https://www.ibm.com/news/ca/en/2014/04/14/p838123w96898u59.html. (Accessed 2.21.18).

IBM, 2018. Green Horizon [Online]. Available from: https://www.research.ibm.com/labs/china/greenhorizon.html. (Accessed 2.21.18).

Jackson, S., Bobrow, S., 2015. Standards and/as Innovation: Protocols, Creativity, and Interactive Systems Development in Ecology. Innovation in Theories & Products, CHI https://doi.org/10.1145/2702123.2702564. 2015, April 18–23 2015, 1769–1778.

Jasanoff, S., 2003. Technologies of Humility: citizen participation in governing science. Minerva 41 (3), 223–244. Retrieved from. http://www.jstor.org.ezproxy.library.ubc.ca/stable/41821248.

Jepson, P., Ladle, R.J., 2015. Nature apps: waiting for the revolution. Ambio 44 (8), 827–832. https://doi.org/10.1007/s13280-015-0739-4.

Jian, H., Liao, J., Fan, X., Xue, Z., 2017. Augmented virtual environment: fusion of realtime video and 3D models in the Digital Earth system. Int. J. Digit. Earth 10 (12), 1–20. https://doi.org/10.1080/17538947.2017.1306126.

Joppa, L.N., 2017. The case for technology investments in the environment. Nature 552 (7685), 325–328.

Khan, K.A., Akhter, G., Ahmad, Z., 2013. Integrated geoscience databanks for interactive analysis and visualization. Int. J. Digit. Earth 6 (2), 41–49. https://doi.org/10.1080/17538947.2011.638990.

Khatami, R., Mountrakis, G., Stehman, S.V., 2016. A meta-analysis of remote sensing research on supervised pixel-based land-cover image classification processes: general guidelines for practitioners and future research. Remote Sens. Environ. 177, 89–100. https://doi.org/10.1016/j.rse.2016.02.028.

Kinchy, A., Jalbert, K., Lyons, J., 2014. What is volunteer water monitoring good for? Fracking and the plural logics of participatory science. In: Frickel, S., Hess, D.J. (Eds.), Fields of

Knowledge: Science, Politics and Publics in the Neoliberal Age (Political Power and Social Theory, Volume 27). Emerald Group Publishing Limited, Bingley, pp. 259–289. https://doi.org/10.1108/S0198–871920140000027017.

Kitchin, R., 2014. Big data, new epistemologies and paradigm shifts. Big Data Soc. 1 (1), 1–12. https://doi.org/10.1177/2053951714528481.

Kitchin, R., McArdle, G., 2016. What makes Big Data, Big Data? Exploring the ontological characteristics of 26 datasets. Big Data Soc. 3 (1), 1–10. https://doi.org/10.1177/2053951716631130.

Kluza, D.A., Vieglais, D.A., Andreasen, J.K., Peterson, A.T., 2007. Sudden oak death: geographic risk estimates and predictions of origins. Plant Pathol. 56 (4), 580–587. https://doi.org/10.1111/j.365–3059.2007.01602.x.

Komninos, N., 2002. Intelligent Cities: Innovation, Knowledge Systems, and Digital Spaces. Spon Press, New York, London.

Koomey, J.G., Matthews, H.S., Williams, E., 2013. Smart everything: will intelligent systems reduce resource use? Ann. Rev. Environ. Resour. 38 (1), 311–343. https://doi.org/10.1146/annurev-environ-021512–110549.

Kratz, T.K., Arzberger, P., Benson, B.J., Chiu, C.Y., Chiu, K., Ding, L., Fountain, T., Hamilton, D., Hanson, P.C., Hu, Y.H., Lin, F.P., 2006. Toward a global lake ecological observatory network. Publ. Karelian Inst. 145, 51–63. Available from: http://users.sdsc.edu/~sameer/pubs/Gleon_Kratz_etal_2006.pdf (Accessed 2.21.18).

Kumar, P., Morawska, L., Martani, C., Biskos, G., Neophytou, M., Di Sabatino, S., Bell, M., Norford, L., Britter, R., 2015. The rise of low-cost sensing for managing air pollution in cities. Environ. Int. 75, 199–205. https://doi.org/10.1016/j.envint.2014.11.019.

Flow [Online]. Available from: https://flow.plumelabs.com/ (Accessed 2.21.18).

Landau, D.M., 2016. Bees With Backpacks: Keeping the Hive Alive [Online]. Available from: I.Q. Intel. https://iq.intel.com/bees-with-backpacks-keep-the-hive-alive/.

Lave, R., 2015. The future of environmental expertise. Ann. Assoc. Am. Geogr. 105 (2), 244–252. https://doi.org/10.1080/00045608.2014.988099.

Lehman, J., 2016. A sea of potential: the politics of global ocean observations. Political Geogr. Q. 55, 113–123. https://doi.org/10.1016/j.polgeo.2016.09.006.

Leszczynski, A., 2016. Speculative futures: cities, data, and governance beyond smart urbanism. Environ. Plan. A 48 (9), 1691–1708. https://doi.org/10.1177/0308518X16651445.

Li, S., Chen, J., 2017. Supporting future earth with global geospatial information. Int. J. Digit. Earth 10 (4), 325–327. https://doi.org/10.1080/17538947.2016.1275832.

Lindenmayer, D.B., Liekns, G.E., 2010. The science and application of ecological monitoring. Biol. Conserv. 143 (6), 1317–1328. https://doi.org/10.1016/j.biocon.2010.02.013.

Lindenmayer, D.B., Likens, G.E., Franklin, J.F., 2017. Earth Observation Networks (EONs): finding the right balance. Trends Ecol. Evol. (Amst.) 33 (1), 1–3. https://doi.org/10.1016/j.tree.2017.10.008.

Little, A.S., Needle, C.L., Hilborn, R., Holland, D.S., Marshall, C.T., 2015. Real-time spatial management approaches to reduce bycatch and discards: experiences from Europe and the United States. Fish Fisher 16 (4), 576–602. https://doi.org/10.1111/faf.12080.

LOOKNorth, 2012. Laser-based Greenhouse Gas Monitoring [Online]. Available from: https://www.looknorth.org/LaserGHGMonitoring (Accessed 2.21.18).

Mahony, M., 2014. The predictive state: science, territory and the future of the Indian climate. Soc. Stud. Sci. 44 (1), 109–133. https://doi.org/10.1177/0306312713501407.

Mairota, P., Cafarelli, B., Didham, R.K., Lovergine, F.P., Lucas, R.M., Nagendra, H., Rocchini, D., Tarantino, C., 2015. Challenges and opportunities in harnessing satellite remote-sensing for biodiversity monitoring. Ecol. Inform. 30, 207–214. https://doi.org/10.1016/j.ecoinf.2015.08.006.

Malm, A., 2016. Fossil Capital: the Rise of Steam Power and the Roots of Global Warming. Verso Books, London, New York.

Manzella, V., Gaz, C., Vitaletti, A., Masi, E., Santopolo, L., Mancuso, S., Salazar, D., De Las Heras, J.J., 2013. Plants as sensing devices: the PLEASED experience. In: Proceedings of the 11th

ACM Conference on Embedded Networked Sensor Systems. ACM, Rome. Available from: https://dl.acm.org/citation.cfm?id=2517403 (Accessed 2.21.18).

Marvin, S., Luque-Ayala, A., McFarlane, C. (Eds.), 2015. Smart Urbanism: Utopian Vision or False Dawn? Routledge, Oxford, New York.

Matabos, M., Hoeberechts, M., Doya, C., Aguzzi, J., Nephin, J., Reimchen, T. E., Leaver, S., Marx, R. M., Branzan Albu, A., Fier, R., Fernandez-Arcaya, U., 2017. Expert, crowd, students or algorithm: who holds the key to deep-sea imagery 'big data' processing? Methods Ecol. Evol. 8 (8), 996–1004. https://doi.org/10.1111/2041–210X.12746.

McGuinness, D. L., Pinheiro da Silva, P., Patton, E. W., Chastain, K., 2014. Semantic escience for ecosystem understanding and monitoring: the Jefferson project case study. AGU Fall Meeting Abstracts [Online]. American Geophysical Union, San Francisco. Available from: http://adsabs.harvard.edu/abs/2014AGUFMIN21B3712M (Accessed 2.21.18).

Michener, W. K., 2015. Ecological data sharing. Ecol. Inform. 29, 33–44. https://doi.org/10.1016/j.ecoinf.2015.06.010.

Michener, W. K., Jones, M. B., 2012. Ecoinformatics: supporting ecology as a data-intensive science. Trends Ecol. Evol. (Amst.) 27 (2), 85–93. https://doi.org/10.1016/j.tree.2011.11.016.

Michener, W. K., Brunt, J. W., Helly, J. J., Kirchner, T. B., Stafford, S. G., 1997. Nongeospatial metadata for the ecological sciences. Ecol. Appl. 7 (1), 330–342. https://doi.org/10.2307/2269427.

Millie, D. F., Weckman, G. R., Fahnenstiel, G. L., Carrick, H. J., Ardjmand, E., Young, W. A., Sayers, M. J., Shuchman, R. A., 2014. Using artificial intelligence for CyanoHAB niche modeling: discovery and visualization of Microcystis-environmental associations within western Lake Erie. Can. J. Fish. Aquat. Sci. 71 (11), 1642–1654. https://doi.org/10.1139/cjfas-2013–0654.

Momoh, J. A., 2012. Smart Grid: Fundamentals of Design and Analysis. Wiley, Hoboken.

Mooney, P., Corcoran, P., 2014. Has OpenStreetMap a role in Digital Earth applications? Int. J. Digit. Earth 7 (7), 534–553. https://doi.org/10.1080/17538947.2013.781688.

Moore, D., 2015a. B.C. Ocean Observation Project Useful for Oil Industry: Report. The Globe and Mail. 23 June. Available from: https://www.theglobeandmail.com/news/british-columbia/bc-ocean-observation-project-useful-for-oil-industry-report/article25069515/. (Accessed 2.21.18).

Moore, J. W., 2015b. Capitalism in the Web of Life: Ecology and the Accumulation of [|211] Capital. Verso, Brooklyn; London.

Morris, B. D., White, E. P., 2013. The EcoData retriever: improving access to existing ecological data. PLoS ONE 8 (6), e65848. https://doi.org/10.1371/journal.pone.0065848.

Murai, S., 2010. Can we predict earthquakes with GPS data? Int. J. Digit. Earth 3 (1), 83–90. https://doi.org/10.1080/17538940903548438.

National Research Council, 2007. Successful Response Starts With a Map: Improving Geospatial Support for Disaster Management. National Academies Press, Washington D. C.

Natural Resources Defence Council (NRDC), 2012. What you need to know about E-waste and New York City's intro. 104-A, the Electronics Collection, Recycling, and Reuse Act [Online]. Available from: https://www.nrdc.org/sites/default/files/ny104A.pdf (Accessed 2.21.18).

Natural Resources Defense Council (NRDC), 2015. Home Idle Load: Devices Wasting Huge Amount of Energy When Not in Use [Online]. Available from: https://www.nrdc.org/sites/default/files/home-idle-load-IP.pdf (Accessed 2.21.18).

O'Brien, M., Costa, D., Servilla, M., 2016. Ensuring the quality of data packages in the LTER network data management system. Ecol. Inform. 36, 237–246. https://doi.org/10.1016/j.ecoinf.2016.08.001.

Ocean Observatories Initiative, 2016. Cyberinfrastructure Technology [Online]. Available from: http://oceanobservatories.org/cyberinfrastructure-technology/ (Accessed 2.21.18).

ONC, 2014. Towards a "Smarter" BC Coast [Online]. Available from: http://www.oceannetworks.ca/towards-smarter-bc-coast. (Accessed 2.21.18).

Ostrom, E., 1990. Governing the Commons: The Evolution of Institutions for Collective Action. Cambridge University Press, Cambridge.

Panzacchi, M., Van Moorter, B., Strand, O., Saerens, M., Kivimäki, I., St Clair, C.C., Herfindal, I., Boitani, L., 2016. Predicting the continuum between corridors and barriers to animal movements using step selection functions and randomized shortest paths. J. Anim. Ecol. 85 (1), 32–42. https://doi.org/10.1111/1365–2656.12386.

Parenti, C., 2012. Tropic of Chaos: Climate Change and the New Geography of Violence. Nation Books, New York.

Paul, K., 2015. The Plan to Create a 'Fitbit for the Oceans' [Online]. Motherboard. Available from: https://motherboard.vice.com/en_us/article/z4maxy/the-plan-tocreate-a-fitbit-for-the-oceans (Accessed 2.21.18).

Pearson, E., Tindle, H., Ferguson, M., Ryan, J., Litchfield, C., 2016. Can we tweet, post, and share our way to a more sustainable society? A review of the current contributions and future potential of #Socialmediaforsustainability. Ann. Rev. Environ. Resour. 41, 363–397. https://doi.org/10.1177/2053951716652914.

Pettorelli, N., Laurance, W. F., O'Brien, T. G., Wegmann, M., Nagendra, H., Turner, W., 2014. Satellite remote sensing for applied ecologists: opportunities and challenges. J. Appl. Ecol. 51 (4), 839–848. https://doi.org/10.1111/1365–2664.12261.

PLEASED, 2017. Plants Employed As Sensing Devices [Online]. Available from: https://cordis.europa.eu/project/rcn/103686_en.html (Accessed 2.21.18).

Porter, J. H., 2010. A brief history of data sharing in the US Long Term Ecological Research Network. Bull. Ecol. Soc. Am. 91 (1), 14–20. https://doi.org/10.1890/0012-9623-91.1.14.

Pötzsch, H., 2015. The emergence of iBorder: bordering bodies, networks, and machines. Environ. Plan. D: Soc. Space 33 (1), 101–118. https://doi.org/10.1068/d14050p.

Ratti, C., Frenchman, D., Pulselli, R. M., Williams, S., 2006. Mobile landscapes: using location data from cell phones for urban analysis. Environ. Plan. B: Plan. Des. 33 (5), 727–748. https://doi.org/10.1068/b32047.

Reed, M. G., Bruyneel, S., 2010. Rescaling environmental governance, rethinking the state: a three-dimensional review. Prog. Hum. Geogr. 34 (5), 646–653.

Reichman, O. J., Jones, M. B., Schildhauer, M. P., 2011. Challenges and opportunities of open data in ecology. Science 331 (6018), 703–705. https://doi.org/10.1126/science.1197962.

Resch, B., Schulz, B., Mittlboeck, M., Heistracher, T., 2014. Pervasive geo-security – a lightweight triple-A approach to securing distributed geo-service infrastructures. Int. J. Digit. Earth 7 (5), 373–390. https://doi.org/10.1080/17538947.2012.674562.

Ritts, M., 2017. Amplifying environmental politics: ocean noise. Antipode 49 (5), 1406–1426. https://doi.org/10.1111/anti.12341.

Robertson, M. M., Wainwright, J. D., 2013. The value of nature to the state. Ann. Assoc. Am. Geogr. 103 (4), 890–905. https://doi.org/10.1080/00045608.2013.765772.

Rocchini, D., Hernández-Stefanoni, J. L., He, K. S., 2015. Advancing species diversity estimate by remotely sensed proxies: a conceptual review. Ecol. Inform. 25, 22–28. https://doi.org/10.1016/j.ecoinf.2014.10.006.

Rocchini, D., Petras, V., Petrasova, A., Horning, N., Furtkevicova, L., Neteler, M., Leutner, B., Wegmann, M., 2017. Open-access and open-source for remote sensing training in ecology. Ecol. Inform. 40, 57–61. https://doi.org/10.1016/j.ecoinf.2017.05.004.

Roche, D. G., Kruuk, L. E., Lanfear, R., Binning, S. A., 2015. Public data archiving in ecology and evolution: how well are we doing? PLoS Biol. 13 (11), 1–12. https://doi.org/10.1371/journal.pbio.1002295.

Rockström, J., Falkenmark, M., Karlberg, L., Hoff, H., Rost, S., Gerten, D., 2009. Future water availability for global food production: the potential of green water for increasing resilience to global change. Water Resour. Res. 45 (7), 1–16. https://doi.org/10.1029/2007WR006767.

Rykiel, E. J., 1989. Artificial intelligence and expert systems in ecology and natural resource management. Ecol. Model. 46 (1–2), 3–8. https://doi.org/10.1016/0304–3800(89)90066–5.

Sandbrook, C., 2015. The social implications of using drones for biodiversity conservation. Ambio 44 (4), 636–647. https://doi.org/10.1007/s13280-015-0714-0.

Save the Elephants, 2017. Geo-fencing [Online]. Available from: http://www.savetheelephants.org/project/geo-fencing/. (Accessed 2.21.18).

Schwab, K., 2017. The Fourth Industrial Revolution. Crown Business, New York.

Shelton, T., Zook, M., Wiig, A., 2015. The 'actually existing smart city.' Cambridge J. Reg. Econ. Soc. 8, 13–25. https://doi.org/10.1093/cjres/rsu026.

Sieber, R., 2006. Public participation geographic information systems: a literature review and framework.Ann.Assoc.Am.Geogr. 96 (3), 491–507.https://doi.org/10.1111/j.1467–8306.2006.00702.x.

Smith, J. W., 2015. Immersive virtual environment technology to supplement environmental perception, preference and behavior research: a review with applications. Int. J. Environ. Res. Public Health 12 (9), 11486–11505.

Snyder, E. G., Watkins, T. H., Solomon, P. A., Thoma, E. D., Williams, R. W., Hagler, G. S., Shelow, D., Hindin, D. A., Kilaru, V. J., Preuss, P. W., 2013. The changing paradigm of air pollution monitoring. Environ. Sci. Technol. 47 (20), 11369–11377. https://doi.org/10.1021/es4022602.

Specht, A., Guru, S., Houghton, L., Keniger, L., Driver, P., Ritchie, E. G., Lai, K., Treloar, A., 2015. Data management challenges in analysis and synthesis in the ecosystem sciences. Sci. Total Environ. 534, 144–158. https://doi.org/10.1016/j.scitotenv.2015.03.092.

Starosielski, N., 2015. The Undersea Network. Duke University Press, Durham.

Strangway, D., 2013. May 9: Committing Science, and Other Letters to the Editor. 9 May. Available from: The Globe and Mail (Accessed 2.21.18). https://www.theglobeandmail.com/globe-debate/letters/may-9-committing-science-and-otherletters-to-the-editor/article11795542/.

Swanstrom, E., 2016. Animal, Vegetable, Digital: Experiments in New Media Aesthetics and Environmental Poetics. University of Alabama Press, Tuscaloosa.

Thatcher, J., 2014. Big data, big questions: living on fumes: digital footprints, data fumes, and the limitations of spatial big data. Int. J. Commun. 8 (2014), 1765–1783. Available from: http://ijoc.org/index.php/ijoc/article/view/2174.

Turner, W., Rondinini, C., Pettorelli, N., Mora, B., Leidner, A. K., Szantoi, Z., Buchanan, G., Dech, S., Dwyer, J., Herold, M., Koh, L. P., 2015. Free and open-access satellite data are key to biodiversity conservation. Biol. Conserv. 182, 173–176. https://doi.org/10.1016/j.biocon.2014.11.048.

Verburg, P. H., Dearing, J. A., Dyke, J. G., van der Leeuw, S., Seitzinger, S., Steffen, W., Syvitski, J., 2016. Methods and approaches to modelling the Anthropocene. Glob. Environ. Chang. 39, 328–340. https://doi.org/10.1016/j.gloenvcha.2015.08.007.

Volk, C. J., Lucero, Y., Barnas, K., 2014. Why is data sharing in collaborative natural resource efforts so hard and what can we do to improve it? Environ. Manag. 53 (5), 883–893. https://doi.org/10.1007/s00267-014-0258-2.

Walker, B., Holling, C. S., Carpenter, S., Kinzig, A., 2004. Resilience, adaptability and transformability in social-ecological systems. Ecol. Soc. 9 (2), 5. https://doi.org/10.5751/ES-00650–090205.

Walsh, S. J., Mena, C. F., 2013. Science and Conservation in the Galápagos Islands. Springer, New York, London. https://doi.org/10.1007/978-1-4614-5794-7.

Welle Donker, F., van Loenen, B., 2017. How to assess the success of the open data ecosystem? Int. J. Digit. Earth 10 (3), 284–306. https://doi.org/10.1080/17538947.2016.1224938.

Whale Alert, 2012. Whale Alert: Reducing Lethal Whale Ship-strikes Worldwide [Online]. Available from. (Accessed 2.21.18). http://www.whalealert.org/.

White, P., 2016. Public Transport: Its Planning, Management and Operation. Taylor & Francis, Oxford, New York.

Whitman, G., Pain, R., 2012. Going with the flow: participatory action research and river catchment management. EGU General Assembly Conference Abstracts Vol. 14. EGU, Vienna, pp. 9280. [Online], Available from: http://adsabs.harvard.edu/abs/2012EGUGA..14.9280W (Accessed 2.21.18).

Wilson, M. W., 2014. Continuous connectivity, handheld computers, and mobile spatial knowledge. Environ. Plan. D: Soc. Space 32 (3), 535–555. https://doi.org/10.1068/d14112.

Wood, D. M., 2013. What is global surveillance? Towards a relational political economy of the global surveillant assemblage. Geoforum 49, 317–326. https://doi.org/10.1016/j.geoforum.2013.07.001.

Woodward, G., Gray, C., Baird, D.J., 2013. Biomonitoring for the 21st century: new perspectives in an age of globalisation and emerging environmental threats. Limnetica 32 (2), 159–174. https://doi.org/10.1068/d14112.

Zald, H.S.J., Wulder, M.A., White, J.C., Hilker, T., Hermosilla, T., Hobart, G.W., Coops, N.C., 2016. Integrating landsat pixel composites and change metrics with lidar plots to predictively map forest structure and aboveground biomass in Saskatchewan, Canada. Remote Sens. Environ. 176, 188–201. https://doi.org/10.1016/j.rse.2016.01.015.

Zandbergen, D., 2017. "We are sensemakers": the (anti-)politics of smart city co-creation. Public Cult. 29 (3_83), 539–562. https://doi.org/10.1215/08992363–3869596.

Zhang, Q., Pierce, F., 2013. Agricultural Automation: Fundamentals and Practices. CRC Press, London.

Zickuhr, K., 2013. Who's Not Online and Why. Pew Research Center, Washington, D.C.

Zimmerman, A.S., 2008. New knowledge from old data: the role of standards in the sharing and reuse of ecological data. Sci. Technol. Human Values 33 (5), 631–652. https://doi.org/10.1177/0162243907306704.

Zoological Society of London, 2018. n.d. Conservation Technology – Instant Detect. [Online]. Available from: https://www.zsl.org/conservation-initiatives/conservation-technology/instant-detect (Accessed 2.21.18).

Zyl, T.Lv., Simonis, I., McFerren, G., 2009. The sensor web: systems of sensor systems. Int. J. Digit. Earth 2 (1), 16–30.

III WIE DATEN, ALGORITHMEN UND SENSOREN AGIEREN

CYBORG GEOGRAPHIES:
TOWARDS HYBRID EPISTEMOLOGIES

Matthew W. Wilson

As a mode of critique, the cyborg is often separated from its role as a figuration. This article reviews Donna Haraway's cyborg theory to restate the importance of the cyborg *as a figuration* in critical methodology. Figuration is about opening knowledge-making practices to interrogation. I argue that the cyborg enables this inquiry through epistemological hybridization. To do so, cyborg figurations not only adopt a language of being or becoming, but narrate this language in the production of knowledges, to know hybridly. The epistemological hybridization of the cyborg includes four strategies: witnessing, situating, diffracting and acquiring. These are modes of knowing in cyborg geographies. To underline the importance of this use of cyborg theory, I review selected geographic literatures in naturecultures and technosciences, to demonstrate how geographers cite the cyborg. My analysis suggests these literatures emphasize an ontological hybridity that leaves underconsidered the epistemological hybridization at work in cyborg figuration. To take up the cyborg in this way is to place at risk our narrations, to re-make these geographies as hybrid, political work.
Keywords: critical methodology; cyborg figuration; hybridity; natureculture; technoscience

INTRODUCTION

Cyborgs can be figures for living within contradictions, attentive to the naturecultures of mundane practices, opposed to the dire myths of self-birthing, embracing mortality as the condition for life, and alert to the emergent historical hybridities actually populating the world at all its contingent scales. (Haraway 2003a, 11)

Cyborg geographies enact hybrid ways of knowing. This article argues that the cyborg's frequent citation as a literal marker for machinic-organic life has clouded the role of the cyborg *as a figuration*. While geographic literatures have cited the cyborg to signal an ontological hybridity, the epistemological hybridity of cyborg figuration has been less explored. I take this argument up to articulate a renewed critical methodology in geographies of naturecultures and technosciences, as these are the domains of cyborgean inhabitation. It is a call for greater specificity of the cyborg as an artifact of feminist critiques of science – a specification that actually broadens its use. Haraway's premising of the cyborg as a machinic-organic hybrid points to its more ontological usage. However, the larger purpose of this hybridization is to know differently our relationships with nature and technology – a partial knowing that requires both ontological and epistemological hybridity. Ontological hybridity is about contingent beings and about forms of becoming that challenge dualist narratives, like human/machine, nature/society and the virtual/real. Geographies of naturecultures and technoscience have each interrogated these kinds of [|500] hybridities (Kitchin 1998; Whatmore 2002; Swyngedouw 1996; Schuurman 2002).

However, to not engage the cyborg as an epistemological hybrid is to be inattentive to the partial and contingent practices of knowledges-in-the-making.

As the epigraph by Haraway alludes, cyborgs are about both *living* within and remaining *attentive to* the contradictions of technoscience and naturecultures. The cyborg is therefore a simultaneous being/becoming and knowing/seeing conduit through which to conduct critical study. Human geographers have only partially made use of this conduit, for example in studies of identity in cyberspace and of urban ecologies. I argue for more attention to cyborg epistemologies in these spaces to further ground our critical projects in their study. I situate this proposal in research that explores boundaries and boundary-makings, such as work in naturecultural geographies that challenges scholarly convention in studying the city and the wild, and technoscientific geographies that explore the contingencies of technological and cultural production. How do we narrate the production of knowledge in these geographic subfields and what is the role of cyborg theory in these narrations? To address this question, I propose a re-reading of cyborg theory, such that narrations of knowledge-in-the-making are conceptualized as a witnessing, situating, acquiring and diffracting – epistemological hybrids of the cyborg.

I begin by revisiting the work of Haraway, from cyborgs to her recent writings on companion species, to demonstrate how figuration works to *do* subjectivities and knowledge productions differently. Indeed, her work has influenced geographic study, namely in two directions: to bolster feminist critiques of the production of spatial knowledges and as a series of jumping-off points for studies of technologically-mediated spaces and human–animal relations. This article is framed as primarily a critique of the latter and an extended contribution of the former. Following this review of the cyborg as a figuration, I develop techniques for researching and writing these geographies. In the final section, I demonstrate my argument in the study of technoscientific and naturecultural space-times, by surveying the cyborg concept in selected geographic literatures. It is important to recognize how hybrid bodies are made in the process of these studies, as risky knowledge-making endeavors are inevitably messy and rife with boundary-crossings. This recognition involves an incessant questioning of *how* we know, *how* we theorize. If the 'cyborg' is left to stand for only the hybridity of being, then how do we engage the cyborg's political project of working knowledges and risky subjectivities? I argue that this engagement emerges through epistemological hybridity – by placing knowledge-making actions within the messy and risky realm of creative, strategic, fallible encounters and by becoming historically aware of the everydayness of our technological adaptations.

CYBORG FIGURATIONS

Over two decades after she offered her 'manifesto for cyborgs', Haraway's reaches into the metaphorical and the figurative remain a rich source for critical engagement. The cyborg is both a site and sighting for boundary crossings, framing the tension for this article. The cyborg can seem to be an academic trend and while its use stretches across popular culture, cultural critique and technological innovation, it is a particular process of critique and critical engagement that deserves further consideration. Here, I emphasize figuration and the cyborg as an example of figuration, to consider its role in writing critique. What is at stake is how we know what we narrate, in projects that research the multiplicities of bodily representations through innovations like cyberspace, urban ecologies, GIS and bioengineering. These are innovations enacted through hybridity; I argue that cyborgs are writing devices to narrate these hybridities. [|501]

Figuration is Haraway's overarching approach in critique, while recognizing Prins' (1995) argument that is it impossible to distill a methodological agenda in her research. Figuration is her aid in narration. Just as authors provide figures to illustrate arguments, figurations illustrate worlds. Haraway chooses figurations that have 'real' meaning and then reclaims their purposes in critique, and in this sense, figurations trope. Examples include the cyborg, gene, brain, chip, database, ecosystem, race, bomb, simian, species and fetus (Haraway 1991a, 1997, 2003a). As reclamations, these figurations act as entry points. Cyborg figurations walk worlds and as Shields (2006) suggests, the cyborg shares tendencies with *flâneur*. Haraway (2000, 138) describes such figurations as stem cells, '[o]ut of each one you can unpack an entire world'. Figurations are a kind of radical personification – an inhabiting of figures with the purpose of narrating (Gane and Haraway 2006). The point of doing so is:

> to make a difference in the world, to cast our lot for some ways of life and not others. To do that, one must be in the action, be finite and dirty, not transcendent and clean. Knowledge-making technologies, including crafting subject positions and ways of inhabiting such positions, must be made relentlessly visible and open to critical intervention. (Haraway 1997, 36)

The making of knowledge is the action of figuration, to open it to a radical visibility. It is this visibility that enables intervention through the un-working and re-working of knowledge production – to inspire an always partial storytelling of (post)modernity. These figurations seek to move beyond polemics and the either – or jousting of certain feminisms, specifically identity politics, by entering (in order to undo) their dualistic fields of operation.

Figuration is neither entirely figurative nor literal; its political prowess lies in its ambiguity. Figurations transcend rationalities and invoke multiplicity, but motivate a kind of objectivity through embodied perspective. This is described as the inhabiting of performativities (Haraway 1997, 179). Figurations are about arrangement, as a series of arguments or the composition of an image. In this sense, figurations are deeply spatial, as they are representative. Indeed, Haraway (1997, 11) invokes a mapping sense of figuration:

> We inhabit and are inhabited by such figures that map universes of knowledge, practice and power. To read such maps with mixed and differential literacies and without the totality, appropriations, apocalyptic disasters, comedic resolutions, and salvation histories of secularized Christian realism is the task of the mutated modest witness.

Figurations map. However, these are maps of contingency and relationality. In other words, figurations form geographies, to inhabit them. That figuring is a matter of inhabitation, and that this inhabitation enables a critical visioning is the epistemological rooting of this sort of ontological messing. In a poststructural vein, Haraway is interested in what gives these figures their particular shapes and what challenges permeate their shaping. This is a renewed storytelling – of re-situating these knowledges in ways that may contradict their usual moorings. In this sense, figurations both map and dis-map with their enrolling of 'mixed and differential literacies'.

The cyborg (short for cybernetic organism) is an image being continually drawn, fabricated, figured since its 'birth' in the 1960s – as the technoscientific processes of (post)modernity enable these images/imaginings. The cyborg figuration emerges from Haraway's need to tell certain truths about scientific processes. She crafts a position (a site/citation) from which and within which to objectively narrate. The cyborg is thus a material-semiotic entity, employed as a figuration in Haraway's critique of military-industrial relationships with science and technology. As a narrative device, the cyborg is [|502] composed of complicated and contradictory associations: of technologies and biologies, virtualities and physicalities, discursivities and materialities. It is complicit in generative projects of difference. The cyborg begins, after all, as the 'cyborg enemy' (Harvey and Haraway 1995, 514) – an enemy that needs to be reclaimed, or queered, into new possibilities. To engage in generative projects of difference, the cyborg advances a re-writing of the narratives about military–industrial relationships with science and technology. However, not all cyborgs tell these particular stories (cf. Gray 1995, 2000; Balsamo 1996; Stone 1995; Foster 2005; Halberstam and Livingston 1995).

Her interest in the cyborg is detailed in her 'manifesto for cyborgs', reprinted in a collection of essays titled *Simians, Cyborgs, and Women* (Haraway 1985, 1991). In this manifesto, Haraway introduces cyborgs as transgressing three boundaries: between human and animal, organism and machine, and the physical and the non-physical. She situates these transgressions in the ubiquity of electronics and their embeddedness in various practices, organizations, industries and militaries. It is this pervasiveness of the microelectronic that marks the potential for a cyborg manifesto, that in these moments when 'the difference between machine and organism is thoroughly blurred' we can recognize 'totalizing theory is a major mistake' and can take 'responsibility for the social relations of science and technology' (Haraway 1991, 165, 181). Here, Haraway is addressing the feminisms and Marxisms of the 1980s as they come aground in the massive movements of capital around the development of communication and biological technologies in Silicon Valley, California. By insisting on the heterogeneity forced by our microelectronic and bioengineered present, Haraway sees the political and ethical potential for hybrid subjects – that in these moments of intense diversification of economies emerge multiple kinds of subjects, resistive and contradictory. The micro(electronic) (bio)politics of the cyborg makes it a trickster in its opposition to grand narratives of progress, domination and emancipation. The cyborg project illuminates, for instance, the heterogeneity of gendered identity and insists on the construction of its supposed naturalness. Cyborg vision thus 'sees' an ontological hybridization premised on hybrid epistemologies. Without such epistemological and ontological visioning, critiques of knowledge practices remain routine and lack the riskiness of embedded narration. Routine critiques of knowledge practices are those that lend themselves too easily to determinisms and constructionisms, the slippage of 'the- machine-made-me-do-it' and the convenience of relative perspective.

From this initial manifesto, cyborg figuration grows into an entire book project, allowing Haraway (1997) to explore this kind of storytelling, of working within figurations. Two major parts make up the project: semantics and pragmatics. The first part emphasizes a meeting between a post-gendered post-human, FemaleMan© and a technically-'enhanced' mouse, OncoMouse™: the former an elaboration of a science fiction character, the latter the first patented animal, 'developed' by Dupont to harbor cancerous cells. Here, Haraway exercises her figuration's strength as a narrative device, to place in conversation literary fiction about post-gendered identity and genetically-altered/infused, cancer-growing rodents. The second part considers how the cyborg, again as a writing device, embodies a troubling of boundaries, between the technical and political. Here, she discusses the gene, race and the fetus as a few of several stem cells in which she places responsibility for the legitimating knowledge systems of the world. These stem cells illustrate her call to a particular, embodied witnessing of scientific practices: figurations. Here, Haraway draws feminists to the practices of science and technology, to challenge reactions against objectivity and fiction, and to complicate feminist concerns with reflexivity. [|503]

In her *Companion Species Manifesto*, Haraway (2003a) introduces a figuration to interrogate human–nonhuman relationships: the companion species, specifically the dog. Companion species are about historicizing our relationships with animals, as mediation for biotechnology's colonization of the genome. Haraway (2003b, 56) is intrigued by dogs as they are beings that are not-us; this figuration enables a worrying of the nature/culture binary, as 'dogs are neither nature nor culture, not both/and, not neither/nor, but something else'. These narrative devices are about *entering into* these histories, by writing their associations. Through inhabiting the narrative and exceeding 'the maze of dualisms', the cyborg insists not on a 'common language, but of a powerful infidel heteroglossia' (Haraway 1991a, 181). These kinds of figurations require work, Haraway (1999) argues, and working hybridities are those that are exposed and are made vulnerable, 'where epistemological and ontological risk define the name of the game'. To provide an example of working hybrids, Haraway (1999) examines a series of reports produced by the Scientific Panel of the government of British Columbia, Canada, to address conflict surrounding forestry practices on Vancouver Island. The alliances formed represent worked knowledges wherein the entities participating put themselves at risk, to challenge what it means (and why it matters) to have sustainable forest communities. As a method, if it could be one, the cyborg works knowledge-making enterprises, but the question remains of *how* to practice this critical methodology.

I have underlined the cyborg's role as a figuration: as a narrative device, to embed and craft associations, to historicize differently. The purpose is to enter into these storytellings, to make a mess of fact/fiction, subject/object and mind/body. This sort of work opens up human geography to new political geographies of contingency, relationality and difference within semiotic *and* material border-lands. The cyborg embodies these spaces, as a hybrid, to practice the production of knowledges. Hybridity is thus the means and ends to this knowledge production – a kind of working hybridity, where subjectivities are re-made in boundary crossings. Working hybrids invoke multiplicity, contingency and blurred, unraveling bound-aries between body and machine. They produce worked knowledges. Our relation-ships with microelectronics and dogs are indicative of working hybrids and worked knowledges – where all entities are altered in the process of association, where the line of association itself is blurred into near invisibility. These alterations, I argue, have two dimensions, ontological hybridity and epistemological hybridity, the former having been the more convenient usage of the cyborg, the latter an un-derutilized resource in critical geographic research. By not remaining attentive to the epistemological hybridity of the cyborg, we lose the critical politics of figura-tion – to make knowledge-in-the-making a visible practice.

EPISTEMOLOGICAL HYBRIDITY AS STRATEGIES

> The richness of the cyborg concept allows us to negotiate a multiplicity of spaces and prac-tices simultaneously and in so doing *develop epistemological strategies* ... (Gandy 2005, 40, emphasis added)

The resourcefulness of the cyborg stems, I argue, from its epistemological hybridity and the risk that comes with knowledge co-productions. The citation of the cyborg as an ontological hybrid – as a troubling of ontology – can mask this resourceful-ness. To challenge this masking, I suggest how the cyborg figuration enacts episte-mological strategies, as proposed by Gandy above. Figurations invoke multiple ways of being/becoming and knowing/seeing; as such, they are both epistemologi-cal *and* ontological. In this section, I develop the epistemological strategies of cy-borg figuration, [|504] as it is these strategies that I argue have been subsumed in our fascination with ontological hybridity.

The cyborg has been taken up to mean and signal a litany of cultural production and critique. I advocate a return to cyborg theory, to recover the 'epistemological subtlety and political prescience' of the Figure (Gandy 2005, 28). The purpose of these risky, working hybrids is to not only provide a language of being or becoming, but to narrate this language in the co-production of knowledges. If ontological hy-bridity is concerned with what it means to *be* hybrid, I suggest that epistemological hybridity considers what it means to *know* hybridly. Here, I propose a cyborg geog-raphy that is attentive to these ways of knowing.

To *know* hybridly, I argue that cyborg figurations take up the language of wit-nessing, situating, diffracting and acquiring. I have distilled these strategies from Haraway's writings of the cyborg and of companion species, from the lab to the kennel. It is a language taken up elsewhere in geography as feminist epistemologies (Rose 1997; Cope 2002; England 1994; Katz 1994; McDowell 1992; Lawson 1995). I refine this language of epistemological strategy to speak to the figuration of the cyborg. In arguing that these modes of epistemological hybridity are strategies of cyborg figuration, I broaden what is potentially enabled in the use of the cyborg. This is a re-activation of the cyborg, to intervene in narratives of knowledge production, to challenge their knowledges-in-the-making. These four interventions should be read as epistemological strategies in cyborg geographies.

Witnessing

The cyborg emerges from a need to witness: to observe, to provide an account and to be present. By placing the cyborg within the strategy of witnessing, I underline the critical impetus for this figuration. For Haraway, the title of her self-help manual is the fictive e-mail address of such a witnessing, *Modest_Witness@Second_Millennium* (1997). Her cyborg is paradoxically a witness situated in modesty and yet challenges the kinds of modest witnessing (observable truths) of science. Here, Haraway recounts female and male modesty, to draw certain distinctions. 'Female modesty was of the body; the new masculine virtue had to be of the mind' (1997, 30). Her modest witness was to be simultaneously of 'the self, biased, opaque', just as it was also transparent and objective (1997, 32). Witnessing was to be an embodied act of providing an account. The paradox of the modest witness, of being both objective and subjective, is inhabited in order to narrate the encounters of technoscience and natureculture. Haraway's historical irreverence continues as we read her e-mail address. Second Millennium situates this witnessing, making visible that our time is literally situated in Christian salvation history – the second millennium of Christ's birth. Here, Haraway calls on the language of witnessing to historicize Science's co-implication with the salvation narrative. To understand this witnessing, Haraway stresses the need to historically situate, to 'know those worlds' (1997, 37) in which our subject–object relations are situated and to realize the fiction 'we are forced to live ... whether or not we fit that story' (1997, 43). In doing so, Haraway grounds/embodies these narrations as a witnessing that is simultaneously partial and yet objective. She writes of a witnessing, that is 'seeing; attesting; ... a collective, limited practice that depends on the constructed and never finished credibility of those who do it' (1997, 267).

Haraway's delight in this kind of cyborgean witnessing allows her to challenge reactions against vision. Instead of avoiding or revoking the concept of vision, she seeks to rework the concept, to insist on a kind of 'seeing' that is necessarily partial – but no less a fact. This is a witnessing distinguished from relativism. By witnessing, we open up the [|505] practices of knowing – uncloaked from scientific rationalisms. Figurations do this work of witnessing – acting as a pivot to draw in the various contingencies and contradictions of knowledge-making practices. Witnessing is a visioning of the various enactions and positionings of knowledge-in-the-making. Geographers need this cyborgean witness to be attentive to a multiple situatedness, not from the single perspective of Author, but from the appendaged collection of authors-in-the-making.

Situating

Situating knowledges is a second epistemological strategy in cyborg figuration. The cyborg is witness to such situatedness – to counter, Haraway (1997, 188) argues, 'a leap out of the marked body and into a conquering gaze from nowhere'. However, the concept of situated knowledges does not indicate that our claims need to be grounded, or put in place. This is not a simple geography of perspective. Haraway seeks to clarify this mis-reading:

> … it is very important to understand that 'situatedness' doesn't necessarily mean place … Sometimes people read 'Situated Knowledges' in a way that seems to me a little flat; i.e., to mean merely what your identifying marks are and literally where you are. 'Situated' in this sense means only to be in one place. Whereas what I mean to emphasize is the *situatedness* of situated. In other words it is a way to get at the multiple modes of embedding that are about place and space in the manner in which geographers draw that distinction. (2000, 71; emphasis original)

Similarly, Gillian Rose (1997) has taken up this concern about situated knowledges in geography. Rose suggests that Haraway's situated knowledges are bound up with vision. Situated knowledges are, as Rose (1997, 308) writes, a 'siting [that] is intimately involved in sighting'. Situating knowledges requires powerful figuration and imagery; it is a tool for visioning difference. Cyborgs are sites from which to witness this 'situatedness of situated'. Witnessing and situating are co-dependent practices in cyborg geographies. To inquire about technoscientific and naturecultural encounters, geographers must inhabit figurations to 'see' and 'place', witness and situate, the multiplicity of relations that make our cyborg geographies. As figurations, cyborgs witness the various knowledge practices that constitute objects and subjects and the differences that are made – to situate, call attention to, the work that places or endows them with a geography. As a hybrid epistemological strategy, our recognition that the 'geography is elsewhere' for these figurations, is about their multiple and often contradictory placings (Haraway 1991b). To be attentive to this cyborgean situatedness, geographers have the responsibility to place these knowledges-in-the-making, not with some reified, exacted place, but as a placing – an objective, yet contingent, collusion of objects, subjects and spaces.

Diffracting

Cyborg geographies adopt a politics of diffraction. In opposition to an epistemology of reflexivity, diffraction is resistive to reflections. The point is to make a difference. Haraway (2000, 102) works the notion of reflexivity in feminist methodology, to oppose repeating the 'Sacred Image of the Same'; instead, diffraction is a recording of the 'history of interaction, interference, reinforcement, [and] difference'. Haraway uses the science of optics to draw a distinction between reflection and diffraction. The passage of light through a crystal separates light into its individuated bands; this sort of diffraction is about recording these various passages. Whereas reflections enable the mirror-images of [|506] ourselves elsewhere, diffractions work the image, to change the figuration, to alter the politics and to construct knowledges differently. Hers is a different optics of politics, a 'pattern [that] does not map where differences appear, but rather maps where the effects of difference appear' (Haraway 1992, 300). As witnessing is about 'seeing' and situating is about placings, diffracting is about changing knowledges, reconstructing knowledge practices such that alternative understandings of these knowledges emerge.

Diffraction works to tell new stories of technosciences and naturecultures – doing so requires not the mirror-image of reflexivity, but a visual metaphor based on difference and the enacting of differences. Diffraction, then, takes up various accounts, witnessed and situated, in order to radically alter them. It is therefore not enough to reflect on one's co-implication in knowledge practices; rather, it is our responsibility to diffract, to document the difference generated by such knowledge practices. Diffraction is the mantra of the unbeliever, to resist incredulously our accepted experiences about knowledge-in-the-making, to enable different explanations within differing geographies.

Acquiring

Cyborg geographies engage in acquisitions to know hybridly. Acquiring, as an epistemological strategy, asks us to take risks in building working alliances to further interrogate naturecultures and technosciences. These risks could involve learning from the observed, taking up their discourses, to diffract, to alter these knowledge productions. That material-semiotic entities acquire each other and 'make each other up, in the flesh', is the kind of ontological and epistemological risk present in working hybrids (Haraway 2003a, 2–3; 1999). Beyond essentialized alliances, acquiring, as an epistemological and political strategy, is to 'remain accountable to each other' (Penley and Ross 1991, 4). Further, to acquire is to become vulnerable to alternative, even contradictory, discourses – doing so enables a kind of hybridizing diffraction that messes knowledge practices based in reflection, extraction and synthesis.

While Haraway works the concept of acquiring after the bulk of her cyborg project was published, I suggest that acquiring is an epistemological strategy of cyborg geographies. The cyborg, after all, is an acquiring figuration – drawing in multiple, incongruous projects, such as projects of destruction and domination as well as projects for the enhancement of (non)human life. The pervasiveness of contemporary boundary crossings requires our permanent availability to hybrid ways of knowing; we have the opportunity and responsibility to acquire and become open to change. This, I believe, is complemented by the kinds of openness – of co-productions – that permeates Massey's (2005, 148) conceptualization of space as 'contemporaneous multiplicity' and as 'under construction'.

Acquiring is one tactic for new imaginations of political responsibility in scholarly endeavors. It is a strategy for knowing hybridly – to allow the unknown and the alternatively known to inhabit our ways of knowing, to alter them permanently.

Certainly, the discipline of geography is haunted by its legacy of acquisition. What I am suggesting here is not a return to those troubling acquisitions, but an ethic of making knowledges by working those hybrid encounters which place us at risk – to acquire one another in an enacting of responsible collaboration.

In this section, I have discussed a series of knowing practices – epistemological strategies – to consider the cyborg *as a figuration*. In the following section, I demonstrate how the cyborg's role as a figuration is lost in geographic literatures on naturecultures and technosciences. Haraway's attentiveness to the spaces of knowledge production (the corporate laboratory, the genome archive and the kennel) should give geographers pause [|507] to critically consider how our cyborg geographies are timed and spaced in important ways, and *how* we know what we narrate. By attending to these spaces for hybrid ways of knowing, we embrace the messiness of our boundary-work and remain responsible to the entities that populate our space-times – human, nonhuman, posthuman, cyborg, etc.

WORKING KNOWLEDGES IN NATURECULTURE AND TECHNOSCIENCE

In the geographies of natureculture and technoscience, there is a tendency to understate (or miss altogether) the epistemological hybridity of the cyborg figuration and instead connote ontological, categorical instability. I demonstrate that this is a missed opportunity to know hybridly the relations and adjacencies of knowledges-in-the-making. To take seriously cyborg figuration, geographers must expand notions of hybridity beyond being or becoming. The inclination towards ontological hybridity might emerge from human geographers' concerns about the centrality of the 'human'. These concerns unravel the monoglossia of the human sciences, to decentralize what we imagine *to be*. Accordingly, hybridity is invoked to draw in other kinds of entities (animals and the computer), to destabilize notions of being and becoming – to ontologically hybridize. The cyborg often signals this ontological hybridity in geographic literatures, drawing upon the more common understanding of the concept: the 'hybrid of machine and organism' (Haraway 1991a, 149). By pushing back on this citation, I argue that geographers need to go further in their engagement here. I examine their engagement in selected geographic literatures about naturecultures and technoscience, and argue that within each the citation of the cyborg works less as a figuration for hybridly knowing and more as a signal for hybrid beings/becomings.

Cyborgs in naturecultural geographies

The discomfort with the centrality of the human in human geography leads Sarah Whatmore (2004) to prefer the 'more-than-human' concept, as opposed to the posthuman. Here, Whatmore problematizes how temporality is invoked in posthumanism (see also Braun 2004). She reflects on her earlier work in *Hybrid Geographies*, a treatise on the production of naturecultural knowledges:

> Using various devices to push hybridity back in time, I sought to demonstrate that whether one works through the long practised intimacies between human and plant communities or the skills configured between bodies and tools, one never arrives at a time/place when the human was not a work in progress. (Whatmore 2004, 1361)

The question of 'the human', as a work of 'practised intimacies', remains. Whatmore (2002) attempts to move beyond dualisms and, for her, hybrid geographies are spaces wherein dualisms like human/nonhuman and nature/culture are untenable. By thinking hybridity in this way, as the impossibility of binary thinking, Whatmore destabilizes nature–society traditions in geography – to demonstrate how this way of interrogating 'nature' intimates the 'social'. Here, Whatmore historicizes urban relationships to the wild, from Roman uses of animals in the gaming arena to scientific inventories and animal management. She tacks between embodied, while partial, accounts of natures–societies and critiques of science through actor-network theory, and seeks to 'practice geography as a craft', to demonstrate the centrality of 'wild(er)ness' to the social (Whatmore 2002, 3–4). While Whatmore's investigatory motivations are insightful, I push back on her interpretation of cyborg figuration, as her reading illustrates how the cyborg has been limited to an ontological hybridity. I examine Whatmore and Eric Swyngedouw's citation [|508] of the cyborg, to ask what work it performs in their naturecultural geographies, to consider: what is enabled by the cyborg in naturecultural geographies?

Naturecultural geographies are concerned more broadly with the nature-society tradition of the discipline and hybridity is indeed one conceptual tool for problematizing this perspective. Energized by debates on the social construction of nature (Demeritt 2002), geographers have troubled the boundaries constituted by the nature-society tradition (Gerber 1997), thereby exploring the relationships between the natural and the urban (Swyngedouw 1996; Gandy 2005; Castree 2003; Braun 2005) and between animals and humans (Wolch and Emel 1998; Whatmore 2002; Philo and Wilbert 2000). Responsibilities and connectivities are at stake in these debates – responsibilities to multiple ways of living and concepts of life and connectivities to those other, constitutive entities.

Whatmore (2002) exercises caution in her use of the cyborg, finding the cyborg a useful ontological figure, while being less enamored with the potential of the cyborg, epistemologically. She draws on the cyborg to illustrate the hybridity that was always present in our relationships with the 'wild'. She recognizes the disruptive potential of the cyborg (1997), to de-purify the natural and the social. This de-purification works in an ontological sense. Further, she finds the cyborg less capable of expressing the material corporeality of nature–society connections.

> [A]lthough Haraway's account of hybridity successfully disrupts the purification of nature and society and relegation of 'nonhumans' to a world of objects, it is less helpful in trying to 'flesh out' the 'material' dimensions of the practices and technologies of connectivity that make the communicability of experience across difference, and hence the constitution of ethical community, possible. (Whatmore 1997, 47)

That there is a tension in the concept of the cyborg as per Whatmore's reading is clear. Considering the cyborg concept limited by a 'one-plus-one' logic, Whatmore (2002, 165) instead suggests a hybridity 'defined less by its departure from patterns of being that went before than with how it articulates the fluxes of becoming that complicate the spacings-timings of social life'. For Whatmore (2002, 187n16), cyborgs are 'couplings ... [where] difference is prefigured in the alterity of already constituted kinds'.

The cyborg figuration, for Whatmore, is not a site of inhabitation. This inhabitation is necessary to take seriously the role of witnessing, situating, diffracting and acquiring. Cyborgs, in Whatmore's reading, are simply the possibility of becoming one from multiple beings/things. They operate less as figurations, accordingly they are not inhabited or points of entry or narrative devices. We are left, then, with a concept of the cyborg that is anemic, unable to take risks, to see (witness) and place (situate) differently, to fold in (acquire) and alter (diffract) knowledges-in-the-making. This is a cyborg-in-passing, a relic of 1980s cultural production.

While Eric Swyngedouw uses the cyborg as an entry point into historical-material analysis of the urban, I want to push his use of the cyborg – to enact the kinds of risks that come in narrating urbanizations. He writes of the 'city as cyborg' to mark urban processes around water as intimately linked to bodily arrangements as well as regional and global relationships (Swyngedouw 1996, 80). More specifically, his entry point is a cup of water, to examine the connectivities between the urban and nature. By doing so, he emphasizes the city as a hybrid, telling stories of 'its people and the powerful socio-ecological processes that produce the urban and its spaces of privilege and exclusion' (Swyngedouw 1996, 67). The cup of water symbolizes – figures – his entry into multiple discussions of urbanizations: [|509]

> of participation and marginality; of rats and bankers; of water-borne disease and speculation in water-industry related futures and options; of chemical, physical and biological reactions and transformations; of the global hydrological cycle and global warming; of the capital, machinations and strategies of dam builders; of urban land developers; of the knowledges of the engineers; of the passage from river to urban reservoir. (Swyngedouw 1996, 67)

By analyzing water in this way, Swyngedouw demonstrates the city as a hybrid – partially composed of the relationships mentioned in the preceding quote.

Swyngedouw's use of cyborg and hybrid are nearly interchangeable, both invoking a composition of various complementary and contradictory elements. He employs cyborgs to package a multiplicity of natural–urban environmental productions (Swyngedouw 1996, 1999, 2006; Swyngedouw and Kaika 2000). As Bakker and Bridge (2006, 17) point out, this invocation of the cyborg serves to emphasize the production of hybrids, or the 'process of hybridization'. Similar to Whatmore's reading of the cyborg as a 'coupling', Swyngedouw cites the cyborg to emphasize the productive combinations within the urban, referenced in his quotation above. By taking up the cyborg as a figuration, I suggest that the processes of hybridization could be opened, through the epistemological strategies of the cyborg. It is Swyngedouw's (1996, 1999) cup of water that needs this witnessing – to remain partial and open to multiple and risky narratives about political, cultural and economic norma-tivities as well as micro resistances and inconsistencies.

My selection of naturecultural geographies by Whatmore and Swyngedouw demonstrate the tension around usage of the cyborg concept – in the former, a read-ing of the cyborg not as a figuration, but as a coupling and in the latter, a reading of the cyborg as emergent, assuming a pre-cyborgean condition, emphasizing the pro-duction of hybrids. This has demonstrated a need for a resuscitation of the cyborg citation, to recognize its potential for witnessing the situatedness of urbanization and urban study, to recognize their co-implicated discourses. What would it mean to write the cyborg city, where the objects of analysis illustrate the differences pro-duced by their study, in an always-incompleted project of working knowledges and inconclusive evidences?

Cyborgs in technoscientific geographies

Technoscientific geographies have the potential to enroll cyborg figuration, to wit-ness, situate, diffract and acquire the multiplicities of subject-objects in space-times. However, I argue that citations of the cyborg in these selected studies of cy-berspace and geographic information systems associate the cyborg with a narrow ontological hybridity. Without sufficient attention to hybrid epistemologies, these technoscientific geographies miss the opportunity to make knowledges of differ-ence. What remains is a technoscience of the same – a kind of inquiry that leaves technological knowledge production unchallenged and furthers a project of techno-logical advancement by the few. To write cyborg geographies of technoscience, geographers must foreground the cobbled-togetherness of technoscientific prac-tice – to elaborate their messy inceptions and risky encounters.

Technoscience indicates an alternative telling of (post)modernisms, wherein the productive tension between science and technology serves to exceed these very dis-tinctions, including mind and body, subject and object, human and nonhuman, nature and society (Haraway 1997, 3). Technoscience is Haraway's emphatic rejoinder to scientific rationalism. The cyborg figure is the narrator of this rejoinder, to 'bring the technical and the political back into realignment so that questions about possible livable worlds, lie visibly at the heart of our best science' (Haraway 1997, 39). Within geography, a limited literature explores the geographies of technoscience, to distin-guish among [|510] historical geographies of science and the histories of the geo-graphical sciences (Powell 2007). The geography of technoscience is concerned with the production of scientific and technical knowledges, particularly how these productions constitute spatial relationships among nature, society and technology. This research examines the practices of statisticians (Barnes 1998), high-energy physicists (Jons 2006), transgenic food production (McAfee 2008) and the gender-ing of office technologies (Boyer and England 2008). Although geographies of cy-berspace and critical geographic information systems have not been cast explicitly as technoscientific study, here I consider how two researchers have drawn upon tech-noscientific critique – specifically that of the cyborg. I extend this critical lens to Rob

Kitchin and Nadine Schuurman's use of the cyborg in technoscientific geographies, to discuss the absence of the cyborg as a figuration involved in hybridly knowing.

The permeation of cyberspace into everyday life is what Rob Kitchin (1998, 394) terms 'cyborging'. Identity is multiply produced in cyberspace and cyborging, according to Kitchin (1998, 394), describes the 'merging of nature with technology, as humans and computers coalesce'. Kitchin and Kneale (2001) also enroll the concept of cyborging to discuss cyberpunk fiction. Kitchin's usage of the cyborg concept is a marker for hybrid identification. Cyborging, for him, is a process of unification through merger and coalescence – the becoming of one, identifiable subject. Cyborging is a writing device to invoke hybridity, through analyses of lives lived online and literary fiction. This device, when used in cyberspace, according to Kitchin, enables a user to actively create identity, *to cyborg*. Cyborging is this process whereby '[u]sers literally become the authors of their lives' (Kitchin 1998, 394). It is this hybridity-in-the-making that draws Kitchin (1998, 395) to the cyborg concept, where cyberspace subjects 'play' with 'fantasies, … othernesses, … [and] crossdressing'. Cyborging in cyberspace is about enacting hybrid identities in virtual and imaginative geographies. However, I stress that this notion of 'cyborging' is limited to an ontological dimension. To enroll cyborg figuration is to witness and situate these productions of cyber-identities, beyond a recognition of their made-up becomings and towards a critical visuality. Cyborging, as I alternatively read it, is not only coalescence, but also the always-unfinished project of attesting to the ways of knowing self and other in the network.

Similarly, Nadine Schuurman (2002, 2004) calls for 'writing the cyborg', arguing for increased use of GIS by women and underrepresented groups (Schuurman 2002, 261). To 'write the cyborg', or 'perform the cyborg' as Kwan (2002, 276) has stated, is to invoke the cyborg as a process. Schuurman draws upon Haraway's notion of cyborgs in the construction/use of geographic information technologies, countering critiques of its potential surveillance capability (Kwan 2002). For Schuurman the prospect of GIS in the hands of the surveilled is a reworking of the technology in a cyborgian tone. Schuurman seeks to challenge the masculine inception of the technology, by actively re-rendering the technology from a feminist perspective operated for/with female/other bodies, described as 'strength in numbers' (Schuurman 2002, 261). Her 'feminist cyborg' seeks to 'make GIS and geography a more equitable place not only for women but for many underrepresented and less powerful groups' (Schuurman 2002, 261). Like cyborging in Kitchin's (1998) review of cyberspace research, the point is to actively constitute the possibility of hybrid becomings. Schuurman (2004, 1337) traces the concept of the cyborg:

> The cyborg of the 20th century was an amalgamation of technology and humanity. Using a computer to write, having a locator chip installed in your dog's ear, or programming military-industrial applications all warranted the designation. Any confluence of silicon with animal or human behaviour and presto: a cyborg. Cyborgs of the 20th century had less to do with data [|511] than with silicon. The very fact that they incorporated computing was enough to earn the designation 'cyborg' … 'New' cyborgs are, however, more than metal and flesh; they come to life in the presence of data.

Schuurman argues that twenty-first century cyborgs are not necessarily made of microelectronics, but data. This use of the cyborg symbolizes a hybridity of being – of being/becoming more-than-human in an intermeshing of data and electronics. Our interactions with computing technologies designates our *being* cyborg. Further Schuurman (2002) advocates that the political challenge is to enable a feminist cyborg, by emphasizing the role of marginalized populations in the production of GIS knowledges. While one aspect of the feminist critique of science was to advocate the placement of more women in science positions (and to make more visible those women who are scientists), Haraway instead proposes situated knowledges – so as to avoid essentializing women's role in (scientific) knowledge productions. The role of gender is multiply interpreted in cyborg figurations, according to Haraway (2003a; see also Wajcman 2004) and Schuurman 'writing the cyborg' is indicative of this tension. Haraway's (2003a, 47) political potency was to use the cyborg to resist militaristic, 'man-in-space' projects, while also narrating their implicatedness in technoscientific agendas. Schuurman's use of the cyborg illustrates for her the complicated arrangement of humans and technology, without further asking what kinds of knowledges are made in these arrangements and how these knowledges may be made differently. Instead, Schuurman's citation of the cyborg figuration is to mark bodies *as* cyborgs and does not inhabit the cyborg as a strategy in narrating knowledges-in-the-making.

The cyborg, as in the selected literatures above, often references a being/becoming hybrid – emphasizing the ontological connotation of the concept. I argue that an opportunity has been missed in this citation. Beyond 'writing the cyborg', 'performing the cyborg', or 'cyborging' (each constitutes hybrid ways of *being*), *knowing* hybridly requires an inhabiting of the figuration of the cyborg – to see the relationships and connectivities of naturecultural and technoscientific practice. To know hybridly, I argue for a return to the knowing practices of the cyborg, based on witnessing, situating, diffracting and acquiring.

CONCLUSIONS

We know, from our bodies and from our machines, that tension is a great source of pleasure and power. May cyborg, and this *Handbook*, help you enjoy both and go beyond dualistic epistemologies to the epistemology of cyborg: thesis, antithesis, synthesis, prosthesis. And again. (Gray, Mentor, and Figueroa-Sarriera 1995, 13)

What would it mean to introduce oneself as a 'cyborg geographer', in the same sense that we introduce ourselves as human geographers? How do we complicate our own proclivities toward the 'human'? As cyborg geographers, we are responsible for being attentive to the partial and contingent practices of knowledges-in-the-making. This responsibility is two-part. First, we must recognize knowledge-making actions as creative, sometimes strategic and often fallible encounters. Second, we must be historically aware of our multiple adaptations in these actions. As Gray, Mentor, and Figueroa-Sarriera (1995) provoke a taking-up of their handbook in the above quotation, they encourage the use of the cyborg *as an epistemology*.

This signaling to the cyborg as a device, to be enrolled and invoked, parallels my insistence that figuration of the cyborg requires an inhabitation, to be *and* know hybridly. Cyborg geographers use figurations to fulfill their responsibilities. [|512]

In the midst of geographers' explorations of the cyborg in terms of an ontological boundary messing, I have argued that the cyborg's potential as an epistemological hybrid is underconsidered. Therefore, I have taken a more specified understanding of this hybrid epistemology as one of figuration. This is not to say that ontological hybridity is not important for cyborg geographers; indeed, knowing hybridly is steeped in being and becoming hybrid. However, I argue that what is at stake in knowing hybridly is *how* we know what we narrate. Cyborg geographers are interested in this question of how we know, but not in lieu of creating action. As hybrid beings interested in how we know, cyborg geographers emerge from their embedded narrations to warn against determinisms and constructionisms, to flag critiques that have become too routine. This is a responsible action – a responsibility fulfilled by recognizing our implications in knowledge-making-actions. The recognition of our implications enables the writing of these geographies of contingency, relationality and difference. This writing becomes the creative action of the cyborg geographer. To practice this writing, I have suggested four strategies: witnessing, situating, acquiring and diffracting. That these ways of knowing are always an attempt to complete an incompletable whole is the name of the game. Partiality *and* objectivity are strange bedfellows. And yet the cyborg geographer enacts their coordination.

Following Haraway and limited geographic literatures that cite the cyborg, I have proposed witnessing, situating, diffracting and acquiring as methodological endeavors, to know hybridly. By knowing hybridly through these strategies, knowledge practices become grounded throughout narratives about naturecultural and technoscientific phenomena. While these strategies are indeed co-dependent on the ontological hybridity of the cyborg concept, I highlight these to both return to the cyborg as a theoretical concept and to ground this concept as a figuration. I propose cyborg geographies as a call to take these hybrids seriously, to recognize that hybridly knowing is bound up in becoming and being hybrid. At stake is our action to know – to not only recognize differences produced in techno-nature-culture worlds, but to inhabit these spaces of difference and become responsible to their vulnerabilities. Hybridly knowing is knowing-at-risk.

That the cyborg is foremost a figuration in Haraway's work frames my insistence on close readings of geography's dabbling with the cyborg, to consider the geographies of the cyborg and the potential for further critique in a cyborgian tone. As a figuration, the cyborg is a writing technology. By invoking the cyborg in this sense of knowing differently, we also conjure certain hybridities of knowing. These hybridities live within and through our writings. There is work to be done in the cyborg geographies of natureculture and technoscience. This is a call for cyborg geographers to:

-   inhabit the spaces where the human and nonhuman are constituted and narrate the conditions that established this entry point;

- resist even temporary stabilizations of urban study, by recognizing urbanization as an always-incompleted project of working knowledges;
- foreground the messy and risky spaces of technoscientific practice;
- articulate the moments in which self and other are known in the virtual; and
- question the knowledges made through human–technology assemblages, so as to create the possibility that they may be made differently.

To conceptualize this work, I have suggested a re-reading of the cyborg figuration, to inspire some strangeness amid the popular familiarity of the cyborg. I have found it helpful to consider Massey's re-conceptualizations of spaces as taking a similar approach. [|513] Massey seeks to disrupt our narratives about space and poses a series of challenges to geographers: how to narrate, how to spatialize, how to mis/represent, how to situate and position, as well as how to (not) obfuscate, how to seize/cease production/representation, how to complicate, etc. Cyborg geographies forward this disruption. It is a political project of knowledge-in-the-making, of knowing hybridly – of finding coeval kinship in knowledge endeavors.

Acknowledgements

I thank Vicky Lawson and my writing workshop colleagues, Elise Bowditch, Todd Faubion, Juan Galvis, Stephen Young, Dena Aufseeser, Srinivas Chokkakula and Tish Lopez for their encouraging, careful reviews. I also thank others who read various drafts of this article or who were supportive of the ideas contained within it, Sarah Elwood, Bonnie Kaserman, Michael Brown and Tim Nyerges. Finally, I sincerely appreciate the useful and caring feedback from Robyn Longhurst and three anonymous referees. I, of course, take final responsibility for these ideas-in-the-making.

Matthew W. Wilson received his PhD from the Department of Geography at the University of Washington and is currently Assistant Professor of Geography at Ball State University. His research is situated across political, feminist and urban geography as well as science and technoculture studies, interfacing these with the more specified field of 'critical geographic information systems'. He is interested in how geographic information technologies enable particular neighborhood assessment endeavors and how these kinds of geocoding activities mobilize notions of 'quality-of-life' and 'sustainability'. His research concentrates at the intersections of several phenomena, namely the energies with which nonprofit and community organizations approach neighborhood quality-of-life issues, the increased role that geographic information technologies have in addressing this kind of indicator work, as well as the increased geocoding of city spaces more generally.

REFERENCES

Bakker, Karen, and Gavin Bridge. 2006. Material worlds? Resource geographies and the 'matter of nature'. *Progress in Human Geography* 30, no. 1: 5–27.

Balsamo, Anne Marie. 1996. *Technologies of the gendered body: Reading cyborg women*. Durham, NC: Duke University Press.

Barnes, Trevor J. 1998. A history of regression: Actors, networks, machines, and numbers. *Environment and Planning A* 30: 203–23.

Boyer, Kate, and Kim England. 2008. Gender, work and technology in the information workplace: From typewriters to ATMs. *Social and Cultural Geography* 9, no. 3: 241–56.

Braun, Bruce. 2004. Querying posthumanisms. *Geoforum* 35: 269–73.

–. 2005. Environmental issues: Writing a more-than-human urban geography. *Progress in Human Geography* 29, no. 5: 635–50.

Castree, Noel. 2003. Environmental issues: Relational ontologies and hybrid politics. *Progress in Human Geography* 27, no. 2: 203–11.

Cope, Meghan. 2002. Feminist epistemology in geography. In *Feminist geography in practice: Research and methods*, ed. P. Moss, 43–56. Oxford, UK: Blackwell Publishers.

Demeritt, David. 2002. What is the 'social construction of nature'? A typology and sympathetic critique. *Progress in Human Geography* 26, no. 6: 767–90.

England, Kim. 1994. Getting personal: Reflexivity, positionality and feminist research. *The Professional Geographer* 46, no. 1: 80–9.

Foster, Thomas. 2005. The souls of cyberfolk: Posthumanism as vernacular theory, Electronic mediations; v. 13. Minneapolis: University of Minnesota Press.

Gandy, Matthew. 2005. Cyborg urbanization: Complexity and monstrosity in the contemporary city. *International Journal of Urban and Regional Research* 29, no. 1: 26–49. [|514]

Gane, Nicholas, and Donna Jeanne Haraway. 2006. When we have never been human, what is to be done? Interview with Donna Haraway. *Theory, Culture and Society* 23, nos. 7–8: 135–58.

Gerber, Judith. 1997. Beyond dualism – The social construction of nature and the natural and social construction of human beings. *Progress in Human Geography* 21, no. 1: 1–17.

Gray, Chris Hables. 1995. *The cyborg handbook*. New York: Routledge.

–. 2000. *Cyborg citizen: Politics in the posthuman age*. New York: Routledge.

Gray, Chris Hables, Steven Mentor, and Heidi J. Figueroa-Sarriera. 1995. Cyborgology: Constructing the knowledge of cybernetic organisms. In *The cyborg handbook*, ed. C.H. Gray, 1–14. New York: Routledge.

Halberstam, Judith, and Ira Livingston, eds. 1995. *Posthuman bodies, unnatural acts*. Bloomington: Indiana University Press.

Haraway, Donna Jeanne. 1985. Manifesto for cyborgs: Science, technology, and socialist feminism in the 1980s. *Socialist Review* 80: 65–108.

–. 1991a. *Simians, cyborgs, and women: The reinvention of nature*. New York: Routledge.

–. 1991b. The actors are cyborg, nature is coyote, and the geography is elsewhere: Postscript to 'Cyborgs at large'. In *Technoculture*, ed. C. Penley and A. Ross, 21–6. Minneapolis: University of Minnesota.

–. 1992. The promises of monsters: A regenerative politics for inappropriate/d others. In *Cultural studies*, ed. L. Grossberg, C. Nelson, and P.A. Treichler, 295–337. New York: Routledge.

–. 1997. *Modest_Witness@Second_Millennium. FemaleManq_Meets_OncoMousee: Feminism and technoscience*. New York: Routledge.

–. 1999. Knowledges and the question of alliances. In *Knowledges and the question of alliances: A conversation with Nancy Hartsock, Donna Haraway, and David Harvey*. Seattle: University of Washington.

–. 2000. *How like a leaf: An interview with Thyrza Nichols Goodeve*. New York: Routledge.

–. 2003a. *The companion species manifesto: Dogs, people, and significant otherness*. Chicago: Prickly Paradigm Press.

–. 2003b. Interview with Donna Haraway. Participants: Randi Markussen, Finn Olesen, and Nina Lykke. In *Chasing technoscience: Matrix for materiality*, ed. D. Ihde and E. Selinger, 47–57. Bloomington: Indiana University Press.

Harvey, David, and Donna Jeanne Haraway. 1995. Nature, politics, and possibilities: A debate and discussion with David Harvey and Donna Haraway. *Environment and Planning D: Society and Space* 13: 507–27.

Jons, Heike. 2006. Dynamic hybrids and the geographies of technoscience: Discussing conceptual resources beyond the human/non-human binary. *Social and Cultural Geography* 7, no. 4: 559–80.

Katz, Cindi. 1994. Playing the field: Questions of fieldwork in geography. *The Professional Geographer* 46, no. 1: 67–72.

Kitchin, Robert M. 1998. Towards geographies of cyberspace. *Progress in Human Geography* 22, no. 3: 385–406.

Kitchin, Robert M., and James Kneale. 2001. Science fiction or future fact? Exploring imaginative geographies of the new millennium. *Progress in Human Geography* 25, no. 1: 19–35.

Kwan, Mei-Po. 2002. Is GIS for women? Reflections on the critical discourse in the 1990s. *Gender, Place and Culture* 9, no. 3: 271–9.

Lawson, Victoria. 1995. The politics of difference: Examining the quantitative/qualitative dualism in post-structuralist feminist research. *The Professional Geographer* 47, no. 4: 449–57.

Massey, Doreen. 2005. *For space*. London and Thousand Oaks, CA: Sage.

McAfee, Kathleen. 2008. Beyond techno-science: Transgenic maize in the fight over Mexico's future. *Geoforum* 39: 148–60.

McDowell, Linda. 1992. Doing gender: Feminism, feminists and research methods in human geography. *Transactions of the IBG* 17, no. 4: 399–416.

Penley, Constance, and Andrew Ross. 1991. Cyborgs at large: Interview with Donna Haraway. In *Technoculture*, ed. C. Penley and A. Ross, 1–20. Minneapolis: University of Minnesota Press.

Philo, Chris, and Chris Wilbert, eds. 2000. *Animal spaces, beastly places: New geographies of human-animal relations, Critical geographies*. London: Routledge. [|515]

Powell, Richard C. 2007. Geographies of science: Histories, localities, practices, futures. *Progress in Human Geography* 31, no. 3: 309–29.

Prins, Baukje. 1995. The ethics of hybrid subjects: Feminist constructivism according to Donna Haraway. *Science, Technology and Human Values* 20, no. 3: 352–67.

Rose, Gillian. 1997. Situating knowledges: Positionality, reflexivities and other tactics. *Progress in Human Geography* 21, no. 3: 305–20.

Schuurman, Nadine. 2002. Women and technology in geography: A cyborg manifesto for GIS. *The Canadian Geographer* 46, no. 3: 258–65.

–. 2004. Databases and bodies: A cyborg update. *Environment and Planning A* 36: 1337–40.

Shields, Rob. 2006. Flânerie for cyborgs. *Theory, Culture and Society* 23, nos. 7–8: 209–20.

Stone, Allucquère Rosanne. 1995. *The war of desire and technology at the close of the mechanical age*. Cambridge, MA: MIT Press.

Swyngedouw, Erik. 1996. The city as a hybrid: On nature, society, and cyborg urbanisation. *Capitalism, Nature, Socialism* 7, no. 25: 65–80.

–. 1999. Modernity and hybridity: Nature, regeneracionismo, and the production of the Spanish waterscape, 1890–1930. *Annals of the Association of American Geographers* 89, no. 3: 443–65.

–. 2006. Circulations and metabolism: (Hybrid) natures and (cyborg) cities. *Science as Culture* 15, no. 2: 105–21.

Swyngedouw, Erik, and Maria Kaika. 2000. Producing geography: Nature, society, and cyborg geographies. In *Launching Greek geography on the eastern EU border*, ed. L. Leontidou, 53–74. Mytiilin, Greece: University of the Aegean.

Wajcman, Judy. 2004. *TechnoFeminism*. Cambridge: Polity.

Whatmore, Sarah. 1997. Dissecting the autonomous self: Hybrid cartographies for a relational eth-
    ics. *Environment and Planning D: Society and Space* 15: 37–53.
–. 2002. *Hybrid geographies: Natures, cultures, spaces*. London and Thousand Oaks, CA: Sage.
–. 2004. Humanism's excess: Some thoughts on the 'post-human/ist' agenda. *Environment and
    Planning A* 36: 1360–3.
Wolch, Jennifer R., and Jody Emel, eds. 1998. *Animal geographies: Place, politics, and identity in
    the nature-culture borderlands*. London: Verso.

Abstract Translation
Geografías *cyborg*: acercándose a las epistemologías híbridas
Como una forma de crítica, el *cyborg* es muchas veces separado de su rol como figuración. Este
artículo da una reseña de la teoría del *cyborg* de Donna Haraway para reafirmar la importancia
del *cyborg como una figuración* en la metodología crítica. La figuración se trata de abrir las
prácticas de producción de conocimiento a la interrogación. Sostengo que el *cyborg* facilita
este interrogatorio a través de la hibridización espistemológica. Para hacerlo, las figuraciones
cyborg no sólo asumen un lenguaje de ser o volverse, sino que narran este lenguaje en la
producción de saberes, para saber híbridamente. La hibridización epistemológica del *cyborg*
incluye cuatro estrategias: presenciar, situar, difractar y adquirir. Estas son formas de saber en
geografías *cyborg*. Para subrayar la importancia de este uso de la teoría del *cyborg*, hago una
revisión de una selección de literaturas geográficas sobre naturaleza-culturas y las tecnocien-
cias para demostrar cómo los geógrafos citan al *cyborg*. Mi análisis sugiere que estas literaturas
remarcan una hibridez ontológica que deja poco estudiada la hibridización epistemológica in-
volucrada en la figuración *cyborg*. Entender el *cyborg* de esta forma significa poner en riesgo
nuestros relatos, rehacer estas geografías como trabajo político, híbrido.
Palabras clave: metodología crítica; figuración *cyborg*; hibridez; naturaleza-cultura; tecno-
ciencia [|516]

半机械人的地理：走向混杂的认识论

半机械人作为一个批判的模式往往与它负起的比喻角色有所分隔。这篇文章综述了唐娜.哈拉维的半机械
人理论来重申半机械人在批判方法学作为比喻的重要性。比喻是为了质问种种营造知识的实践。我认为
，该半机械人能够通过认识论的混杂来促使这个调查。要做到这一点，半机械人的比喻不止要采用成为
或即将成为的言语来描述，该语言也应讲述知识的生产，以便得到混杂的认知。半机械人的认识论的混
杂包括了目睹，情境，衍射和获取四个战略。这些是半机械人的地理中认知模式。为了强调半机械人理
论在该运用方面的重要性，我审查了特定关于自然的文化以及技术科学的地理研究来展示地理学家如何
引用半机械人。我的分析表明，这些研究强调了本体论的混杂，因而忽略了认识论的混杂在半机械人的
比喻中所扮演的重要角色。以这种方式来对待半机械人，我们把论述摆在危险的处境，并重新塑造半机
械人的地理为混杂式，政治化的工作。

关键词：批判方法学，半机械人的比喻　混杂　自然的文化，技术科学

THINKING CRITICALLY ABOUT
AND RESEARCHING ALGORITHMS

Rob Kitchin

Abstract: More and more aspects of our everyday lives are being mediated, augmented, produced and regulated by software-enabled technologies. Software is fundamentally composed of algorithms: sets of defined steps structured to process instructions/data to produce an output. This paper synthesises and extends emerging critical thinking about algorithms and considers how best to research them in practice. Four main arguments are developed. First, there is a pressing need to focus critical and empirical attention on algorithms and the work that they do given their increasing importance in shaping social and economic life. Second, algorithms can be conceived in a number of ways – technically, computationally, mathematically, politically, culturally, economically, contextually, materially, philosophically, ethically – but are best understood as being contingent, ontogenetic and performative in nature, and embedded in wider socio-technical assemblages. Third, there are three main challenges that hinder research about algorithms (gaining access to their formulation; they are heterogeneous and embedded in wider systems; their work unfolds contextually and contingently), which require practical and epistemological attention. Fourth, the constitution and work of algorithms can be empirically studied in a number of ways, each of which has strengths and weaknesses that need to be systematically evaluated. Six methodological approaches designed to produce insights into the nature and work of algorithms are critically appraised. It is contended that these methods are best used in combination in order to help overcome epistemological and practical challenges.
Keywords: Algorithm; code; epistemology; methodology; research

INTRODUCTION: WHY STUDY ALGORITHMS?

The era of ubiquitous computing and big data is now firmly established, with more and more aspects of our everyday lives – play, consumption, work, travel, communication, domestic tasks, security, etc. – being mediated, augmented, produced and regulated by digital devices and networked systems powered by software (Greenfield, 2006; Kitchin & Dodge, 2011; Manovich, 2013; Steiner, 2012). Software is fundamentally composed of algorithms – sets of defined steps structured to process instructions/data to produce an output – with all digital technologies thus constituting 'algorithm machines' (Gillespie, 2014a). These 'algorithm machines' enable extensive and complex tasks to be tackled that would be all but impossible by hand or analogue machines. They can perform millions [|15] of operations per second; minimise human error and bias in how a task is performed; and can significantly reduce costs and increase turnover and profit through automation and creating new services/products (Kitchin & Dodge, 2011). As such, dozens of key sets of algorithms are shaping everyday practices and tasks, including those that perform search, secure encrypted exchange, recommendation, pattern recognition, data compression, auto-correction, routing, predicting, profiling, simulation and optimisation (MacCormick, 2013).

As Diakopoulos (2013, p. 2) argues: 'We're living in a world now where algorithms adjudicate more and more consequential decisions in our lives. ... Algorithms, driven by vast troves of data, are the new power brokers in society.' Steiner (2012, p. 214) thus contends:

> algorithms already have control of your money market funds, your stocks, and your retirement accounts. They'll soon decide who you talk to on phone calls; they will control the music that reaches your radio; they will decide your chances of getting lifesaving organs transplant; and for millions of people, algorithms will make perhaps the largest decision of in their life: choosing a spouse.

Similarly, Lenglet (2011), MacKenzie (2014), Arnoldi (2016), Pasquale (2015) document how algorithms have deeply and pervasively restructured how all aspects of the finance sector operate, from how funds are traded to how credit agencies assess risk and sort customers. Amoore (2006, 2009) details how algorithms are used to assess security risks in the 'war on terror' through the profiling passengers and citizens. With respect to the creation of Wikipedia, Geiger (2014, p. 345) notes how algorithms 'help create new articles, edit existing articles, enforce rules and standards, patrol for spam and vandalism, and generally work to support encyclopaedic or administrative work.' Likewise, Anderson (2011) details how algorithms are playing an increasingly important role in producing content and mediating the relationships between journalists, audiences, newsrooms and media products.

In whatever domain algorithms are deployed they appear to be having disruptive and transformative effect, both to how that domain is organised and operates, and to the labour market associated with it. Steiner (2012) provides numerous examples of how algorithms and computation have led to widespread job losses in some industries through automation. He concludes

> programmers now scout new industries for soft spots where algorithms might render old paradigms extinct, and in the process make mountains of money ... Determining the next field to be invaded by bots [automated algorithms] is the sum of two simple functions: the potential to disrupt plus the reward for disruption. (Steiner, 2012, pp. 6, 119)

Such conclusions have led a number of commentators to argue that we are now entering an era of widespread algorithmic governance, wherein algorithms will play an everincreasing role in the exercise of power, a means through which to automate the disciplining and controlling of societies and to increase the efficiency of capital accumulation. However, Diakopoulos (2013, p. 2, original emphasis) warns that: 'What we generally lack as a public is *clarity about how algorithms exercise their power over us.*' Such clarity is absent because although algorithms are imbued with the power to act upon data and make consequential decisions (such as to issue fines or block travel or approve a loan) they are largely black boxed and beyond query or question. What is at stake [|16] then with the rise of 'algorithm machines' is new forms of algorithmic power that are reshaping how social and economic systems work.

In response, over the past decade or so, a growing number of scholars have started to focus critical attention on software code and algorithms, drawing on and contributing to science and technology studies, new media studies and software studies, in order to unpack the nature of algorithms and their power and work. Their analyses typically take one of three forms: a detailed case study of a single algorithm, or class of algorithms, to examine the nature of algorithms more generally (e. g., Bucher, 2012; Geiger, 2014; Mackenzie, 2007; Montfort et al., 2012); a detailed examination of the use of algorithms in one domain, such as journalism (Anderson, 2011), security (Amoore, 2006, 2009) or finance (Pasquale, 2014, 2015); or a more general, critical account of algorithms, their nature and how they perform work (e. g., Cox, 2013; Gillespie, 2014a, 2014b; Seaver, 2013).

This paper synthesises, critiques and extends these studies. Divided into two main sections – thinking critically about and researching algorithms – the paper makes four key arguments. First, as already noted, there is a pressing need to focus critical and empirical attention on algorithms and the work that they do in the world. Second, it is most productive to conceive of algorithms as being contingent, ontogenetic, performative in nature and embedded in wider socio-technical assemblages. Third, there are three main challenges that hinder research about algorithms (gaining access to their formulation; they are heterogeneous and embedded in wider systems; their work unfolds contextually and contingently), which require practical and epistemological attention. Fourth, the constitution and work of algorithms can be empirically studied in a number of ways, each of which has strengths and weaknesses that need to be systematically evaluated. With respect to the latter, the paper provides a critical appraisal of six methodological approaches that might profitably be used to produce insights into the nature and work of algorithms.

THINKING CRITICALLY ABOUT ALGORITHMS

While an algorithm is commonly understood as a set of defined steps to produce particular outputs it is important to note that this is somewhat of a simplification. What constitutes an algorithm has changed over time and they can be thought about in a number of ways: technically, computationally, mathematically, politically, culturally, economically, contextually, materially, philosophically, ethically and so on.

Miyazaki (2012) traces the term 'algorithm' to twelfth-century Spain when the scripts of the Arabian mathematician Muḥammad ibn Mūsā al-Khwārizmī were translated into Latin. These scripts describe methods of addition, subtraction, multiplication and division using numbers. Thereafter, 'algorism' meant 'the specific step-by-step method of performing written elementary arithmetic' (Miyazaki, 2012, p. 2) and 'came to describe any method of systematic or automatic calculation' (Steiner, 2012, p. 55). By the mid-twentieth century and the development of scientific computation and early high level programming languages, such as Algol 58 and its derivatives (short for ALGOrithmic Language), an algorithm was understood to be a set of defined steps that if followed in the correct order will computationally process input (instructions and/or data) to produce a desired outcome (Miyazaki, 2012).

From a computational and programming perspective an 'Algorithm = Logic + Control'; where the logic is the problem domain-specific component and specifies the abstract [|17] formulation and expression of a solution (what is to be done) and the control component is the problem-solving strategy and the instructions for processing the logic under different scenarios (how it should be done) (Kowalski, 1979). The efficiency of an algorithm can be enhanced by either refining the logic component or by improving the control over its use, including altering data structures (input) to improve efficiency (Kowalski, 1979). As reasoned logic, the formulation of an algorithm is, in theory at least, independent of programming languages and the machines that execute them; 'it has an autonomous existence independent of "implementation details"' (Goffey, 2008, p. 15).

Some ideas explicitly take the form of an algorithm. Mathematical formulae, for example, are expressed as precise algorithms in the form of equations. In other cases problems have to be abstracted and structured into a set of instructions (pseudo-code) which can then be coded (Goffey, 2008). A computer programme structures lots of relatively simple algorithms together to form large, often complex, recursive decision trees (Neyland, 2015; Steiner, 2012). The methods of guiding and calculating decisions are largely based on Boolean logic (e. g., if this, then that) and the mathematical formulae and equations of calculus, graph theory and probability theory. Coding thus consists of two key translation challenges centred on producing algorithms. The first is translating a task or problem into a structured formula with an appropriate rule set (pseudo-code). The second is translating this pseudo-code into source code that when compiled will perform the task or solve the problem. Both translations can be challenging, requiring the precise definition of what a task/problem is (logic), then breaking that down into a precise set of instructions, factoring in any contingencies such as how the algorithm should perform under different conditions (control). The consequences of mistranslating the problem and/or solution are erroneous outcomes and random uncertainties (Drucker, 2013).

The processes of translation are often portrayed as technical, benign and commonsensical. This is how algorithms are mostly presented by computer scientists and technology companies: that they are 'purely formal beings of reason' (Goffey, 2008, p. 16). Thus, as Seaver (2013) notes, in computer science texts the focus is centred on how to design an algorithm, determine its efficiency and prove its optimality from a purely technical perspective. If there is discussion of the work algorithms do in real-world contexts this concentrates on how algorithms function in practice to perform a specific task. In other words, algorithms are understood 'to be strictly rational concerns, marrying the certainties of mathematics with the objectivity of technology' (Seaver, 2013, p. 2). 'Other knowledge about algorithms – such as their applications, effects, and circulation – is strictly out of frame' (Seaver, 2013, pp. 1–2). As are the complex set of decision-making processes and practices, and the wider assemblage of systems of thought, finance, politics, legal codes and regulations, materialities and infrastructures, institutions, inter-personal relations, which shape their production (Kitchin, 2014).

Far from being objective, impartial, reliable and legitimate, critical scholars argue that algorithms possess none of these qualities except as carefully crafted fictions (Gillespie, 2014a). As Montfort et al. (2012, p. 3) note, '[c]ode is not purely abstract and mathematical; it has significant social, political, and aesthetic dimensions,' inherently framed and shaped by all kinds of decisions, politics, ideology and the materialities of hardware and infrastructure that enact its instruction. Whilst programmers might seek to maintain a high degree of mechanical objectivity – being distant, detached and impartial in how [|18] they work and thus acting independent of local customs, culture, knowledge and context (Porter, 1995) – in the process of translating a task or process or calculation into an algorithm they can never fully escape these. Nor can they escape factors such as available resources and the choice and quality of training data; requirements relating to standards, protocols and the law; and choices and conditionalities relating to hardware, platforms, bandwidth and languages (Diakopoulos, 2013; Drucker, 2013; Kitchin & Dodge, 2011; Neyland, 2015). In reality then, a great deal of expertise, judgement, choice and constraints are exercised in producing algorithms (Gillespie, 2014a). Moreover, algorithms are created for purposes that are often far from neutral: to create value and capital; to nudge behaviour and structure preferences in a certain way; and to identify, sort and classify people.

At the same time, 'programming is ... a live process of engagement between thinking with and working on materials and the problem space that emerges' (Fuller, 2008, p. 10) and it 'is not a dry technical exercise but an exploration of aesthetic, material, and formal qualities' (Montfort et al., 2012, p. 266). In other words, creating an algorithm unfolds in context through processes such as trial and error, play, collaboration, discussion and negotiation. They are ontogenetic in nature (always in a state of becoming), teased into being: edited, revised, deleted and restarted, shared with others, passing through multiple iterations stretched out over time and space (Kitchin & Dodge, 2011). As a result, they are always somewhat uncertain, provisional and messy fragile accomplishments (Gillespie, 2014a; Neyland, 2015). And such practices are complemented by many others, such as researching the concept, selecting and cleaning data, tuning parameters, selling the idea and product, building coding teams, raising finance and so on. These practices are framed by systems of thought and forms of knowledge, modes of political economy, organisational and institutional cultures and politics, governmentalities and legalities, subjectivities and communities. As Seaver (2013, p. 10) notes, 'algorithmic systems are not standalone little boxes, but massive, networked ones with hundreds of hands reaching into them, tweaking and tuning, swapping out parts and experimenting with new arrangements.'

Creating algorithms thus sits at the 'intersection of dozens of ... social and material practices' that are culturally, historically and institutionally situated (Montfort et al., 2012, p. 262; Napoli, 2013; Takhteyev, 2012). As such, as Mackenzie (2007, p. 93) argues treating algorithms simply 'as a general expression of mental effort, or, perhaps even more abstractly, as process of abstraction, is to lose track of proximities and relationalities that algorithms articulate.' Algorithms cannot be divorced from the conditions under which they are developed and deployed (Geiger, 2014). What this means is that algorithms need to be understood as relational, contingent, contextual in nature, framed within the wider context of their socio-technical

assemblage. From this perspective, 'algorithm' is one element in a broader apparatus which means it can never be understood as a technical, objective, impartial form
of knowledge or mode of operation.

Beyond thinking critically about the nature of algorithms, there is also a need
to consider their work, effects and power. Just as algorithms are not neutral, impartial expressions of knowledge, their work is not impassive and apolitical. Algorithms search, collate, sort, categorise, group, match, analyse, profile, model, simulate, visualise and regulate people, processes and places. They shape how we understand the world and they do work in and make the world through their execution as
software, with profound consequences (Kitchin & Dodge, 2011). In this sense, they
are profoundly performative [|19] as they cause things to happen (Mackenzie &
Vurdubakis, 2011). And while the creators of these algorithms might argue that they
'replace, displace, or reduce the role of biased or self-serving intermediaries' and
remove subjectivity from decision-making, computation often deepens and acceler-
ates processes of sorting, classifying and differentially treating, and reifying traditional pathologies, rather than reforming them (Pasquale, 2014, p. 5).

Far from being neutral in nature, algorithms construct and implement regimes
of power and knowledge (Kushner, 2013) and their use has normative implications
(Anderson, 2011). Algorithms are used to seduce, coerce, discipline, regulate and
control: to guide and reshape how people, animals and objects interact with and
pass through various systems. This is the same for systems designed to empower,
entertain and enlighten, as they are also predicated on defined rule-sets about how
a system behaves at any one time and situation. Algorithms thus claim and express
algorithmic authority (Shirky, 2009) or algorithmic governance (Beer, 2009; Musiani, 2013), often through what Dodge and Kitchin (2007) term 'automated management' (decision-making processes that are automated, automatic and autonomous; outside of human oversight). The consequence for Lash (2007) is that society
now has a new rule set to live by to complement constitutive and regulative rules:
algorithmic, generative rules. He explains that such rules are embedded within
computation, an expression of 'power through the algorithm'; they are 'virtuals that
generate a whole variety of actuals. They are compressed and hidden and we do not
encounter them in the way that we encounter constitutive and regulative rules. ...
They are ... pathways through which capitalist power works' (Lash, 2007, p. 71).

It should be noted, however, that the effects of algorithms or their power is not
always linear or always predictable for three reasons. First, algorithms act as part
of a wider network of relations which mediate and refract their work, for example,
poor input data will lead to weak outcomes (Goffey, 2008; Pasquale, 2014). Second, the performance of algorithms can have side effects and unintended consequences, and left unattended or unsupervised they can perform unanticipated acts
(Steiner, 2012). Third, algorithms can have biases or make mistakes due to bugs
or miscoding (Diakopoulos, 2013; Drucker, 2013). Moreover, once computation
is made public it undergoes a process of domestication, with users embedding the
technology in their lives in all kinds of alternative ways and using it for different means, or resisting, subverting and reworking the algorithms' intent (consider
the ways in which users try to game Google's PageRank algorithm). In this sense,

algorithms are not just what programmers create, or the effects they create based on certain input, they are also what users make of them on a daily basis (Gillespie, 2014a).

Steiner's (2012, p. 218) solution to living with the power of algorithms is to suggest that we '[g]et friendly with bots.' He argues that the way to thrive in the algorithmic future is to learn to 'build, maintain, and improve upon code and algorithms,' as if knowing how to produce algorithms protects oneself from their diverse and pernicious effects across multiple domains. Instead, I would argue, there is a need to focus more critical attention on the production, deployment and effects of algorithms in order to understand and contest the various ways that they can overtly and covertly shape life chances. However, such a programme of research is not as straightforward as one might hope, as the next section details. [|20]

RESEARCHING ALGORITHMS

The logical way to flesh out our understanding of algorithms and the work they do in the world is to conduct detailed empirical research centrally focused on algorithms. Such research could approach algorithms from a number of perspectives:

> a technical approach that studies algorithms as computer science; a sociological approach that studies algorithms as the product of interactions among programmers and designers; a legal approach that studies algorithms as a figure and agent in law; a philosophical approach that studies the ethics of algorithms, (Barocas, Hood, & Ziewitz, 2013, p. 3)

and a code/software studies' perspective that studies the politics and power embedded in algorithms, their framing within a wider socio-technical assemblage and how they reshape particular domains. There are a number of methodological approaches that can be used to operationalise such research, six of which are critically appraised below. Before doing so, however, it is important to acknowledge that there are three significant challenges to researching algorithms that require epistemological and practical attention.

CHALLENGES

Access/black boxed

Many of the most important algorithms that people encounter on a regular basis and which (re)shape how they perform tasks or the services they receive are created in environments that are not open to scrutiny and their source code is hidden inside impenetrable executable files. Coding often happens in private settings, such as within companies or state agencies, and it can be difficult to negotiate access to coding teams to observe them work, interview programmers or analyse the source code they produce. This is unsurprising since it is often a company's algorithms that provide it with a competitive advantage and they are reluctant to expose their intellectual property even with non-disclosure agreements in place. They also want to limit the ability of users to game the algorithm to unfairly gain a competitive edge. Access is a little easier in the case of open-source programming teams and open-source programmes through repositories such as Github, but while they provide access to much code, this is limited in scope and does not include key proprietary algorithms that might be of more interest with respect to holding forms of algorithmic governance to account.

Heterogeneous and embedded

If access is gained, algorithms, as Seaver (2013) notes, are rarely straightforward to deconstruct. Within code algorithms are usually woven together with hundreds of other algorithms to create algorithmic systems. It is the workings of these algorithmic systems that we are mostly interested in, not specific algorithms, many of which are quite benign and procedural. Algorithmic systems are most often 'works of collective authorship, made, maintained, and revised by many people with different goals at different times' (Seaver, 2013, p. 10). They can consist of original formulations mashed together with those sourced from code libraries, including stock algorithms that are re-used in multiple instances. Moreover, they are embedded within complex socio-technical assemblages made up of a heterogeneous set of relations including potentially thousands of individuals, data sets, objects, apparatus, elements, protocols, standards, laws, etc. that frame their development. [|21] Their construction, therefore, is often quite messy, full of 'flux, revisability, and negotiation' (p. 10), making unpacking the logic and rationality behind their formulation difficult in practice. Indeed, it is unlikely that any one programmer has a complete understanding of a system, especially large, complex ones that are built by many teams of programmers, some of whom may be distributed all over the planet or may have only had sight of smaller outsourced segments. Getting access to a credit rating agency's algorithmic system then might give an insight into its formula for assessing and sorting individuals, its underlying logics and principles, and how it was created and works in practice, but will not necessarily provide full transparency as to its full reasoning, workings or the choices made in its construction (Bucher, 2012; Chun, 2011).

Ontogenetic, performative and contingent

As well as being heterogeneous and embedded, algorithms are rarely fixed in form and their work in practice unfolds in multifarious ways. As such, algorithms need to be recognised as being ontogenetic, performative and contingent: that is, they are never fixed in nature, but are emergent and constantly unfolding. In cases where an algorithm is static, for example, in firmware that is not patched, its work unfolds contextually, reactive to input, interaction and situation. In other cases, algorithms and their instantiation in code are often being refined, reworked, extended and patched, iterating through various versions (Miyazaki, 2012). Companies such as Google and Facebook might be live running dozens of different versions of an algorithm to assess their relative merits, with no guarantee that the version a user interacts with at one moment in time is the same as five seconds later. In some cases, the code has been programmed to evolve, re-writing its algorithms as it observes, experiments and learns independently of its creators (Steiner, 2012).

Similarly, many algorithms are designed to be reactive and mutable to inputs. As Bucher (2012) notes, Facebook's EdgeRank algorithm (that determines what posts and in what order are fed into each users' timeline) does not act from above in a static, fixed manner, but rather works in concert with the each individual user, ordering posts dependent on how one interacts with 'friends.' Its parameters then are contextually weighted and fluid. In other cases, randomness might be built into an algorithm's design meaning its outcomes can never be perfectly predicted. What this means is that the outcomes for users inputting the same data might vary for contextual reasons (e. g., Mahnke and Uprichard (2014) examined Google's auto-complete search algorithm by typing in the same terms from two locations and comparing the results, finding differences in the suggestions the algorithm gave), and the same algorithms might be being used in quite varied and mutable ways (e. g., for work or for play). Examining one version of an algorithm will then provide a snapshot reading that fails to acknowledge or account for the mutable and often multiple natures of algorithms and their work (Bucher, 2012).

Algorithms then are often 'out of control' in the sense that their outcomes are sometimes not easily anticipated, producing unexpected effects in terms of their work in the world (Mackenzie, 2005). As such, understanding the work and effects of algorithms needs to be sensitive to their contextual, contingent unfolding across situation, time and space. What this means in practice is that single or limited engagements with algorithms cannot be simply extrapolated to all cases and that a set of comparative case studies [|22] need to be employed, or a series of experiments performed with the same algorithm operating under different conditions.

APPROACHES

Keeping in mind these challenges, this final section critically appraises six methodological approaches for researching algorithms that I believe present the most promise for shedding light on the nature and workings of algorithms, their embedding in socio-technical systems, their effects and power, and dealing with and overcoming the difficulties of gaining access to source code. Each approach has its strengths and drawbacks and their use is not mutually exclusive. Indeed, I would argue that there would be much to be gained by using two or more of the approaches in combination to compensate for the drawbacks of employing them in isolation. Nor are they the only possible approaches, with ethnomethodologies, surveys and historical analysis using archives and oral histories offering other possible avenues of analysis and insight.

Examining pseudo-code/source code

Perhaps the most obvious way to try and understand an algorithm is to examine its pseudo-code (how a task or puzzle is translated into a model or recipe) and/or its construction in source code. There are three ways in which this can be undertaken in practice. The first is to carefully deconstruct the pseudo-code and/or source code, teasing apart the rule set to determine how the algorithm works to translate input to produce an outcome (Krysa & Sedek, 2008). In practice this means carefully sifting through documentation, code and programmer comments, tracing out how the algorithm works to process data and calculate outcomes, and decoding the translation process undertaken to construct the algorithm. The second is to map out a genealogy of how an algorithm mutates and evolves over time as it is tweaked and rewritten across different versions of code. For example, one might deconstruct how an algorithm is re-scripted in multiple instantiations of a programme within a code library such as github. Such a genealogy would reveal how thinking with respect to a problem is refined and transformed with respect to how the algorithm/code performs 'in the wild' and in relation to new technologies, situations and contexts (such as new platforms or regulations being introduced). The third is to examine how the same task is translated into various software languages and how it runs across different platforms. This is an approach used by Montfort et al. (2012) in their exploration of the '10 PRINT' algorithm, where they scripted code to perform the same task in multiple languages and ran it on different hardware, and also tweaked the parameters, to observe the specific contingencies and affordances this introduced.

While these methods do offer the promise of providing valuable insights into the ways in which algorithms are built, how power is vested in them through their various parameters and rules, and how they process data in abstract and material terms to complete a task, there are three significant issues with their deployment. First, as noted by Chandra (2013), deconstructing and tracing how an algorithm is constructed in code and mutates over time is not straightforward. Code often takes the form of a 'Big Ball of Mud': '[a] haphazardly structured, sprawling, sloppy, duct-tape and bailing wire, spaghetti code jungle' (Foote & Yoder, 1997; cited in Chandra, 2013, p. 126). Even those that have produced it can find it very difficult to unpack its algorithms and routines; those unfamiliar with its [|23] development can often find that the ball of mud remains just that. Second, it requires that the researcher is both an expert in the domain to which the algorithm refers and possesses sufficient skill and knowledge as a programmer that they can make sense of a 'Big Ball of Mud'; a pairing that few social scientists and humanities scholars possess. Third, these approaches largely decontextualise the algorithm from its wider socio-technical assemblage and its use.

Reflexively producing code

A related approach is to conduct auto-ethnographies of translating tasks into pseudo-code and the practices of producing algorithms in code. Here, rather than studying an algorithm created by others, a researcher reflects on and critically interrogates their own experiences of translating and formulating an algorithm. This would include an analysis of not only the practices of exploring and translating a task, originating and developing ideas, writing and revising code, but also how these practices are situated within and shaped by wider socio-technical factors such as regulatory and legal frameworks, form of knowledge, institutional arrangements, financial terms and conditions, and anticipated users and market. Ziewitz (2011) employed this kind of approach to reflect on producing a random routing algorithm for directing a walking path through a city, reflecting on the ontological uncertainty in the task itself (that there is often an ontological gerrymandering effect at work as the task itself is re-thought and re-defined while the process of producing an algorithm is undertaken), and the messy, contingent process of creating the rule set and parameters in practice and how these also kept shifting through deferred accountability. Similarly, Ullman (1997) uses such an approach to consider the practices of developing software and how this changed over her career.

While this approach will provide useful insights into how algorithms are created, it also has a couple of limitations. The first is the inherent subjectivities involved in doing an auto-ethnography and the difficulties of detaching oneself and gaining critical distance to be able to give clear insight into what is unfolding. Moreover, there is the possibility that in seeking to be reflexive what would usually take place is inflected in unknown ways. Further, it excludes any non-representational, unconscious acts from analysis. Second, one generally wants to study algorithms and code that have real concrete effects on peoples' everyday lives, such as those used in algorithmic governance. One way to try and achieve this is to contribute to open-source projects where the code is incorporated into products that others use, or to seek access to a commercial project as a programmer (on an overt, approved basis with non-disclosure agreements in place). The benefit here is that the method can be complemented with the sixth approach set out below, examining and reflecting on the relationship between the production of an algorithm and any associated ambitions and expectations vis-à-vis how it actually does work in the world.

Reverse engineering

In cases where the code remains black boxed, a researcher interested in the algorithm at the heart of its workings is left with the option of trying to reverse engineer the compiled software. Diakopoulos (2013, p. 13) explains that '[r]everse engineering is the process of articulating the specifications of a system through a rigorous examination drawing on domain knowledge, observation, and deduction to unearth a model of how that system [|24] works.' While software producers might desire their products to remain opaque, each programme inherently has two openings that enable lines of enquiry: input and output. By examining what data are fed into an algorithm and what output is produced it is possible to start to reverse engineer how the recipe of the algorithm is composed (how it weights and preferences some criteria) and what it does.

The main way this is attempted is by using carefully selected dummy data and seeing what is outputted under different scenarios. For example, researchers might search Google using the same terms on multiple computers in multiple jurisdictions to get a sense of how its PageRank algorithm is constructed and works in practice (Mahnke & Uprichard, 2014), or they might experiment with posting and interacting with posts on Facebook to try and determine how its EdgeRank algorithm positions and prioritises posts in user time lines (Bucher, 2012), or they might use proxy servers and feed dummy user profiles into e-commerce systems to see how prices might vary across users and locales (*Wall Street Journal*, detailed in Diakopoulos, 2013). One can also get a sense of an algorithm by 'looking closely at how information must be oriented to face them, how it is made algorithm-ready'; how the input data are delineated in terms of what input variables are sought and structured, and the associated meta-data (Gillespie, 2014a). Another possibility is to follow debates on online forums by users about how they perceive an algorithm works or has changed, or interview marketers, media strategists, and public relations firms that seek to game an algorithm to optimise an outcome for a client (Bucher, 2012).

While reverse engineering can give some indication of the factors and conditions embedded into an algorithm, they generally cannot do so with any specificity (Seaver, 2013). As such, they usually only provide fuzzy glimpses of how an algorithm works in practice but not its actual constitution (Diakopoulos, 2013). One solution to try and enhance clarity has been to employ bots, which posing as users, can more systematically engage with a system, running dummy data and interactions. However, as Seaver (2013) notes, many proprietary systems are aware that many people are seeking to determine and game their algorithm, and thus seek to identify and block bot users.

Interviewing designers or conducting an ethnography of a coding team

While deconstructing or reverse engineering code might provide some insights into the workings of an algorithm, they provide little more than conjecture as to the intent of the algorithm designers, and examining that and how and why an algorithm was produced requires a different approach. Interviewing designers and coders, or conducting an ethnography of a coding team, provides a means of uncovering the story behind the production of an algorithm and to interrogate its purpose and assumptions.

In the first case, respondents are questioned as to how they framed objectives, created pseudo-code and translated this into code, and quizzed about design decisions and choices with respect to languages and technologies, practices, influences, constraints, debates within a team or with clients, institutional politics and major changes in direction over time (Diakopoulos, 2013; MacKenzie, 2014; Mager, 2012).

In the second case, a researcher seeks to spend time within a coding team, either observing the work of the coders, discussing it with them, and attending associated events such as team meetings, or working in situ as part of the team, taking an active role in producing code. An example of the former is Rosenberg's (2007) study of one company's attempt to produce a new product conducted over a three-year period in which he was given full access to the company, including [|25] observing and talking to coders, and having access to team chat rooms and phone conferences. An example of the latter is Takhteyev's (2012) study of an open-source coding project in Rio de Janeiro where he actively worked on developing the code, as well as taking part in the social life of the team. In both cases, Rosenberg and Takhteyev generate much insight into the contingent, relational and contextual way in which algorithms and software are produced, though in neither case are the specificities of algorithms and their work unpacked and detailed.

Unpacking the full socio-technical assemblage of algorithms

As already noted, algorithms are not formulated or do not work in isolation, but form part of a technological stack that includes infrastructure/hardware, code platforms, data and interfaces, and are framed and conditions by forms of knowledge, legalities, governmentalities, institutions, marketplaces, finance and so on. A wider understanding of algorithms then requires their full socio-technical assemblage to be examined, including an analysis of the reasons for subjecting the system to the logic of computation in the first place. Examining algorithms without considering their wider assemblage is, as Geiger (2014) argues, like considering a law without reference to the debate for its introduction, legal institutions, infrastructures such as courts, implementers such as the police, and the operating and business practices of the legal profession. It also risks fetishising the algorithm and code at the expense of the rest of the assemblage (Chun, 2011).

Interviews and ethnographies of coding projects, and the wider institutional apparatus surrounding them (e. g., management and institutional collaboration), start to produce such knowledge, but they need to supplemented with other approaches, such as a discursive analysis of company documents, promotional/industry material, procurement tenders and legal and standards frameworks; attending trade fairs and other inter-company interactions; examining the practices, structures and behaviour of institutions; and documenting the biographies of key actors and the histories of projects (Montfort et al., 2012; Napoli, 2013). Such a discursive analysis will also help to reveal how algorithms are imagined and narrated, illuminate the discourses surrounding and promoting them, and how they are understood by those that create and promote them. Gaining access to such a wider range of elements, and being able to gather data and interlink them to be able to unpack a socio-technical assemblage, is no easy task but it is manageable as a large case study, especially if undertaken by a research team rather than a single individual.

Examining how algorithms do work in the world

Given that algorithms do active work in the world it is important not only to focus on the construction of algorithms, and their production within a wider assemblage, but also to examine how they are deployed within different domains to perform a multitude of tasks. This cannot be simply denoted from an examination of the algorithm/code alone for two reasons. First, what an algorithm is designed to do in theory and what it actually does in practice do not always correspond due to a lack of refinement, miscodings, errors and bugs. Second, algorithms perform in context – in collaboration with data, technologies, people, etc. under varying conditions – and therefore their effects unfold in contingent and relational ways, producing localised and situated outcomes. When users employ an algorithm, say for play or work, they are not simply playing or working in conjunction with the algorithm, rather they are 'learning, internalizing, and becoming intimate with' it [|26] (Galloway, 2006, p. 90); how they behave is subtly reshaped through the engagement, but at the same time what the algorithm does is conditional on the input it receives from the user. We can therefore only know how algorithms make a different to everyday life by observing their work in the world under different conditions.

One way to undertake such research is to conduct ethnographies of how people engage with and are conditioned by algorithmic systems and how such systems reshape how organisations conduct their endeavours and are structured (e. g., Lenglet, 2011). It would also explore the ways in which people resist, subvert and transgress against the work of algorithms, and re-purpose and re-deploy them for purposes they were not originally intended. For example, examining the ways in which various mobile and web applications were re-purposed in the aftermath of the Haiti earthquake to coordinate disaster response, remap the nation and provide donations (Al-Akkad et al., 2013). Such research requires detailed observation and interviews focused on the use of particular systems and technologies by different populations and within different scenarios, and how individuals interfaced with the algorithm through software, including their assessments as to their intentions, sense of what is occurring and associated consequences, tactics of engagement, feelings, concerns and so on. In cases where an algorithm is black boxed, such research is also likely to shed some light on the constitution of the algorithm itself.

CONCLUSION

On an average day, people around the world come into contact with hundreds of algorithms embedded into the software that operates communications, utilities and transport infrastructure, and powers all kinds of digital devices used for work, play and consumption. These algorithms have disruptive and transformative effect, reconfiguring how systems operate, enacting new forms of algorithmic governance and enabling new forms of capital accumulation. Yet, despite their increasing pervasiveness, utility and the power vested in them to act in autonomous, automatic and automated ways, to date there has been limited critical attention paid to algorithms in contrast to the vast literature that approaches algorithms from a more technical perspective. This imbalance in how algorithms are thought about and intellectually engaged with is perhaps somewhat surprising given what is at stake in a computationally rich world. As such, there is a pressing need for critical attention across the social sciences and humanities to be focused on algorithms and forms of algorithmic governance. The contribution of this paper to this endeavour has been to: advance an understanding of algorithms as contingent, ontogenetic, performative in nature and embedded in wider socio-technical assemblages; to detail the epistemological and practical challenges facing algorithm scholars; and to critically appraise six promising methodological options to empirically research and make sense of algorithms. It is apparent from the studies conducted to date that there is a range of different ways of making sense of algorithms and the intention of the paper has not been to foreclose this diversity, but rather to encourage synthesis, comparison and evaluation of different positions and to create new ones. Indeed, the more angles taken to uncover and debate the nature and work of algorithms the better we will come to know them.

Likewise, the six approaches appraised were selected because I believe they hold the most promise in exposing how algorithms are constructed, how they work within socio- [|27] technical assemblages and how they perform actions and make a difference in particular domains, but they are by no means the only approaches that might be profitably pursued. My contention is, given each approach's varying strengths and weaknesses, that how they reveal the nature and work of algorithms needs to be systematically evaluated through methodologically focused research. Studies that have access to the pseudo-code, code and coders may well be the most illuminating, though they still face a number of challenges, such as deciphering how the algorithm works in practice. Moreover, there is a need to assess: (1) how they might be profitably used in conjunction with each other to overcome epistemological and practical challenges; (2) what other methods might be beneficially deployed in order to better understand the nature, production and use of algorithms? With respect to the latter, such methods might include ethnomethodologies, surveys, historical analysis using archives and oral histories, and comparative case studies. As such, while the approaches and foci I have detailed provide a useful starting set that others can apply, critique, refine and extend, there are others that can potentially emerge as critical research and thinking on algorithms develops and matures.

Acknowledgements

Many thanks to Tracey Lauriault, Sung-Yueh Perng and the referees for comments on earlier versions of this paper.

Disclosure statement

No potential conflict of interest was reported by the author.

Funding

The research for this paper was funded by a European Research Council Advanced Investigator award [ERC-2012-AdG-323636-SOFTCITY].

Rob Kitchin is a professor and ERC Advanced Investigator at the National University of Ireland Maynooth. He is currently a principal investigator on the Programmable City project, the Digital Repository of Ireland, the All-Island Research Observatory and the Dublin Dashboard. [email: rob.kitchin@nuim.ie]

REFERENCES

Al-Akkad, A., Ramirez, L., Denef, S., Boden, A., Wood, L., Buscher, M., & Zimmermann, A. (2013). *'Reconstructing normality': The use of infrastructure leftovers in crisis situations as inspiration for the design of resilient technology.* Proceedings of the 25th Australian Computer-Human Interaction Conference: Augmentation, Application, Innovation, Collaboration (pp. 457–466). New York, NY: ACM. Retrieved October 16, 2014, from http://dl.acm.org/citation.cfm?doid= 2541016.2541051

Amoore, L. (2006). Biometric borders: Governing mobilities in the war on terror. *Political Geography, 25*, 336–351. [|28]

Amoore, L. (2009). Algorithmic war: Everyday geographies of the war on terror. *Antipode, 41*, 49–69.

Anderson C. W. (2011). Deliberative, agonistic, and algorithmic audiences: Journalism's vision of its public in an age of audience. *Journal of Communication, 5*, 529–547.

Arnoldi, J. (2016). Computer algorithms, market manipulation and the institutionalization of high frequency trading. *Theory, Culture & Society, 33*(1), 29–52.

Barocas, S., Hood, S., & Ziewitz, M. (2013). Governing algorithms: A provocation piece. Retrieved October 16, 2014, from http://papers.ssrn.com/sol3/papers.cfm?abstract_id=2245322

Beer, D. (2009). Power through the algorithm? Participatory Web cultures and the technological unconscious. *New Media and Society, 11*(6), 985–1002.

Bucher, T. (2012). 'Want to be on the top?' Algorithmic power and the threat of invisibility on Facebook. *New Media and Society, 14*(7), 1164–1180.

Chandra, V. (2013). *Geek sublime: Writing fiction, coding software.* London: Faber.

Chun, W. H. K. (2011). *Programmed visions.* Cambridge: MIT Press.

Cox, G. (2013). *Speaking code: Coding as aesthetic and political expression.* Cambridge: MIT Press.

Diakopoulos, N. (2013). *Algorithmic accountability reporting: On the investigation of black boxes.* A Tow/Knight Brief. Tow Center for Digital Journalism, Columbia Journalism School. Retrieved August 21, 2014, from http://towcenter.org/algorithmic-accountability-2/

Dodge, M., & Kitchin, R. (2007). The automatic management of drivers and driving spaces. *Geoforum, 38*(2), 264–275.

Drucker, J. (2013). Performative materiality and theoretical approaches to interface. *Digital Humanities Quarterly, 7*(1). Retrieved June 5, 2014, from http://www.digitalhumanities.org/dhq/vol/7/1/000143/000143.html

Foote, B., & Yoder, J. (1997). Big Ball of Mud. *Pattern Languages of Program Design, 4*, 654–692.

Fuller, M. (2008). Introduction. In M. Fuller (Ed.), *Software studies – A lexicon* (pp. 1–14). Cambridge: MIT Press.

Galloway, A. R. (2006). *Gaming: Essays on algorithmic culture.* Minneapolis: University of Minnesota Press.

Geiger, S. R. (2014). Bots, bespoke, code and the materiality of software platforms. *Information, Communication & Society*, *17*(3), 342–356.

Gillespie, T. (2014a). The relevance of algorithms. In T. Gillespie, P. J. Boczkowski, & K. A. Foot (Eds.), *Media technologies: Essays on communication, materiality, and society* (pp. 167–193). Cambridge: MIT Press.

Gillespie, T. (2014b, June 25). Algorithm [draft] [#digitalkeyword]. *Culture Digitally*. Retrieved October 16, 2014, from http://culturedigitally.org/2014/06/algorithm-draft-digitalkeyword/

Goffey, A. (2008). Algorithm. In M. Fuller (Ed.), *Software studies – A lexicon* (pp. 15–20). Cambridge: MIT Press.

Greenfield, A. (2006). *Everyware: The dawning age of ubiquitous computing*. Boston, MA: New Riders.

Kitchin, R. (2014). *The data revolution: Big data, open data, data infrastructures and their consequences*. London: Sage.

Kitchin, R., & Dodge, M. (2011). *Code/space: Software and everyday life*. Cambridge: MIT Press.

Kowalski, R. (1979). Algorithm = Logic + Control. *Communications of the ACM*, *22*(7), 424–436.

Krysa, J., & Sedek, G. (2008). Source code. In M. Fuller (Ed.), *Software studies – A lexicon* (pp. 236–242). Cambridge: MIT Press.

Kushner, S. (2013). The freelance translation machine: Algorithmic culture and the invisible industry. *New Media & Society*, *15*(8), 1241–1258.

Lash, S. (2007). Power after hegemony: Cultural studies in mutation. *Theory, Culture & Society*, *24*(3), 55–78.

Lenglet, M. (2011). Conflicting codes and codings: How algorithmic trading is reshaping financial regulation. *Theory, Culture & Society*, *28*(6), 44–66.

MacCormick, J. (2013). *Nine algorithms that changed the future: The ingenious ideas that drive today's computers*. Princeton, NJ: Princeton University Press. [|29]

Mackenzie, A. (2005). The performativity of code: Software and cultures of circulation. *Theory, Culture & Society*, *22*(1), 71–92.

Mackenzie, A. (2007). Protocols and the irreducible traces of embodiment: The Viterbi algorithm and the mosaic of machine time. In R. Hassan & R. E. Purser (Eds.), *24/7: Time and temporality in the network society* (pp. 89–106). Stanford, CA: Stanford University Press.

Mackenzie, A., & Vurdubakis, T. (2011). Code and codings in Crisis: Signification, performativity and excess. *Theory, Culture & Society*, *28*(6), 3–23.

MacKenzie, D. (2014). *A sociology of algorithms: High-frequency trading and the shaping of markets*. Working paper, University of Edinburgh. Retrieved July 6, 2015, from http://www.sps.ed.ac.uk/__data/assets/pdf_file/0004/156298/Algorithms25.pdf

Mager, A. (2012). Algorithmic ideology: How capitalist society shapes search engines. *Information, Communication, & Society*, *15*(5), 769–787.

Mahnke, M., & Uprichard, E. (2014). Algorithming the algorithm. In R. König & M. Rasch (Eds.), *Society of the query reader: Reflections on web search* (pp. 256–270). Amsterdam: Institute of Network Cultures.

Manovich, L. (2013). *Software takes control*. New York, NY: Bloomsbury.

Miyazaki, S. (2012). Algorhythmics: Understanding micro-temporality in computational cultures. *Computational Culture*, Issue 2. Retrieved June 25, 2014, from http://computationalculture.net/article/algorhythmics-understanding-micro-temporality-in-computational-cultures

Montfort, N., Baudoin, P., Bell, J., Bogost, I., Douglass, J., Marino, M. C., ... Vawter, N. (2012). *10 PRINT CHR$ (205.5 + RND (1)): GOTO 10*. Cambridge: MIT Press.

Musiani, F. (2013). Governance by algorithms. *Internet Policy Review*, *2*(3). Retrieved October 7, 2014, from http://policyreview.info/articles/analysis/governance-algorithms

Napoli, P. M. (2013, May). *The algorithm as institution: Toward a theoretical framework for automated media production and consumption*. Paper presented at the Media in Transition Conference, Massachusetts Institute of Technology, Cambridge, MA. Retrieved from ssrn.com/abstract = 2260923

Neyland, D. (2015). On organizing algorithms. *Theory, Culture & Society, 32*(1), 119–132.

Pasquale, F. (2014). *The emperor's new codes: Reputation and search algorithms in the finance sector*. Draft for discussion at the NYU 'Governing Algorithms' conference. Retrieved October 16, 2014, from http://governingalgorithms.org/wp-content/uploads/2013/05/2-paper-pasquale.pdf

Pasquale, F. (2015). *The black box society: The secret algorithms that control money and information*. Cambridge, MA: Harvard University Press.

Porter, T.M. (1995). *Trust in numbers: The pursuit of objectivity in science and public life*. Princeton, NJ: Princeton University Press.

Rosenberg, S. (2007). *Dreaming in code: Two dozen programmers, three years, 4,732 bugs, and one quest for transcendent software*. New York: Three Rivers Press.

Seaver, N. (2013). *Knowing Algorithms*. Media in Transition 8, Cambridge, MA. Retrieved August 21, 2014, from http://nickseaver.net/papers/seaverMiT8.pdf

Shirky, C. (2009). *A speculative post on the idea of algorithmic authority*. Shirky.com. Retrieved October 7, 2014, from http://www.shirky.com/weblog/2009/11/a-speculative-post-on-the-idea of-algorithmic-authority/

Steiner, C. (2012). *Automate this: How algorithms took over our markets, our jobs, and the world*. New York, NY: Portfolio.

Takhteyev, Y. (2012). *Coding places: Software practice in a South American City*. Cambridge: MIT Press.

Ullman, E. (1997). *Close to the machine*. San Francisco, CA: City Lights Books.

Ziewitz, M. (2011, September 29). *How to think about an algorithm? Notes from a not quite random walk*. Discussion paper for Symposium on 'Knowledge Machines between Freedom and Control'. Retrieved August 21, 2014, from http://ziewitz.org/papers/ziewitz_algorithm.pdf

PROGRAMMIEREN VON UMGEBUNGEN – ENVIRONMENTALITÄT UND CITIZEN SENSING IN DER SMARTEN STADT

Jennifer Gabrys

Übersetzt von Michael Schmidt, überarbeitet von Christoph Engemann und Florian Sprenger; für die vorliegende Version mit Genehmigung der Autorin gekürzt und überarbeitet von Anke Strüver und Tabea Bork-Hüffer.*

EINLEITUNG: SMARTE UND NACHHALTIGE STÄDTE

Städte, die von computergestützten Prozessen durchdrungen und transformiert werden, sind offenbar Objekte ständiger Neuerfindung. Während Informationstechniken oder kybernetisch geplante Städte bereits seit den 1960er Jahren entstehen,[1] tauchen Vorhaben für vernetzte oder computergesteuerte Städte seit den 1980er Jahren regelmäßig in Stadtentwicklungsplänen auf.[2] Von Entwürfen modellierbarer urbaner Architektur bis hin zur Vorstellung der Stadt als einer Zone für technologisch angeregtes ökonomisches Wachstum, haben digitale Entwürfe urbane Räume

---

* „While every effort has been made to ensure that the contents of this publication are factually correct, neither the authors nor the publisher accepts, and they hereby expressly exclude to the fullest extent permissible under applicable law, any and all liability arising from the contents published in this Article, including, without limitation, from any errors, omissions, inaccuracies in original or following translation, or for any consequences arising therefrom. Nothing in this notice shall exclude liability which may not be excluded by law. Approved product information should be reviewed before prescribing any subject medications." – „Während alle Anstrengungen unternommen wurden, dass der Inhalt dieser Veröffentlichung sachlich korrekt ist, übernehmen die Autoren und der Verlag keine Gewähr für die Richtigkeit, die Genauigkeit und die Vollständigkeit der Angaben in der hier abgedruckten Übersetzung."

1 Vgl. Archigram: A Guide to Archigram, 1961–74, London: Academi Editions 1994, und Forrester, Jay Wright: Urban Dynamics, Cambridge, MA: MIT Press 1969.

2 Vgl. Batty, Michael: „The computable city", in: International Planning Studies 2 (1995), S. 155–173; Castells, Manuel: The Informational City: Information Technology, Economic Restructuring, and the Urban-Regional Process, Oxford: Blackwell 1989; Droege, Peter (Hg.): Intelligent Environments: Spatial Aspects of the Information Revolution, Amsterdam: Elsevier 1997; Gabrys, Jennifer: „Cite Multimedia: noise and contamination in the information city", Sitzungsprotokolle der Konferenz „Visual Knowledges" an der Universität Edinburgh, 17.–20. September 2003, www_jennifergabrys.net/wp-content/uploads/2003/09/Gabrys_InfoCity-VKnowledges. pdf, vom 11. Juli 2015; Graham, Stephen/Marvin, Simon: Splintering Urbanism: Networked Infrastructures, Technological Mobilities and the Urban Condition, London: Routledge 2001; Mitchell, William J.: City of Bits. Leben in der Stadt des 21. Jahrhunderts, Basel u.a.: Birkhäuser 1996.

in vernetzte, dezentralisierte und flexible Stätten der Kapitalakkumulation und des urbanen Erlebens verwandelt.

Jüngere, kommerziell motivierte Entwürfe für „smarte Städte" konzentrieren sich darauf, wie eine vernetzte Urbanität im Verbund mit partizipatorischen Medien zu „grüneren" oder effizienteren Städten führen könnten, die zugleich Maschinen für ökonomisches Wachstum sein sollen. Die Anhänger der smarten Stadt plädieren üblicherweise für die Unausweichlichkeit dieser Entwicklungen, indem sie auf die Trends der zunehmenden Urbanisierung verweisen. Städte, so argumentieren die Befürworter der Smart City, seien zwar Zentren von ökonomischem Wachstum und Innovationen, doch sie führen auch zu einem erheblichen Verbrauch von Ressourcen und zu Treibhausgasemissionen, weshalb sie als entscheidende Zonen der Umsetzung von Initiativen für Nachhaltigkeit betrachtet werden müssten. In diesen Vorhaben werden heruntergekommene oder noch zu errichtende Infrastrukturen zur Grundlage der Entwicklung smarter Städte. Diese werden als hübsch verpackte Möglichkeit präsentiert, solche verallgemeinerten Herausforderungen zu bewältigen und letztlich dafür zu sorgen, dass Städte – sowohl anzupassende als auch neu errichtete – in Zukunft nachhaltiger und effizienter sein werden als je zuvor.

Von digitalen Technologien und den mit ihnen einhergehenden Imaginationen durchdrungene Städte sind zwar keine neue Entwicklung, doch dass mit ihrer Umsetzung unter dem Deckmantel der Smartness Richtlinien für Nachhaltigkeit gewonnen werden sollen, stellt eine jüngere Taktik zur Forcierung des digitalen Wandels dar. In vielen Entwürfen smarter Städte synchronisieren Computertechnologien urbane Prozesse mit vorhandenen oder neuen Infrastrukturen, um die effiziente Nutzung von Ressourcen, die Verbreitung von Dienstleistungen und die Teilhabe am urbanen Leben zu verbessern. Digitale Technologien und speziell das Ubiquitous Computing werden in diesen Entwürfen immer dann thematisiert, wenn man verdeutlichen will, wie nachhaltige Stadtplanung umgesetzt werden könnte. Doch die Schnittstelle von smarter und nachhaltiger Stadtplanung ist ein Forschungsgebiet, das erst noch im Detail untersucht werden muss, insbesondere im Zusammenhang mit der Frage, welche Modalitäten einer umgebungsbezogenen urbanen Bürgerschaft in der smarten Stadt hervortreten oder eliminiert werden.[3]

Dieser Aufsatz befasst sich mit dem Entstehen der smarten Stadt als nachhaltiger Stadt anhand einer Fallstudie, nämlich dem vom MIT und dem Netzwerktechnikhersteller Cisco im Rahmen der Initiative Connected Urban Development (CUD) entwickelten Projekt Connected Sustainable Cities (CSC).

Dieses Projekt umfasst Designvorschläge, die zwischen 2007 und 2008 von William Mitchell und Federico Casalegno am MIT Mobile Experience Lab in Zusammenarbeit mit CUD entwickelt wurden. Die CUD-Initiative wurde 2006 als Reaktion auf Bill Clintons „Global Initiative" zur Auseinandersetzung mit dem

---

3   Da der Begriff *environment* mit *Umwelt* nur ungenügend übersetzt ist, wird im Folgenden abhängig vom Kontext entweder *environment* im Englischen belassen, *Umgebung* verwendet oder aber dort, wo eindeutig vom Umweltschutz (*environmentalism*) die Rede ist, der Begriff *Umwelt* verwendet. Die hier vorliegende Übersetzung unterscheidet sich daher teilweise auch von der deutschen Erstveröffentlichung von 2015.

Klimawandel initiiert. In Zusammenarbeit mit acht Städten auf der ganzen Welt – von San Francisco bis Madrid, von Seoul bis Hamburg – lief die CUD-Initiative bis 2010; daraus ging Ciscos derzeit laufendes Project Smart + Connected Communities hervor, das weiterhin Pläne für smarte Städte produziert, die von der aktuellen Entwicklung der Smart City Songdo in Südkorea bis hin zu Vorschlägen für ein „nachhaltiges San Francisco fürs 21. Jahrhundert" reichen.[4]

Indem ich diesen Designvorschlag mit Smart City-Projekten kontrastiere, die in ihren Entwicklungsplänen Nachhaltigkeit einbeziehen, möchte ich untersuchen, wie dieses spekulative und frühe Smart City-Projekt nachhaltigere und effizientere Urbanitäten zu erreichen versucht. Die in diesen Projekten vorgeschlagenen Szenarien des Ubiquitous Computings sollen, so der Plan des CSC-Projekts, auf existierende wie auf hypothetische Städte angewandt werden. Der Entwurf weist starke Ähnlichkeiten mit den Entwicklungen anderer smarter Städte auf und kann durch seine Verbindung zu Cisco, einem der ersten Entwickler von Netzwerkarchitekturen für Städte, als eine einflussreiche Demonstration von Vorstellungen smarter Städte gelten. Die vom CUD-Projekt entwickelten Instrumente bestehen aus Planungsdokumenten, White Papers, Öko-Werkzeugkisten, Multimediavorführungen und spekulativen Entwürfen, welche die Entwicklung der smarten Stadt leiten sollen.[5] Als wichtiger, aber vielleicht übersehener Teil solcher Vorhaben spielen diese Designs, Narrative und Dokumente eine bedeutsame Rolle bei der Umdeutung der smarten Stadt zur nachhaltigen Stadt. Vor allem jedoch versteht dieser Aufsatz die Vorschläge nicht einfach als *diskursive* Darstellungen von Städten, sondern als Elemente eines urbanen kalkulatorischen *Dispositivs* oder *Apparats*,[6] der die materiell-politischen Zusammenhänge von spekulativen Designs, technologischen Vorstellungswelten, Stadtentwicklungsplänen und demokratischen Engagements mithilfe partizipatorischer Medien und vernetzter Infrastrukturen herstellt. All dies ist Teil vieler gegenwärtiger Stadtentwicklungspläne und -praktiken, selbst wenn sich das Projekt der smarten Stadt nur schwerlich jemals in der anvisierten Form realisieren ließe.

Die vorgeschlagenen Pläne und Designs smarter Städte artikulieren eigene Materialitäten und Räumlichkeiten ebenso wie Formationen von Regierungs- und Verwaltungsmacht. Anhand des von Foucault inspirierten Begriffs der Environmentalität untersuche ich die Art und Weise, wie das CSC-Projekt in und durch Vorschläge für smarte Umgebungen und entsprechende Technologien eine Verteilung solcher Macht sicherstellt.

---

4   Cisco: „Smart + connected communities", www.cisco.com/web/strategy/smart_connected_communities.html, vom 11. Juli 2015.

5   Allein über die Rolle von White Papers im Rahmen der Entwicklung von smarten Städten ließe sich ein eigener Aufsatz schreiben. Von der Industrie, von Universitäten und Behörden erstellte Smart City-White Papers erweisen sich als Schlüsselmedium, in dem die Vorstellungen und die Umsetzung dieser Stadtentwicklungen zirkulieren. Die „Zirkulation" von Politik gehört, wie es Robinson 2011 dargelegt hat, zu der Art und Weise, wie Städte im Rahmen von Projekten zur urbanen Vorstellungswelt vielfach verortete „Anderswos" akkumulieren.

6   Vgl. Foucault, Michel: Schriften in vier Bänden. Dits et Ecrits, Frankfurt a.M.: Suhrkamp 2001–2005, Bd. III, S. 391 ff.

Dieser Aspekt von Foucaults Auseinandersetzung mit Environmentalität steht im Mittelpunkt meiner Ausführungen, um Foucaults unvollendete Überlegungen zu der Frage aufzugreifen und weiterzuentwickeln, wie Technologien des Umgebens als räumliche Modi von Regierungsmacht materiell-politische Verteilungen von Regulation und mögliche Modi von Subjektivierung verändern könnten. Indem ich Foucaults Vorstellung von Environmentalität nicht auf die Produktion von Subjekten beschränke, die von ihren Umgebungen geformt werden, sondern als eine räumlich-materielle Verteilung und Relationalität von Macht durch Umgebungen, Technologien und Lebensweisen verstehe, gehe ich der Frage nach, wie Praktiken und Operationen von Bürgerschaft entstehen, die ein entscheidender Teil der Imaginationen smarter und nachhaltiger Städte geworden sind. Dieses Verständnis von Environmentalität in der smarten Stadt formuliert neu, wer oder was als „Bewohner" zählt. Entsprechend geht es um die Art und Weise, wie Bewohner und Bürger weniger als zu beherrschende individuelle Subjekte, sondern vielmehr durch die Verteilung und das Feedback von Überwachung sowie durch urbane Datenpraktiken in Umgebungen geprägt werden.

Primär soll Nachhaltigkeit in smarten Städten durch effizientere Prozesse und durch verantwortungsvolle Bürger erreicht werden, die an computergestützten Sensor- und Überwachungspraktiken teilhaben. Im Kontext der Entwürfe smarter Städte werden die Bewohner der Stadt zu Sensorknoten oder vielmehr zu Bürgersensoren. Auf diese Weise lässt sich „Citizen Sensing" nicht nur als eine Erhebung von Daten über Bürger, sondern als Modalität von Bürgerschaft verstehen, die durch die Interaktion mit computergestützten Sensortechniken entsteht, welche für Überwachung und Feedback von Umgebungen benutzt werden. In diesem Zusammenhang greife ich die Vorschläge für smarte Städte auf, wie sie im CSC-Projekt entwickelt wurden, um folgende Fragen zu stellen: Welche Implikationen haben die mittels Computern organisierten Verteilungen von Regierungsmacht über die Umgebung [environmental governance], die für bestimmte Funktionalitäten programmiert sind und von privatwirtschaftlichen und staatlichen Akteuren gemanagt werden, welche Städte wie manipulierbare Datensätze behandeln? Wie artikuliert sich Environmentalität im Rahmen von Plänen und Entwicklungen nachhaltiger smarter Städte, wenn Governance durch Umgebungen ausgeübt wird, die computergestützt programmiert sind? Und wenn als Sensoren agierende Bewohner Operateure in urbanen Computersystemen werden, wie können dann Technologien des Umgebens die den Bewohnern angemessenen Praktiken auf eine Reihe von Aktionen beschränken, die durch die Überwachung und Verwaltung von Daten konstituiert werden? Hieße dies etwa, dass Bewohner smarter Städte weniger starre menschliche Subjekte sind, sondern als Operationalisierung von „Citizenship" verstanden werden können, die auf digitale Techniken angewiesen sind, um lebendig zu sein?

DIE NEUGESTALTUNG SMARTER STÄDTE

Wie man der vielschichtigen Literatur über smarte Städte und den auf sie abzielenden Projekten entnehmen kann, gibt es zahlreiche Interpretationen davon, was überhaupt als smarte Stadt gelten darf.[7] Sie könnte etwa mit neuen Medienanwendungen oder automatisierten Infrastrukturen verbunden sein, die mit vernetzten digitalen Sensoren ausgestattet sind; sie könnte sich auf die Konvergenz von Online- und Offlinewelten beziehen oder jene urbanen Erlebnisse umfassen, die durch mobile Geräte ermöglicht werden. Während sich die frühere Forschung zu diesen Fragen auf das Verhältnis zwischen der digitalen und der physischen Stadt oder auf die Art und Weise konzentriert hat, wie „virtuelle" digitale Technologien physische Städte räumlich umgestalten oder darstellen können,[8] befasst man sich inzwischen zunehmend mit den Verfahren, mit denen Städte sowohl durch Software wie auch durch die materiellen Infrastrukturen digitaler Technologien umgebaut und vermarktet werden.[9] Durch das Generieren beträchtlicher Datenmengen, die zum Managen urbaner Prozesse notwendig sind, und durch das direkte Einbetten von Endgeräten in urbane Infrastrukturen und ihre Räume, gestaltet Ubiquitous Computing Städte eher um, als sie zu ersetzen oder virtuell zu repräsentieren.

„Smartness" bezieht sich zwar ganz allgemein auf computergestützte Stadtentwicklung, bezeichnet aber zunehmend urbane Nachhaltigkeitsstrategien, die von der Implementierung des Ubiquitous Computing abhängen, der von Cisco sogenannten „vierten Nutzung".[10] In dem industriellen White Paper „A theory of smart cities" behaupten an der Initiative Smarter Planet beteiligte IBM-Autoren, der Begriff „smart cities" gehe auf „smart growth" zurück, ein Konzept, mit dem Stadtplaner in den späten 1990er Jahren Strategien zur Einschränkung von Zersiedlungseffekten und ineffizienter Ressourcennutzung umschrieben, während man damit später auf Informationstechnologien basierende Infrastrukturen und Prozesse bezeichnete.[11] In staatlichen wie industriellen White Papers ist immer wieder die Rede davon, wie vernetzte Sensortechnologien urbane Prozesse und Ressourcenverbrauch optimieren und effizienter machen sollen, was auch Transport und Verkehr, Gebäudemanagement, Stromversorgung und Industrie umfasst. Durch Sensoren operationalisierte und automatisierte Umgebungen führen zu einer bestimmten

---

7   Vgl. Allwinkle, Sam / Cruickshank, Peter: „Creating smarter cities: an overview", in: Journal of Urban Technology 18 (2011), S. 1–16.

8   Vgl. Lovink, Geert: „Die digitale Stadt – Metapher und Gemeinschaft" in: ders., Dark Fiber – auf den Spuren einer kritischen Internetkultur, Opladen: Leske + Budrich 2004, S. 42–61.

9   Vgl. Ellison, N. / Burrows, R. / Parker, S. (Hg.): „Urban informatics: software, cities and the new cartographies of knowing capitalism", in: Information, Communication and Society 10 (2007), S. 785–960; sowie Graham, Stephen (Hg.): The Cybercities Reader, London: Routledge 2004.

10 Vgl. Elfrink, Wim: „Intelligent Urbanization", Video, http://blogs.cisco.com/news/video/, vom 12. Juli 2015.

11 Vgl. Harrison, Colin / Donelly, Ian Abbott: „A theory of smart cities", Protokolle der 55. Jahrestagung der International Society for the System Sciences (ISSS) 2011, http://journals.isss.org/index.php/proceedings 55th/article/view/1703, vom 11. Juli 2015.

Version von Nachhaltigkeit, für die Effizienz das eigentliche Ziel der Konvergenz von ökonomischem Wachstum mit grünen Zielen darstellt. Tatsächlich werden smarte Städte häufig als erhoffte Quelle zur Generierung beträchtlicher neuer Einnahmen behandelt. In einem von der Rockefeller Foundation finanzierten Report deutet das Institute for the Future an, smarte Städte seien wahrscheinlich „ein Markt mit einem Volumen von vielen Billionen Dollar".[12]

Die derzeitige Welle von geplanten wie bereits laufenden Projekten für smarte und nachhaltige Städte umfasst zahlreiche, auf der ganzen Welt angesiedelte Vorhaben mit ähnlichen Zielen, Entwürfen und Designs. Von Abu Dhabi bis Helsinki wie von Smart Grids in Indien bis zum PlanIT Valley in Portugal sind viele Stadtentwicklungsprojekte auf die Implementierung von vernetzten Sensorumgebungen gerichtet, die mit der Logik von Effizienz und Nachhaltigkeit vermarktet werden. Projekte für smarte Städte werden oft als öffentlich/private Partnerschaften (PPP) zwischen multinationalen Technikkonzernen wie Cisco, IBM und Hewlett Packard zusammen mit Stadtverwaltungen, Universitäten oder Design- und Ingenieurbüros begonnen. Bei diesen Vorhaben geht es etwa um das Nachrüsten urbaner Infrastrukturen wie in New York oder London, um das Entwickeln neuer Städte auf Brachflächen wie im koreanischen Songdo oder in Lake Nona in Florida, sowie um das Verstärken von Netzwerkeinrichtungen in mittleren Großstädten wie Dubuque in Iowa zum Testen vernetzter Sensoranwendungen. Der Schwerpunkt liegt hier auf der Art und Weise, wie Smartness Ausdrucksformen urbaner Nachhaltigkeit beeinflusst. Aber statt sich auf eine starre Definition der smarten Stadt festzulegen, bewegt sich meine Arbeit im Spannungsfeld von den Vermutungen, dass sich an der Art und Weise, wie informatisierte Städte mobilisiert werden, politische und ökonomische Interessen ablesen lassen.[13] Digital informierte Städte können in diesem Sinne Gebilde sein, die sich in ihrer Vorstellungswelt, Implementierung und Erfahrung ständig verändern.[14] Smarte Städte könnten in ihrer urbanen Grundidee zwar als idealtypisch und universal verstanden werden, doch konkrete Projekte gehen letztlich aus den materiell und politisch zufälligen Räumen und Praktiken von Design, Politik und Entwicklung hervor, während sie zugleich spezifische, wenn auch spekulative urbane Lebensweisen in die Welt bringen.

DIE NEUGESTALTUNG SMARTER BEWOHNER

Die in den Entwürfen smarter Städte vorgeschlagenen und weiterentwickelten Computertechnologien sollen zusammen mit den Interaktionen und Praktiken der Stadtbewohner urbane Umgebungen und Prozesse gestalten. Citizen Sensing und

---

12   Townsend, Anthony et al.: A Planet of Civic Laboratories: The Future of Cities, Information, and Inclusion, Palo Alto: Institute for the Future and the Rockefeller Foundation 2010.

13   Hollands, Robert G.: „Will the real smart city please stand up?", in: City 12 (2008), S. 303–320.

14   Mackenzie, Adrian: Wirelessness: Radical Empiricism in Network Cultures, Cambridge, MA: MIT Press 2010.

partizipatorische Plattformen werden in diesen Plänen gefördert, um es den Bewohnern zu ermöglichen, Ereignisse in ihren Umgebungen durch mobile Geräte und Sensortechniken in Echtzeit zu verfolgen. Doch wenn man es Bürgern ermöglicht, ihre eigenen Aktivitäten zu überwachen, macht man aus diesen Bürgern unfreiwillige Sammler und Lieferanten von Daten, die für mehr genutzt werden können als für den Ausgleich des Energieverbrauchs. Es geht auch darum, Energieversorgern und Stadtverwaltungen Details über die Muster des Alltagslebens zu verschaffen. Das Überwachen, das Managen von Daten und das Einspeisen dieser Informationen in die Systeme einer Stadt werden so zu Praktiken, die für Bürgerschaft konstitutiv sind. Bewohnen verwandelt sich damit in das, was ich Citizen Sensing nennen möchte. Es ist manifestiert in Praktiken, die in Reaktion auf computergestützte Umgebungen und in Kommunikation mit diesen Technologien vollzogen werden.

Citizen Sensing als eine Form des Engagements ist sowohl für die entwicklungsbestimmte wie für die kreativ-praktische Beschäftigung mit smarten Städten ein durchgängiger, wenn auch unterschiedlich stark betonter Richtwert. DIY-Projekte schlagen eine Bürgerbeteiligung durch die Nutzung partizipatorischer Medien und Sensortechnologien vor, um den Unterschied zwischen basisdemokratischen und eher groß angelegten Entwicklungen von smarten Städten zu betonen. Zu einem interessanten Zusammenfließen von Imaginationen und Praktiken kommt es beim Aufrüsten von Bürgern, das bis hin zum „Verändern der Subjektivität heutiger Bürgerschaft"[15] reicht, indem es Stadtbewohnern ermöglicht wird, durch die Nutzung von Sensortechnologien mit ihren urbanen Environments zu interagieren. Mit was für einer Subjektivität haben wir es hier zu tun? Könnte man anhand von computergestützten Umgebungen untersuchen, wie (und wo) diese Subjektivität und die Bedeutung von Bürgerschaft verändert werden? Mit anderen Worten: Wenn sich urbane Prozesse und Architekturen durch Ubiquitous Computing wandeln, dessen Ziel Effizienz und Nachhaltigkeit darstellt, wie verwandeln sich dann auch die materielle Politik einer Stadt und die Möglichkeiten für demokratisches Engagement?[16] Mein Interesse an diesen Modalitäten von Citizen Sensing in smarten Städten läuft jedoch nicht darauf hinaus, diese Vorhaben und Projekte als Kontrollinstrumente zu denunzieren, wie es für eine Technikkritik typisch wäre. Vielmehr geht es mir darum, genauer zu verstehen, wie computergestützte Materialisierungen Macht durch urbane Räume und Prozesse verteilen. Oder wie es Foucault dargelegt hat: Statt sich einen machtfreien Raum vorzustellen, kann es produktiver sein, zu untersuchen, wie Macht als eine Möglichkeit verteilt ist, konkrete Praktiken der Verwaltung zu kritisieren, indem man sich vorstellt, wie es möglich sein könnte, nicht so beherrscht zu werden – oder nicht auf diese Weise.[17]

15   Borden, Ed / Greenfield, Adam: „You are the smart city", http://blog.xively.com/2011/06/30/you-are-the-smart-city/, vom 11. Juli 2015.

16   Vgl. Fuller, Matthew / Haque, Usman: „Urban Versioning System 1.0", in: Situated Technologies Pamphlets 2, The Architectural League of New York 2008.

17   Vgl. Foucault, Michel: The Politics of Truth, New York: Semiotexte 1997, S. 44 f.

ENVIRONMENTALITÄT

Ich greife diese Fragen der Transformationen in der Entwicklung, Form und dem Bewohnen von Städten auf, um gründlicher zu analysieren, wie die umgebungsbezogenen Technologien des Ubiquitous Computing urbane Regierungen und Bürgerschaft informieren. Environmentalität ist ein Begriff, mit dem ich diese urbanen Transformationen beschreiben und den ich durch eine Interpretation von Foucaults unvollendeter Erörterung dieses Begriffs in einer seiner letzten Vorlesungen, *Die Geburt der Biopolitik*, aufgreifen und überarbeiten möchte. Foucault signalisiert sein Interesse an Environmentalität und umgebungsbezogenen Technologien, wenn er von einer historischen zu einer eher zeitgenössischen und neoliberalen Betrachtung von Biopolitik im Zusammenhang mit dem Milieu oder der Umgebung als Ort von Regierungsmacht übergeht.[18] In diesem Übergang verliert, so Foucault, das Subjekt oder die Bevölkerung für das Verständnis der Ausübung biopolitischer Techniken an Bedeutung, da Änderungen der Umgebungsbedingungen neue Formen der Regulierung mit sich bringen.[19]

Foucaults Erörterung von Environmentalität geht aus einer Analyse der Kriminalität hervor. An einem Beispiel untersucht er, wie Versuche, den Drogenhandel zu regulieren, sich vielleicht stärker auf die Bedingungen der Sucht ausgewirkt haben als Strategien, die sich gegen individuelle abhängige Konsumenten oder Gruppen abhängiger Konsumenten richten. Foucault identifiziert, weniger erklärend als tastend, einen zunehmenden Trend zu einer umgebungsbezogenen Regierungsmacht, die an die Stelle einer auf das Subjekt oder auf die Bevölkerung gerichteten Regierungsmacht tritt. Er stellt fest:

> Es ist […] die Idee […] einer Gesellschaft, […] in der es keine Einflussnahme auf die Spieler des Spiels, sondern auf die Spielregeln geben würde und in der es schließlich eine Intervention gäbe, die die Individuen nicht innerlich unterwerfen würde, sondern sich auf ihre Umwelt bezöge.[20]

Im Anschluss daran verweist Foucault auf einen erweiterten Begriff von Environmentalität, wobei das Beeinflussen der „Spielregeln" durch das Modulieren und Regulieren von Umgebungen vielleicht eine aktuellere Umschreibung von Gouvernementalität ist und über direkte Versuche hinausgeht, individuelles Verhalten oder die Normen der Bevölkerung zu beherrschen. Das Verhalten mag zwar adressiert oder beherrscht werden, doch die Techniken sind umgebungsbezogen.

Foucault weist am Ende seiner Vorlesung daraufhin, dass er in der nächsten Vorlesung diese Fragen der Regulierung von und über Umgebung ausführlicher untersuchen werde. Allerdings entwickelt er diesen Gedankengang nicht weiter. Stattdessen sind die sechs Seiten, die seinen Denkansatz zur Environmentalität

---

18 Foucault verwendet in seinen Vorlesungen sowohl den Begriff *milieu* wie auch den Begriff *environment*; ihre Unterscheidung ist nicht immer eindeutig.

19 Foucault, Michel: Sicherheit, Territorium, Bevölkerung. Geschichte der Gouvernementalität I, Frankfurt a. M.: Suhrkamp 2004, S. 40 f., und Foucault, Michel: Die Geburt der Biopolitik. Geschichte der Gouvernementalität II, Frankfurt a. M.: Suhrkamp 2006, S. 359 ff.

20 M. Foucault: Die Geburt der Biopolitik, S. 359.

skizzieren, als Fußnote in den Vorlesungen über *Die Geburt der Biopolitik* enthalten.[21] Foucaults Erörterung der Environmentalität stellt also eher eine unbeantwortete Frage als einen theoretischen Leitgedanken dar und reicht von einer historischen Analyse der Herrschaft über Bevölkerung bis zu einer Betrachtung zeitgenössischer Modi von Governance, die zum Zeitpunkt seiner Vorlesung vielleicht gerade erst im Entstehen begriffen waren. Sein spezifisches Konzept der Environmentalität bleibt zwar eine Fußnote zu seiner Erörterung neoliberaler Regierungsmacht, doch es fordert dazu heraus, gründlich über die Auswirkungen der zunehmenden Förderung und Verbreitung von Computertechnologien nachzudenken, mit deren Hilfe urbane Umgebungen verwaltet werden sollen. Auf welche Weise also können Vorschläge zur Entwicklung smarter Städte umgebungsbezogene Modi der Governance artikulieren und gezielt umsetzen, und wie sehen die räumlichen, materiellen und bürgerbezogenen Umrisse dieser Regierung und Verwaltung aus?

Der Gebrauch des Begriffs Environmentalität, den ich hier auf der Basis der Vorlesungen über Biopolitik entwickle und umwandle, unterscheidet sich von der Art und Weise, wie er oft auf der Grundlage von Foucaults früheren Werken verstanden wird – von der Entwicklung umweltbewusster Individuen zur Erhaltung der Wälder in Indien[22] bis zur „grünen Gouvernementalität" von Umweltorganisationen.[23] Mithilfe des Begriffs Environmentalität kann man gewiss auch über die politischen Ziele des Umweltschutzes nachdenken. Dabei gilt es aber im Auge zu behalten, dass Environmentalität und Umweltschutz nicht deckungsgleich sind. Foucaults Analyse von Environmentalität befasst sich nicht direkt mit dem Umweltschutz als solchem, sondern vielmehr mit einem Verständnis von durch die Umgebung operierender Regierungsmacht. Tatsächlich gilt Foucaults Interesse an umgebungsbezogenen Governance-Strategien „der Umwelttechnologie oder der Umweltpsychologie"[24], Räumen, die auch das Design von Überlebenssystemen oder von Erlebniswelten in Shoppingmalls umfassen könnten.[25] Umgebungsbezogene Modi von Governance können genauso gut auch aus dem Scheitern beim Umsetzen von Zielen des Umweltschutzes hervorgehen. Katastrophen wie der Hurrikan Katrina führen, wie Brian Massumi in seiner Analyse der Environmentalität darlegt, zu ausgeprägten Verfahren von krisenorientierter Governance, die im Zusammenhang mit der Ungewissheit des Klimawandels entstehen – einem Zustand von „Krieg und Wetter", der eine Raumpolitik von permanenter Störung und Reaktion in Gang setzt.[26]

21 M. Foucault: Die Geburt der Biopolitik, S. 359–361.

22 Agrawal, Amar Nath: Environmentality: Technologies of Government and the Making of Subjects, Durham/NC: Duke University Press 2005.

23 Luke, Timothy W.: „Environmentality as Green Governmentality", in: Eric Darier (Hg.), Discourses of the Environment, Oxford: Blackwell 1998, S. 121–151.

24 M. Foucault: Die Geburt der Biopolitik, S. 359.

25 Vgl. Anker, Peder: „The closed world of ecological architecture", in: The Journal of Architecture 10 (2005), S. 527–552.

26 Massumi, Brian: „National Enterprise Emergency", in: Theory, Culture and Society 26 (2009), S. 153–185, hier S. 154.

BIOPOLITIK 2.0

Foucaults Erörterung des Begriffs Environmentalität mag zwar verkürzt sein, adressiert aber die Rolle von Technologien des Umgebens als Instrumente der Regierungsmacht und hängt auf vielfache Weise mit seiner beharrlichen Betrachtung der Umgebung als Ort des biopolitischen Managements zusammen. Biopolitik oder die Regierung des Lebens, wie er sie in ihren Ausprägungen des späten 18. und 19. Jahrhunderts analysierte, befasst sich mit den „Beziehungen innerhalb der menschlichen Gattung, dem menschlichen Wesen als Gattung, als Lebewesen und in ihrem Umfeld, ihrem Lebensumfeld [milieu]"[27]. Wenn wir ferner davon ausgehen, dass Biopolitik auch jene Verteilungen von Macht umfasst, die nicht nur das Leben informieren, sondern auch die Art und Weise, wie wir leben, wie werden dann Lebensweisen durch diese besonderen Verteilungen des Umgebens beherrscht? Die Formulierung „Lebensweisen", derer sich Foucault bedient, um biopolitische Arrangements und Verteilungen von Macht zu erörtern, wird von Judith Revel aufgegriffen. Sie legt dar, dass sich das Konzept der Biopolitik nicht ausschließlich mit „Kontrolle" befasse, wie dies vielleicht Interpretationen von Foucaults Frühwerk überbetont haben, sondern sein Hauptaugenmerk auf die räumlich-materiellen Bedingungen und die Verteilungen von Macht richte, die für jede Zeit und jeden Ort charakteristisch sind.[28] „Lebensweisen" oder „gelebtes Leben" sind biopolitische Ansätze, die über das Verständnis von Leben als gegebene biologische Einheit[29] hinausgehen und stattdessen darlegen, dass Lebensweisen verortet und gewachsen sind und durch räumliche und materielle Machtbeziehungen praktiziert werden. Ein solches Konzept bezeichnet kein totalisierendes Machtschema, sondern verweist auf ein Verständnis davon, wie Macht in Lebensweisen entsteht und operiert. So kann man auch Möglichkeiten vorschlagen, wie sich alternative Lebensweisen generieren lassen.

Eine andere Formation der Biopolitik entsteht im Kontext der Environmentalität, da sich Biopolitik in Relation zu einer Umgebung entfaltet, die sich weniger an der Kontrolle der Bevölkerung orientiert, sondern stattdessen durch umgebungsbezogene Modi der Regierung funktioniert. Um die Lebensweisen zu erfassen und zu untersuchen, die im Rahmen des CSC-Vorhabens für smarte Städte entstehen, verwende ich (mit einer gewissen Ironie) den Begriff Biopolitik 2.0, um damit die partizipatorischen oder digitalen Technologien 2.0 zu bezeichnen, die in smarten Städten im Spiel sind. Dies eröffnet die Möglichkeit, spezifische Lebensweisen zu untersuchen, die sich in der smarten Stadt entfalten. Biopolitik 2.0 ist somit ein methodologischer Kunstgriff zur Analyse von Biopolitik als einem historisch gebundenen Konzept, wie dies Foucault bei seiner Entwicklung des Begriffs betont

---

27 Foucault, Michel: In Verteidigung der Gesellschaft, Frankfurt a. M.: Suhrkamp 2001, S. 288.
28 Revel, Judith: „Identity, Nature, Life: Three Biopolitical Deconstructions", in: Theory, Culture and Society 26 (2009), S. 45–54, hier S. 49–52.
29 Diese Interpretation hängt mit Agambens Arbeit über Biopolitik und das nackte Leben zusammen. Agamben, Giorgio: Homo Sacer. Die souveräne Macht und das nackte Leben, Frankfurt a. M: Suhrkamp 2002.

hat. Das 2.0 von Biopolitik erfasst die historische Situation dieses Begriffs, welche
die ungeheure Vermehrung von nutzergenerierten Inhalten durch partizipatorische
digitale Medien einschließt. Dieser entscheidende Teil der Vorstellung davon, wie
smarte Städte funktionieren sollen, schließt auch die *Versionierung* digitaler Technologien durch den Übergang der Computertechnologie von Desktops zu Umgebungen ein,[30] sei es in Form von mobilen digitalen Geräten oder von Sensoren, die
in der urbanen Infrastruktur, in Gegenständen und Netzwerken eingebettet sind –
was auch durch den Begriff *Stadt 2.0* erfasst wird, der als Alternative zur smarten
Stadt in Umlauf ist.

Das biopolitische Milieu generiert materiell-räumliche Arrangements, in denen
und durch die unterschiedliche *Dispositive* oder Apparate operieren. Der Apparat
der computergestützten Stadtplanung lässt sich durch Netzwerke, Techniken und
Machtbeziehungen analysieren, die von der Infrastruktur bis zu Governance und
Planung, Alltagspraktiken, Imaginationen des Urbanen, Architekturen, Ressourcen
und mehr reichen. Doch diese „heterogene Gesamtheit" lässt sich durch die „Natur
der Verbindung" beschreiben, die sich zwischen all diesen Elementen entfaltet.[31] In
seinen Ausführungen zu Biopolitik, Apparat und Milieu erklärt Foucault wieder-
holt, dass die Art und Weise, wie Zusammenhänge hergestellt werden, entscheidend
ist für das Verständnis davon, wie Modi von Governance, Lebensweisen und politische Möglichkeiten entstehen oder aufrechterhalten werden.

Die Eigenschaften des computergestützten Monitorings und der Responsiveness[32] charakterisieren die „Natur der Verbindung" in smarten Umgebungen und
zwischen ihren Bewohnern. Biopolitische Zusammenhänge 2.0 artikulieren das Be-
dürfnis, die wirtschaftliche Entwicklung zu fördern, während gleichzeitig dem dro-
henden Unheil für die Umwelt entgegengetreten wird – Zustände, die durch eine
„dringende Anforderung" charakterisiert sind, welche Foucault als entscheidend
für die historische Situation des Apparats und folglich für die Ausübung von Biopolitik erkennt.[33] In Vorschlägen und Entwürfen für smarte Städte werden Städte als
umweltbezogene, soziale und ökonomische Probleme dargestellt. Die digitale Reorganisation durchzieht urbane Infrastrukturen, wobei die Produktivität erhöht und
zugleich Effizienz erzielt werden soll. Indem ich Foucaults Verständnis davon, wie
Macht umgebungsbezogen und biopolitisch operieren könnte, zusammenfasse, lege
ich den Schwerpunkt auf das Verständnis urbaner Räume und der Bürgerschaft im
Rahmen relationaler oder konnektiver Register, wobei ich mich auf die Computer-
praktiken und -prozesse konzentriere, die Lebensweisen in smarten Städten erneu-
ern und beeinflussen sollen. Während ich diese Aspekte der Überlegungen Foucaults als weniger auf disziplinierte und kontrollierte Individuen oder Populationen

---

30 Hayles, N. Katherine: „RFID: Human Agency and Meaning in information-intensive Environ-
    ments", in: Theory, Culture and Society 26 (2009), S. 47–72.

31 M. Foucault: Schriften in vier Bänden. Bd. III., S. 392.

32 Da der Begriff der *Responsivität* bereits von der Phänomenologie besetzt ist und Reaktions-
    freudigkeit oder Ansprechbarkeit keine treffenden Übersetzungen darstellen, wird hier und im
    Folgenden der Begriff *responsiveness* im Englischen belassen. [Anm. F.5.].

33 M. Foucault: Schriften in vier Bänden, Bd. III., S. 393. Siehe auch Agamben, Giorgio: Was ist
    ein Dispositiv? Berlin: Diaphanes 2008.

konzentriert interpretiere, verlagere ich Environmentalität in einen Raum, in dem sich betrachten lässt, wie sich smarte Städte als nachhaltig qualifizieren – durch ein Neufassen der „Spielregeln".

Wenn hier gesagt wird, dass sich smarte Städte durch eine Analyse der Biopolitik 2.0 verstehen ließen, soll nicht so sehr behauptet werden, dass digitale Technologien schlicht Kontrollinstrumente seien, sondern vielmehr untersucht werden, wie die räumlichen und materiellen Programme, die in Vorschlägen für smarte Städte vorgestellt und implementiert werden, bestimmte Arten von Machtarrangements und Modi von Environmentalität generieren. Stadtbewohner werden damit in spezifische Anforderungen ihrer Bürgerschaft verstrickt. Aber im Rahmen dieser Programme für eine computergestützte Stadtentwicklung werden sich die prozesshaften und praktizierten Lebensweisen, die sich tatsächlich entfalten oder noch entfalten sollen, unweigerlich auf vielfache Weise materialisieren. Die „Spielregeln", die für Foucault von zentraler Bedeutung für Environmentalität sind, müssen vielleicht als eine weniger statische oder deterministische Darstellung dessen, wie Regierungsmacht funktioniert, revidiert werden. Vorschläge für das Design smarter Städte führen auf einer Ebene Vorhaben und Programme ein, wie computergestützte Stadtentwicklungen operieren sollen; doch auf einer anderen Ebene verlaufen Programme nie nach Plan und werden niemals isoliert umgesetzt. Environmentalität könnte dadurch vorangetrieben werden, dass man smarte Städte nicht als Anwendung von Quellcode in einer Command-and-Control-Logik von Herrschaftsraum betrachtet, sondern als vielfache, sich wiederholende und sogar schwankende Materialisierungen vorgestellter und belebter computergestützter Urbanität.

VERNETZTE NACHHALTIGE STÄDTIE

Das CSC-Projekt arbeitet an dieser Schnittstelle von umgebungsbezogenen Modi der Regierungsmacht, Technologien des Umgebens und Nachhaltigkeit, wie sie in smarten Städten operationalisiert werden. Es stellt damit eine Vision von einer nahen Zukunft urbanen Ubiquitous Computing dar, die auf eine zunehmende Nachhaltigkeit ausgerichtet ist. Die Planungsmaterialien des Projekts befürworten die smarte Stadt, weil sie eine entscheidende Rolle für Fragen des Klimawandels und der Ressourcenknappheit spielt, wobei nachhaltige urbane Umwelten durch intelligente digitale Architekturen verwirklicht werden können. Die Designvorschläge und politischen Instrumente des CSC-Projekts, ebenso wie sein zentrales visionäres Dokument – das von William J. Mitchell und Federico Casalegno verfasste Buch *Connected Sustainable Cities* – entwickeln Szenarien für ein Alltagsleben, das erweitert und sogar verändert werde durch smarte Informationstechnologien, die „neue, auf intelligente Weise nachhaltige urbane Lebensmuster fördern" werden.

In den CSC-Designvorschlägen ist die Rede von einer Technologie, die smarte Environments und die Interaktionen zwischen der Stadt und ihren Bürgern weitgehend operationalisiert: Ubiquitous Computing in Form eines „kontinuierlichen, hochauflösenden elektronischen Sensing" durch „Sensoren und Tags", die „auf Gebäuden und in Infrastrukturen montiert, in bewegten Fahrzeugen transportiert, in

Mobilgeräten wie Telefonen integriert und an Produkten befestigt werden". Überall werden Sensoren verteilt, die die urbane Umgebung überwachen. Das unablässige Generieren von Daten liefert „detaillierte Echtzeitbilder" von urbanen Praktiken und Infrastrukturen, die sich managen, synchronisieren und aufteilen lassen, um „die optimale Verteilung knapper Ressourcen" zu fördern.[34] Digitale Sensortechnologien machen urbane Prozesse effizient, wobei Computertechnologien derart in Umgebungen eingebettet sind, dass sie das Management und die Regulierung der Stadt ermöglichen.

Wie bei vielen Vorhaben für solche Städte werden auch die CSC-Städte smart auf mehreren gemeinsamen Ebenen der Intervention, die großenteils auf Produktionssteigerung ausgerichtet sind und zugleich die Effizienz verbessern sollen. Ein Video veranschaulicht dieses Grundprinzip für das Projekt: Seine Kernbereiche sind Plattformen, die entwickelt werden, um den Pendelverkehr, das Recycling in Privathaushalten, die Selbstverwaltung der eigenen CO_2-Bilanz zu unterstützen oder Flexibilität in urbanen Räumen und kollaborative Entscheidungsfindungen zu ermöglichen. All dies sollen Modellbereiche sein, in denen eine verbesserte Effizienz durch digitale Vernetzung und verbesserte Sichtbarkeit von umgebungsbezogenen Daten Ressourcen sparen und Treibhausgasemissionen senken können. Während viele der in diesem Vorhaben vorgestellten Anwendungen bereits in zahlreichen Städten Einzug gehalten haben, von elektronischen Systemen für Leihfahrräder bis zu smarten Messgeräten für den Energieverbrauch, schlägt das Projekt eine weitere koordinierte Verbreitung von Sensortechnologien und Plattformen vor, die urbane Prozesse noch effizienter machen sollen.

Im Video des CUD-Projekts wie im CSC-Designdokument werden Vorhaben für Design und Planung smarter Städte nicht in der Größenordnung eines Masterplans, sondern eher eines Szenarios vorgestellt. Von Curitiba bis Hamburg umfassen die in diesen Designs und politischen Überlegungen angesprochenen Episoden des urbanen Lebens kommunale Dienstleistungen, Instrumente des Umweltmonitorings und spekulative Plattformen, die ein smartes und „nahtlos" automatisiertes Leben ermöglichen sollen. Doch in vielen Fällen vollziehen sich die urbanen Interventionen in einer hypothetischen Stadt. Sie wird so verallgemeinert dargestellt, dass die computergestützten Interventionen zu einer universell gültigen Vorstellung des Alltags führen. In einem Designszenario für das „Management von Privathaushalten" in Madrid werden zahlreiche Möglichkeiten vorgeschlagen, die diese Haushalte effizienter machen sollen. Mobiltelefone kommunizieren per GPS mit sensorbestückten Küchengeräten, sodass ein Abendessen für die ganze Familie mittels kollektiver Abstimmung des geeigneten Orts und der passenden Zeit zubereitet werden kann. Auf ähnliche Weise werden Thermostate mit GPS und Kalendern auf Mobiltelefonen synchronisiert, damit das Haus rechtzeitig vor der Ankunft der Familie angenehm beheizt ist. Die Organisation von Tätigkeiten entwickelt sich durch programmierte und aktivierte Umgebungen, sodass Zeit und Ressourcen am

34 Mitchell, William J. / Casalegno, Federico: Connected Sustainable Cities, Cambridge, MA: MIT Mobile Experience Lab Publishing 2008. http://connectedsustainablecities.com/downloads/con nected_sustainable_cities.pdf, vom 11. Juli 2015.

produktivsten und effizientesten genutzt werden können. Im Madrider Szenario ist
das genaue Überwachen des Verhaltens der Bewohner durch Sensoren und Daten
von wesentlicher Bedeutung, damit eine solche Effizienz erreicht werden kann.
Mithilfe dieser Informationen sollen Umgebungen selbstregulierend operieren und
optimale Leistung erbringen.

Die Effizienzinitiativen des CSC versprechen, „das Management von Städten
zu optimieren", ökologische Fußabdrücke zu verkleinern und „zu verbessern, wie
Menschen urbanes Leben erfahren".[35] Indem smarte Technologien die Orte tägli-
cher Aktivitäten verfolgen, bieten sie die Möglichkeit, dass sich das Abendessen
selbst zubereitet und sich das Haus selbst heizt. Diese „Hilfstechnologien" führen
zu neuen Arrangements von Umgebungen und Lebensweisen: Smarte Thermostate
verbinden sich mit Kalendern, Orten und sogar „Biosignalen des menschlichen
Körpers". „Hauttemperatur und Puls" lassen sich durch Sensoren überwachen, um
für optimale Raumtemperaturen zu sorgen. Auf die gleiche Weise soll eine Kommunikation mit Küchengeräten durch „Toyotas Produktlinie ,Femininity' für Netz-
werkanwendungen" erfolgen. Diese Technologien garantieren, dass der Haushalt
warm, sicher und mit den neuesten Rezepten versorgt wird.[36]

Die Bedeutung des Alltags als Ort der Intervention signalisiert, auf welche
Weise smarte Städte bestimmte Lebensweisen generieren, indem eine „Mikrophy-
sik der Macht" in Alltagsszenarien aus- und eingeübt wird.[37] Governance und Management des urbanen Milieus geschehen nicht durch das Abgrenzen von Territorien, sondern durch das Ermöglichen von Verbindungen und Prozessen der Urbanität im Rahmen computergestützter Modalitäten. Die Handlungen der Bewohner
haben weniger mit der individuellen Ausübung von Rechten und Verantwortlich-
keiten zu tun als mit dem Operationalisieren der kybernetischen Funktionen der
smarten Stadt. Partizipation erfordert eine computergestützte Responsiveness und
ist eher deckungsgleich mit Aktionen des Überwachens und Managens der Beziehungen zur eigenen Umgebung, als dass sie demokratisches Engagement durch Di-
alog und Debatte fördert. Der Bewohner ist ein Kreuzungspunkt von Daten und
damit sowohl ihr Generator wie reagierender Knoten in einem Feedbacksystem.
Das Effizienzprogramm geht davon aus, dass menschliche Teilnehmer innerhalb
eines akzeptablen Handlungsspielraums reagieren werden, damit smarte Städte op-
timal funktionieren. Doch solche Effizienzspielräume werden unweigerlich auch in
menschlichen wie nichtmenschlichen Registern umgesetzt, sodass smarte Fahrrä-
der in Bäche geworfen und Sensing-Geräte gehackt werden können, um Haushalte
zu überwachen oder in sie einzugreifen. Angesichts dieser Seiten der smarten Stadt
stellt sich die Frage, inwieweit fein abgestimmte Lebensweisen tatsächlich gelebt
oder Effizienz- und Produktivitätsprogramme anderen Zwecken als den genannten
zugeführt werden können.

---

35 Ebd., S. 2.
36 Ebd., S. 58 f.
37 Deleuze, Gilles: Unterhandlungen: 1972–1990, Frankfurt a. M.: Suhrkamp 1993, S. 125.

DAS PROGRAMMIEREN VON UMGEBUNGEN

Die Motivationslogik der Nachhaltigkeit, die smarte Technologien implementiert, orientiert sich am Einsparen von Zeit und Ressourcen. Dies wiederum wirkt sich auf Pläne aus, wie smarte Technologien in Alltagsumgebungen eingebettet werden können, um für effizientere Lebensweisen zu sorgen. Die Überwachung wird als eine durch Sensoren ermöglichte Praxis zu einem zentralen Fluchtpunkt der Debatten um die Nachhaltigkeit und Effizienz smarter Städte. Das Sensing, das in der smarten Stadt vollzogen wird, erfordert kontinuierliche Überwachung. Die generierten Sensingdaten sollen die Regulierung urbaner Prozesse innerhalb eines Mensch-Maschine-Kontinuums von Sensing und Agieren ermöglichen, sodass sich „die Reaktionsbereitschaft vernetzter nachhaltiger Städte durch gut informiertes und koordiniertes menschliches Handeln, automatisches Aktivieren von Maschinen und Systemen oder eine Kombination beider erreichen lässt".[38] Menschen mögen sich zwar mit mobilen Geräten und auf digitalen Plattformen an der sensorüberwachten Stadt beteiligen, aber die Koordination aller „manuellen und automatischen" urbanen Prozesse vollzieht sich in programmierten Umgebungen, die den In- und Output von Menschen und Maschinen organisieren. Die „programmierte Stadt" ist als ein spekulatives wie reales Projekt von entscheidender Bedeutung für die laufende Entwicklung von Ubiquitous Computing. Zugleich demonstriert sie, wie kompliziert und unsicher es ist, programmierbare Umgebungen im Konkreten zu realisieren.[39] Das im CSC-Dokument beschriebene Programmieren hat einige Resonanz gefunden, was den architektonischen Sinn für die Gestaltung von Raum für bestimmte Aktivitäten[40] ebenso anzeigt wie die Programme der Stadtentwicklung und -politik sowie das computergestützte Programmieren von Umgebungen. In den Plänen für smarte Städte ist dies eine Möglichkeit, wie sich die „Natur der Verbindung" innerhalb des computergestützten Dispositivs als ein räumliches Arrangement von digitalen Geräten, Software, Städten, Entwicklungsplänen, Bürgern und Praktiken umsetzen lässt.

Die Vorstellung vom Programmieren ist über die spezifische Arbeit des Computers hinaus mit Vorstellungen dessen verknüpft, was eine Umgebung ist und wie sie programmierbar gemacht werden kann. Einige der frühen Imaginationen von Sensorumgebungen spekulieren darüber, wie sich das Alltagsleben mit der Verlagerung

---

38   W. J. Mitchell / F. Casalegno: Connected Sustainable Cities, S. 98.

39   Vgl. Gabrys, Jennifer: „Telepathically urban", in: Alexandra Boutros / Will Straw (Hg.), Circulation and the City: Essays on Urban Culture, Montreal: McGill-Queens University Press 2010, S. 48–63, hier S. 58 f. Dieser Aufsatz ist Teil eines in Bälde erscheinenden Buches mit dem Titel *Program Earth: Environment as Experiment in Sensing Technology*, das sich mit umgebungsbezogenem Sensing und programmierten Umgebungen befasst. Dieser Aufsatz bezieht sich auch auf ein laufendes Forschungsprojekt im Zusammenhang mit digitalen Sensortechnologien und umgebungsbezogener Praxis, „Citizen Sense" (siehe www.citizensense.net).

40   Vgl. Mitchell, William J.: Me++: The Cyborg Self and the Networked City, Cambridge, MA: MIT Press 2003.

des Computers vom Schreibtisch in die Umgebung umwandeln lasse.[41] Während viele dieser Visionen nutzerorientiert sind, beeinflussen Umgebungssensoren auch Vorstellungen davon, wie oder wo das Sensing stattfindet, um weiter verteilte und nichtmenschliche Modalitäten des Sensing einzubeziehen.[42] Die Möglichkeiten des Programmierens von Umgebungen zeigen vielleicht heute an, warum sich das „Milieu" inzwischen am besten als „Environment" verstehen lässt. Dem nach dem Zweiten Weltkrieg aufgekommenen Begriff „Environment" liegt einerseits ein eher kybernetischer Umgang mit Systemen und Ökologie[43] zugrunde und wird andererseits zur Bezeichnung der Environments des Computers verwendet, also den Bedingungen, unter denen Computer sowohl intern als auch extern operieren können.

Forschungen auf dem Gebiet der Software Studies befassen sich inzwischen zunehmend mit der Schnittstelle von Computern und Raum, indem sie prinzipiell davon ausgehen, dass die Computerarbeit – oft in Form von Software oder Quellcode – einen erheblichen Einfluss darauf hat, wie sich räumliche Prozesse entfalten oder gar aufhören zu funktionieren, wenn die Software versagt.[44] Während Software immer stärker räumliche und materielle Prozesse informiert, siedle ich die Performativität von Software innerhalb der (statt oberhalb oder vor den) materiellpolitisch-technischen Operationen des digitalen Dispositivs an. Programmierbarkeit beinhaltet notwendigerweise mehr als das Entfalten von Skripts, die sich in einer diskursiven Architektur von Befehl und Steuerung auf die Welt auswirken. Außerdem lässt sich Software nicht so leicht von der Hardware trennen, die sie aktivieren würde.[45] Vielmehr geht es mir darum, dass Programmierbarkeit darauf verweist, *wie Computerlogiken in materiell-kulturellen Situationen umgesetzt werden*, sogar auf der Ebene von spekulativen Designs oder von Vorstellungen politischer Prozesse. In diesen können digital gestützte Herangehensweisen an dringliche urbane „Probleme" darüber informieren, wie diese Fragen ursprünglich formuliert wurden, *um computergestützt berechnet zu werden*. Zugleich zeigen diese Logiken an, dass real-laufende Programme nicht unbedingt planmäßig funktionieren.

[...]

---

41   Vgl. Weiser, Mark: „The Computer for the 21st Century", in: Scientific American 265 (1991), S. 94–104.

42   Vgl. Gabrys, Jennifer: „Automatic sensation: environmental sensors in the digital city", in: The Senses and Society 2 (2007), S. 189–200. Gabrys, Jennifer: „Sensing an experimental forest: processing environments and distributing relations", in: Computational Culture 2 (2012); sowie K. Hayles: RFID.

43   Vgl. Haraway, Donna: Die Neuerfindung der Natur. Primaten, Cyborgs und Frauen, Frankfurt a.M.: Campus 1995.

44   Vgl. Graham, Stephen: „Software-sorted geographies", in: Progress in Human Geography 29 (2005), S. 562–580; Kitchin, Rob/Dodge, Martin: Code/Space: Software and Everyday Life, Cambridge, MA: MIT Press 2011. Thrift, Nigel/French, Shaun: „The Automatic Production of Space", in: Transactions of the Institute of British Geographers, New Series 27 (2002), S. 309–335.

45   Vgl. Gabrys, Jennifer: Digital Rubbish: A Natural History of Electronics, Ann Arbor, MI: University of Michigan Press 2010 und Kittler, Friedrich: „There is no software", in: CTHEORY www.ctheory.netjarticles.aspx?id=74, vom 11. Juli 2015.

DAS PROGRAMMIEREN VON PARTIZIPATION

Die Infrastrukturen, mit denen die CSC-Vision spielt, bestehen zum Teil aus vorhandenen Netzen und Dienstleistungen, die in smarte Strom-, Transport und Verkehrsnetze sowie smarte Wasserversorgung umgewandelt werden. Sie bestehen aber auch aus partizipatorischen und mobilen Plattformen für das Citizen Sensing, durch die Stadtbewohner ihre Umgebungen überwachen und sich an smarten Systemen beteiligen sollen. Partizipatorische Medien und umgebungsbezogene Geräte ermöglichen diese nachhaltigere Stadt durch Formen der Teilhabe. Die smarten Infrastrukturen und Plattformen des Citizen Sensing des CSC-Projekts ermöglichen Überwachungspraktiken, indem sie Reaktionen strukturieren, die wiederum Alltagspraktiken regulieren oder neu formieren. Nachhaltige Optionen des Nahverkehrs werden durch das Einrichten von kommunalen Initiativen erleichtert,[46] die eine personalisierte Planung von Busrouten, Fahrgemeinschaften und Leihfahrrädern ermöglichen. Energie kann durch smarte Transitsysteme oder Gebäudeoberflächen und mobile Überwachungsgeräte gespart werden. Urbane Räume können auf diese Weise leicht neu konfiguriert oder angepasst werden, um ein vernetztes Arbeiten an jedem Ort und zu jeder Zeit ebenso zu ermöglichen wie die „Intensivierung der urbanen Flächennutzung"[47]. Aktiviert werden diese Praktiken in den Programmen, die in urbane Umgebungen und mobile Geräte eingebettet sind. Digital optimierte Infrastrukturen und entsprechend angepasste Bürger werden als korrespondierende Knoten angesehen, wobei Technologien und Strategien für eine umweltschonende Effizienz sich mit der Bürgerbeteiligung überlappen – und entsprechend „verändertem menschlichem Verhalten"[48].

[…]

Verhaltensüberwachung und Datenproduktion sollen die Basis für vernünftige Entscheidungen als Teil zunehmend nachhaltiger Alltagspraktiken darstellen. Programme für eine optimierte Responsiveness sind wichtig für das Design nachhaltiger Praktiken, wie sie in diesem Entwurf [des CSC Projekts] auftauchen. Damit diese Pläne funktionieren, müssen die Bewohner der Stadt ihre angestammten Rollen spielen, indem sie öffentliche Verkehrssysteme nutzen oder durch ihre ständige Bewegung in urbanen Räumen Energie erzeugen. Umweltbewusste Stadtbürger sind verantwortlich dafür, „informierte, verantwortungsbewusste Entscheidungen" zu treffen.[49] Diese Entwürfe skizzieren ausführlich das Repertoire von Handlungen und Reflexionen, die die smarte Stadt ermöglichen werden und in der das Citizen Sensing ein Ausdruck produktiver Infrastrukturen sein wird: Mitchell und Castalegno betonen die Vorteile von informierter Teilhabe an urbanen Prozessen, die durch partizipatorische Medien und Ubiquitous Computing ermöglicht werde – Technologien, so argumentieren sie, die zu einem erhöhten Verantwortungsbewusstsein

---

46 Connected Urban Development: www.connectedurbandevelopment.org/, vom 11. Juli 2015.
47 Ebd.
48 Ebd.
49 W. J. Mitchell / F. Casalegno: Connected Sustainable Cities, S. 2.

führen werden.[50] Was urbane Bürgerschaft heißt, werde durch diese Umgebungs-technologien verändert. Sie mobilisieren die Bewohner als Agenten der Verarbei-tung kollektiver Umgebungsdaten; Aktivitäten der Bewohner werden dabei zu Er-weiterungen und Ausdrucksformen informationalisierter und effizienter materiell-politischer Praktiken. Bewohnern, die ihre eigenen Konsummuster sowie lokale Umgebungsprozesse digital erfassen, ermöglichen die durch Umgebungstechnolo-gien geschaffenen Bürgerangebote, an der smarten Stadt teilzuhaben.

Die Balance von smarten Systemen mit dem Engagement der Bürger wird üb-licherweise dann thematisiert, wenn man mit den Herausforderungen von Überwa-chung und Kontrolle smarter Städte konfrontiert ist. Wie der zuvor zitierte, von der Rockefeller-Stiftung finanzierte Report darlegt, mögen globale Technologiekon-zerne wie IBM und Cisco ganz andere Ziele haben als „bürgerliche Hacktivisten". Dennoch haben diese beiden Konzerne durchaus Eigeninteressen, an entstehenden Plänen für smarte Städte mitzuwirken.[51] Digitale Technologien erscheinen als Inst-rumente der Befreiung, die es Bürgern gestatten, sich in immer mehr demokrati-schen Aktionen zu engagieren – und dennoch suggerieren das Überwachen und Erfassen von Sensordaten zu nahezu jedem Aspekt urbanen Lebens durch Geräte, die von globalen Technologiekonzernen vertrieben werden, eine neue Dimension der Kontrolle. Aber könnte es nicht sein, dass diese offenkundige Dichotomie zwi-schen Citizen Sensing und smarter Stadt weniger eindeutig ist als es den Anschein hat? In vielerlei Hinsicht ließen sich die partizipatorischen Medien, mit denen wir heute umgehen, bereits als Instrumente eines auf unterschiedliche Weise einge-schränkten politischen Engagements verstehen,[52] während smarte urbane Infra-strukturen in den vorgelegten totalisierenden Visionen nie ganz (wenn überhaupt) greifbar werden.

Der „Sensing Citizen", also der Daten erfassende und sammelnde Bürger, könnte als eine Manifestation des idealen Modus von Bürgerbeteiligung verstanden werden und nicht als jemand, der sich dagegen wehrt, ihr Agent zu sein. Sensing Citizens sind eine notwendige Voraussetzung von smarten Städten – vor allem dort, wo die Smartness der Städte von vornherein vorausgesetzt wird. Stumpfsinnige Bürger in smarten Städten wären eine totalitäre Überzeichnung, da sie als der Über-wachung unterworfene Wesen nicht am Informationsfluss beteiligt wären. Die smarte Stadt wirft weitere Fragen über die Politik der urbanen Exklusion auf, also darüber, wer in der Lage ist, ein partizipierender Bürger in einer Stadt zu sein, die durch den Zugang zu digitalen Geräten in Gang gehalten wird. Doch die partizipa-torische Handlungsbereitschaft, die in Entwürfen smarter Städte eingebettet ist, be-ruft sich nicht auf ein individuelles menschliches Subjekt. Bürgerschaft wird statt-dessen durch Operationen in urbanen Umgebungen artikuliert. Die Pläne des CSC-Projekts enthalten die Möglichkeit, dass angesichts eines möglichen Versagens oder einer Einschränkung menschlicher Responsiveness – etwa eines mangelnden

---

50 Ebd., S. 101.
51 Vgl. A. Townsend et al.: „A planet of civic laboratories".
52 Barney, D.: „Politics and emerging media: the revenge of publicity", in: Global Media Journal
     1 (2008), S, 89–106.

Interesses, sich an der smarten Stadt zu beteiligen – das System eigenständig operieren kann. Aufgrund des möglichen Mangels an „menschlicher Aufmerksamkeit und kognitiver Fähigkeit" ebenso wie aufgrund des Wunsches, „Menschen nicht damit zu belasten, ständig daran zu denken, die Systeme zu kontrollieren, die sie umgeben", könne es, so die Projektautoren, wichtig werden, eine „automatische Aktivierung" anzuwenden. Dies würde zu einem Selbstmanagement urbaner Systeme führen, sodass „Gebäude und Städte sich zu stationären Robotern entwickeln werden"[53]. Ihre Bewohner könnte man als Figuren verstehen, die durch die Environmentalität ihrer Technologien organisiert werden. Doch das Programm der smarten Stadt ist in der Lage, unabhängig zu operieren, indem es Umgebungen erfasst, sie aktiviert und bis zu dem Punkt in sie eingreift, an dem diese Technologien sich über die Bewohner hinwegsetzen, wenn sie nicht gemäß den vorgegebenen Spielregeln mitwirken.

Prozesse, die in den Entwürfen smarter Städte urbane Umgebungen regulieren, erfordern eine solche innere Unterwerfung nicht, da Regierungsmacht in Umgebungen verteilt ist, die wiederum automatische Regulierungsmodi als Standardeinstellungen aufweisen. Hier handelt es sich um eine Version von Biopolitik 2.0, weil es beim Überwachen von Verhalten weniger um die Herrschaft über Individuen oder die Bevölkerung geht, sondern um die Einführung von Umgebungsbedingungen, unter denen an die Umgebung angepasste (und korrekte) Verhaltensmodi entstehen können. Environmentalität erfordert nicht die Erschaffung der von Foucault beschriebenen normativen Subjekte, da die Bewohner von Environments nicht als eigenständige Wesen beherrscht werden müssen; vielmehr ist Environmentalität eine Erweiterung der Handlungen und Kräfte des Automatismus und der Responsiveness, die in Umgebungen eingebettet sind und durch sie ausgeübt werden. Eine solche Situation ließe sich mit Deleuzes Begriff „dividuell" umschreiben. Er bezeichnet die fließende Einheit, die im „Computer"-Zeitalter entstehe.[54] Für Deleuze existiert die Automatisierung gleichzeitig mit einem de-individualisierenden Set von Prozessen, die durch Muster der Responsiveness charakterisiert sind. Sie basieren weniger auf individuellem Engagement als auf korrekter kybernetischer Vernetzung.

Demgegenüber würde ich allerdings behaupten, dass Pläne für smarte Städte nicht so sehr die absolute Eliminierung von Individuen implizieren, da der „Bürger" ein wichtiger Operator in diesen Räumen ist. Vielmehr betätigt sich dieser durch Prozesse, die *Ambividuen* generieren könnten: umgebungsbezogene und geschmeidige urbane Operatoren, die Ausdrucksformen der Anpassung an computerisierte Umgebungen sind. Da das Ambividuum kein kognitives Subjekt ist, artikuliert es die Verteilung von Handlungsknoten in der smarten Stadt. Ambividuen werden nicht auf eine singularisierende Weise demarkiert oder getilgt, sondern sind kontingent und empfänglich für fluktuierende Ereignisse, die mittels Praktiken der Informationsverarbeitung gemanagt werden. Dies entspricht Foucaults Hinweis, ein Merkmal der umgebungsbezogenen Technologien sei die „Errichtung eines

53   W. J. Mitchell / F. Casalegno: Connected Sustainable Cities, S. 98.
54   G. Deleuze: Unterhandlungen, S. 261.

ziemlich lockeren Rahmens um das Individuum herum, damit es überhaupt spielen kann"[55]. Ich möchte aber darauf hinweisen, dass sich das- oder derjenige, das oder der als Ambividuum zählt, nicht auf einen menschlichen Akteur in der smarten Stadt beschränkt, da die Möglichkeiten von Aktionen und Reaktionen quer durch die Handlungsräume zwischen Menschen und Maschinen sowie zwischen Maschinen und Maschinen eröffnen.

CITIZEN SENSING UND SENSING CITIZENS

Ein weiteres Thema, das im Rahmen von smarten Städten und Citizen Sensing auftaucht, ist das Ausmaß, in dem „environmental monitoring" zu handlungsleitenden Daten führt. In den Entwürfen smarter Städte sollen Infrastrukturen als selbstregulierende Umgebungen operieren, doch die Überwachungstechnologien, die in diesen Systemen für Effizienz sorgen sollen, sind offenbar kaum in der Lage, die Bewohner zu effizienterem Handeln anzuleiten. In einem CSC-Szenario, das verschiedene Arten von urbaner, umgebungsbezogener Bürgerschaft zeigen soll, die in der grünen und digitalen Stadt möglich seien, wird vorgeschlagen, dass den Bewohnern von Curitiba verbesserte und synchronisierte Möglichkeiten des Massentransitverkehrs zur Verfügung stehen könnten, wenn sie an den Verkehrsknoten die Luftverschmutzung überwachen und melden. Solche Auswertungen durch Bewohner und ihr Engagement für die Gemeinschaft werden durch die Konnektivität von Informationstechnologien verstärkt. Dank dieser Überwachungs- und Auswertungsmöglichkeiten soll die erhöhte Informationsdichte und Konnektivität positive Veränderungen nach sich ziehen: Man sammle die Luftverschmutzungsdaten, melde sie der zuständigen Behörde, und „Umweltgerechtigkeit" werde realisiert. Diesen Aktivitäten wird universelle Geltung zugesprochen, insofern alle Bewohner sich veranlasst fühlen mögen, die Luftverschmutzung zu kontrollieren, die Daten zu sammeln und sie der zuständigen Behörde zu melden. Die ambividuellen Handlungen, die in diesen Prozessen „codiert" sind, setzen kein bestimmtes Subjekt voraus, da auch ein vollautomatischer Sensor eine derartige Funktion übernehmen kann. Diese Programme der Responsiveness ermöglichen vielmehr eine in alle Richtungen kompatible Prozessierung von Mensch-Maschine- oder Maschine-Maschine-Datenoperationen.

Einen ähnlichen Verlauf stellt man sich typischerweise für selbstregulierende Bürgeraktivitäten vor: Informationen über den Energieverbrauch werden sichtbar gemacht, Korrekturmaßnahmen eingeleitet und das Gleichgewicht des kybernetischen Informationssystems wiederhergestellt. In diesen Szenarien überwachen Technologien Bewohner in ihren Umgebungen, während Bewohner diese Umgebungen auch selbst überwachen. Mit Umgebungsdaten ausgerüstete Bewohner werden als zentrale demokratische Operatoren in diesen Umgebungen verstanden. Doch die der Kybernetik immanente Herrschaft lässt sich vielleicht nicht direkt in

55 M. Foucault: Die Geburt der Biopolitik, S. 361.

die Herrschaft über die Umgebung umsetzen.[56] So kann es durchaus möglich sein, dass gerade die Responsiveness, die es den Bewohnern ermöglicht, Daten zu sammeln, nicht so weit geht, dass sie in der Lage sind, sinnvoll auf der Basis der gesammelten Daten zu handeln, denn das hieße ja, das urbane „System" zu verändern, in dem sie zu leistungsfähigen Operatoren geworden sind. Gleichermaßen führen dominante, wenn auch problematische Narrative über die Nachhaltigkeit anhaltenden Wachstums, wie verbesserte Effizienz und laufende Überwachung, typischerweise nicht zu einer umfassenden Verringerung des Ressourcenverbrauchs oder einer Abfallvermeidung (der aus der Energiediskussion bekannte „Rebound-Effekt"). Strategien der Überwachung und der Effizienz könnten somit verstanden werden als biopolitische Ko-Optierung von Stadtbewohnern und ihren Umgebungen, während die Modi neoliberaler Macht nicht hinterfragt werden müssen, da die Orientierung am Ziel der Nachhaltigkeit als wertvolle Betätigung der Bewohner angesehen wird.

Foucault interessierte sich in seinen Vorlesungen über Biopolitik auch für die Frage, wie sich die Verfahren des Neoliberalismus auf die Regierung und auf die regierten Subjekte auswirken, durch welche die ökonomische Logik der Effizienz die einstigen sozialen oder nichtökonomischen Modalitäten ablöst.[57] Environmentalität bezeichnet die Verteilung von Regierungsmacht in Umgebungen ebenso wie die Aufwertung von Gouvernementalität durch eine Marktlogik, die Effizienz und Produktivität als Leitprinzipien für urbane Lebensweisen anvisiert. Das Individuum wird „gouvernementalisierbar", sodass es als Homo oeconomicus operiert,[58] wobei sich in den hier vorgestellten Beispielen Regierung und Verwaltung als umgebungsbezogene Verteilungen möglicher Reaktionen entfalten, die den Kriterien von Effizienz und maximaler Nützlichkeit gehorchen.

Die Verwandlung von Bewohnern in Daten sammelnde Netzwerkknoten beschränkt die Komplexität bürgerlichen Handelns potenziell auf relativ reduzierte, wenn überhaupt noch durchführbare Handlungsmöglichkeiten. Partizipation in der smarten und nachhaltigen Stadt wird damit instrumentalisiert, und zwar in Bezug auf eine Verbesserung der Umwelt sowohl durch Effizienz wie durch Endgeräte, die Informationen sammeln und verarbeiten. Doch die informationelle und auf Effizienz basierende Methode der Überwachung von Umgebungen wirft mehr Fragen darüber auf, was effektives Handeln ausmacht, als dass sie sie beantwortet. Damit eine solche Instrumentalisierung stattfinden kann, müssen urbane Prozesse und Partizipationsformen, die auf verschiedene Weise auf Nachhaltigkeit abzielen, so programmiert werden, dass sie für eine (computergestützte) Politik zugänglich sind, die diese Herausforderungen zu bearbeiten vermag. Die Modi des Sensing als Überwachung und der Responsiveness, wie sie in vielen auf Sensoren und Smartness fokussierten Entwürfen dargestellt werden, werfen die Frage auf, ob ein Bewohner nicht mehr sein könnte als eine Einheit, die innerhalb von Parametern akzeptabler Responsiveness agieren muss.

56 Vgl. Wiener, Norbert: Kybernetik. Regelung und Nachrichtenübertragung in Lebewesen und Maschine, Reinbek: Rowohlt 1968.
57 Vgl. M. Foucault: Die Geburt der Biopolitik, S. 342 f.
58 Vgl. ebd., S. 349.

ABSCHLIESSENDE ÜBERLEGUNGEN –
VON NETZWERKEN ZU RELAIS, VON PROGRAMMEN
ZU LEBENSWEISEN

Die hier erörterte Vision der smarten und nachhaltigen Stadt stellt sich als eine technische Lösung von politischen und umgebungsbezogenen Problemen dar – eine Vorgehensweise, die als charakteristisch für viele Projekte smarter Städte verstanden werden kann. Während die Entwürfe des CSC- und CUD-Projekts als Design- und Planungsdokumente auf der konzeptuellen Ebene entwickelt werden, sind viele der hier gestellten Fragen der Bürgerüberwachung, der politischen Partizipation und der Organisation des urbanen Lebens relevant für eine allgemeine Betrachtung der zahlreichen Vorhaben, die inzwischen auf diesen Gebieten umgesetzt werden, und zwar sowohl auf der Ebene von kommunalem Engagement als auch auf der Ebene von Stadtplanung und Stadtentwicklungspartnerschaften.[59]

Wie ich dargelegt habe, operieren Entwürfe für nachhaltige smarte Städte mit neuen Modi der Environmentalität ebenso wie mit biopolitischen Konfigurationen von Regierungsmacht, durch die neuartige digitale Dispositive entstehen. Nachdem sich Foucault bereits mit der historischen Spezifität dieser Konzepte und den Ereignissen, auf die sie sich beziehen, befasst hat, ist es an der Zeit, diese Konzepte im Kontext neu entstehender Pläne für smarte Städte zu durchdenken und zu revidieren. Environmentalität, Biopolitik 2.0 und digitale politische Technologien sind Ausdruck der Neuaushandlung von Regierungsmacht und der Operationen, mit denen Bürger mittels programmierter Umgebungen und Technologien geformt werden. In smarten Städten entsteht eine Biopolitik 2.0, die das Programmieren von Umgebungen und ihren Bewohnern erfordert, um sie zu Responsiveness und Effizienz zu erziehen. Ein solches Programmieren erzeugt politische Techniken zur Beherrschung des Alltagslebens, wobei sich urbane Prozesse, bürgerliches Engagement und Governance durch die raum-zeitlichen Netzwerke von Sensoren, Algorithmen, Datenbanken und mobilen Plattformen entfalten. Sie konstituieren die Environments [Umgebungen und Umwelten] smarter Städte.

Die Environmentalität, die durch Entwürfe urbaner Nachhaltigkeit im CSC-Projekt und in vielen ähnlichen Ansätzen entsteht, erfordert das Überwachen, Ökonomisieren und Produzieren einer Vision des digitalisierten ökonomischen Wachstums. Solche smarten Städte ermöglichen Lebensweisen, die auf Ziele der Nachhaltigkeit hin orchestriert werden, welche wiederum durch Produktivität und Effizienz charakterisiert sind. Die Daten, die sich durch diese Praktiken sammeln lassen, stammen aus der Überwachung von Umgebungen. Zugleich erlauben diese Daten umgebungsbezogene Modi von Governance, die nicht ausschließlich in der Jurisdiktion „öffentlicher" Behörden lokalisiert sind, sondern sich auch auf Technologiekonzerne erstrecken können, die urbane Daten besitzen, managen und nutzen.

---

59   Vgl. Europäische Kommission: „Report of the meeting of advisory group. ICT infrastructure for energy-efficient buildings and neighborhoods for carbon-neutral cities", http://ee.europa. eu/information_society/activities/sustainable_growth/docs/smartcities/smart-citiesadv-group_ report.pdf, vom 11. Juli 2015.

Von Google Transit bis zu Cisco TelePresence, HP Halo und Toshiba Femininity fungieren eine Reihe von umgebungsbezogenen Sensor- und Partizipationstechnologien im CSC-Projekt und in anderen Szenarien smarter Städte als Instrumente neoliberaler Governance, die auf staatliche wie nichtstaatliche Akteure angewandt werden.

Ich habe betont, dass sich Foucaults Interesse an Environmentalität im Kontext smarter Städte weiterdenken lässt, um zu erfassen, wie Verteilungen von Macht innerhalb von und durch Environments und ihre Technologien festlegen, wie Bewohner und Bürgerpflichten organisiert werden – anstatt individuelle Subjektivitäten zu prägen. Die „environmentalen" Aspekte der smarten und nachhaltigen Stadt sind nicht abhängig von der Produktion einer umweltbewussten oder ökologischen reflexiven Subjektivität. Die Leistung smarter Urbanität wird nicht durch die Eröffnung neuer Partizipationsmöglichkeiten für demokratisch engagierte Bürger erbracht, sondern vielmehr durch das Begrenzen der Praktiken, die Bürgerschaft konstituieren. Die „Spielregeln" der smarten Stadt artikulieren keine Transformationen, Öffnungen oder Kritiken urbaner Lebensweisen. Vielmehr werden Praktiken effizient gemacht, rationalisiert und auf eine Verbesserung bestehender ökonomischer Prozesse hin ausgerichtet. Und dennoch ließe sich im Rahmen dieses Umgangs mit Environmentalität durch smarte Städte das, was wir als die Regeln oder das Programm der smarten Stadt interpretieren können, weniger als ein deterministisches Codieren von Städten und mehr als etwas verstehen, das sich ungleichmäßig in Praktiken und Ereignissen materialisiert. Während die Designentwürfe ein unbehaglich überzeugendes Argument für das Programm der smarten Stadt liefern, entstehen durch die Zirkulation und Implementierung dieses Programms unweigerlich sehr unterschiedliche smarte Städte.

Doch um Foucaults Vorstellung von Environmentalität noch zu verfeinern, schlage ich vor, das Konzept der „Spielregeln" im Kontext smarter Städte neu zu fassen, nämlich weniger als Regeln, denn als Programme – hier der Responsiveness –, die auf besondere Art und Weise Grenzen setzen und Möglichkeiten eröffnen, die sich aber auch auf unerwartete Weise entfalten, materialisieren oder versagen. Wenn urbane Programme nicht isoliert sind, sondern sich ständig in Veränderung befinden, dann könnte ein neues Verständnis von Environmentalität auch die Art und Weise betreffen, dass und wie Programme nicht nach Plan verlaufen. Umwege und Übergangslösungen könnten auf der Hand liegen. Eine derartige Herangehensweise ist nicht so sehr ein simples Wiedergewinnen menschlichen Widerstands als ein Hinweis darauf, dass diese Programme nicht starr fixiert sind und dass sie in ihrem Entfalten und Operieren unweigerlich neue Praktiken urbaner, environmentaler Bürgerschaft und Lebensweisen aufkommen lassen, wie sie in allen menschlichen und mehr als menschlichen urbanen Verwicklungen entstehen.

Dieser Umgang mit Lebensweisen ist wichtig, wenn man die smarte Stadt nicht schlicht denunzieren möchte, sondern vielmehr vorschlagen möchte, sich mit bestimmten umgebungsbezogenen Modalitäten des Bewohnens – und Möglichkeiten für urbane Kollektive – zu befassen. Subjektivierung, die Deleuze als einen Schlüsselbegriff in Foucaults Werk ausmacht, befasst sich letztlich nicht mit der Produktion starrer Subjekte, sondern vielmehr mit der Möglichkeit, Lebensweisen zu ermitteln, zu kritisieren und sogar zu erschaffen.[60] Projekte smarter Städte erfordern eine Betrachtung – und eine Kritik – der Lebensweisen, die in diesen Plänen und Entwicklungen hervorgebracht und aufrechterhalten werden. Kritik, wie sie in einem Gespräch zwischen Deleuze und Foucault gefasst wird, kann eine bedeutende Möglichkeit sein, mit politischen Engagements zu experimentieren und „Verbindungselemente" zwischen „Aktion der Theorie" und „Aktion der Praxis" zu bilden.[61] Aus dieser Perspektive könnten die in den CSC-Szenarien vorgeschlagenen Lebensweisen als Provokation dazu dienen, zu bedenken, wie sich mit urbanen Vorstellungswelten und Praktiken experimentieren ließe, um nicht auf diese Weise beherrscht zu werden. Wenn wir Biopolitik 2.0 als ein Konzept verstehen, das Lebensweisen bedenkt, die in smarten Städten generiert und gestützt werden, und wenn dieser digitale Apparat in Umgebungen operiert, welche neuen Verbindungselemente für Theorie und Praxis könnten dann in unseren zunehmend computergestützten Städten entstehen?

---

60   G. Deleuze: Unterhandlungen, S. 147–171.
61   M. Foucault: Schriften in vier Bänden, Bd. II, S. 383.

NACHWEIS DER DRUCKORTE

Marc Boeckler (2014)
Digitale Geographien: Neogeographie, Ortsmedien und der Ort der Geographie im digitalen Zeitalter.
Zuerst erschienen in: Geographische Rundschau, 2014, 4–10. Abgedruckt: Überarbeitung der Preprint-Version 2021. Mit freundlicher Genehmigung des Autors.

Georg Glasze (2015)
Neue Kartografien, neue Geografien: Weltbilder im digitalen Zeitalter. Aus Politik und Zeitgeschichte, 41–42/2015, 29–37. Mit freundlicher Genehmigung des Autors.

James Ash / Rob Kitchin / Agnieszka Leszczynski (2018)
Digital turn, digital geographies? Progress in Human Geography, 42 (1), 25–43. doi:10.1177/0309132516664800, reprinted by Permission of SAGE Publications, Ltd.

Lizzie Richardson (2018)
Feminist geographies of digital work. Progress in Human Geography, 42 (2), 244–263. doi:10.1177/0309132516677177, Reprinted by Permission of SAGE Publications, Ltd.

Ayona Datta (2018)
The digital turn in postcolonial urbanism: Smart citizenship in the making of India's 100 smart cities. Transactions of the Institute of British Geographers, 43 (3), 405–419. doi:10.1111/tran.12225. Published by JohnWiley & Sons Ltd on behalf of Royal Geographical Society (with the Institute of British Geographers). Mit freundlicher Genehmigung der Autorin.

Jason C. Young / Renee Lynch / Stanley Boakye-Achampong / Chris Jowaisas / Joel Sam / Bree Norlander
Volunteer geographic information in the Global South: barriers to local implementation of mapping projects across Africa, in: Geo Journal (2020), https://doi.org/10.1007/s10708-020-10184-6

Christian Dorsch / Detlef Kanwischer (2020), Mündigkeit in einer Kultur der Digitalität – Geographische Bildung und „Spatial Citizenship". Zeitschrift für Didaktik der Gesellschaftswissenschaften, 11 (1), 23–40.

Karen Bakker/Max Ritts (2018). Smart Earth: A meta-review and implications for environmental governance. Global Environmental Change, 52, 201–211. doi:https://doi.org/10.1016/j.gloenvcha.2018.07.011

Matthew W. Wilson (2009). Cyborg geographies: towards hybrid epistemologies. Gender, Place & Culture, 16 (5), 499–516. doi:10.1080/09663690903148390, reprinted by permission of the publisher Taylor & Francis Ltd, http://www.tandfonline.com

Rob Kitchin (2017). Thinking critically about and researching algorithms. Information, Communication & Society: The Social Power of Algorithms, 20 (1), 14–29. doi:10.1080/1369118X.2016.1154087, reprinted by permission of the publisher Taylor & Francis Ltd, http://www.tandfonline.com

Jennifer Gabrys, (2015). Programming Environments: Environmentality and Citizen Sensing in the Smart City. Environment and Planning D: Society and Space. 2014;32(1):30–48. doi:10.1068/d16812. Reprinted by permission of Sage Publications, Ltd.
Deutsche Übersetzung aus: Internet der Dinge, edited by Florian Sprenger and Christoph Engemann (Bielefeld: Transcript), 313–342 (Original: doi: 10.1068/d16812)

WEITERFÜHRENDE LITERATUR

Ash, J., Kitchin, R. & Leszczynski, A. (eds.) (2019): *Digital Geographies.* London: SAGE.

Bork-Hüffer, T., Füller, H. & Straube, T. (eds.) (2021): Handbuch Digitale Geographien – Welt, Wissen, Werkzeuge. Stuttgart: UTB.

Kitchin, R. & Dodge, M. (2011): *Code/Space: Software and Everyday Life,* London, Cambridge: MIT Press.

REGISTER

Peter Dirksmeier / Mathis Stock (Hg.)

Urbanität

BASISTEXTE GEOGRAPHIE – BAND 2
201 Seiten mit 4 s/w-Abbildungen und 2 Tabellen
978-3-515-12410-2-0 KARTONIERT

Die Stadt gilt in der wissenschaftlichen Humangeographie als wichtiger Forschungsgegenstand – mit dem Schlüsselbegriff der „Urbanität" versucht man spätestens seit Ende des 19. Jahrhunderts die strukturellen Unterschiede zwischen Stadt und Land begrifflich zu fassen. Der Band umfasst eine Auswahl der wegweisenden Texte des deutschsprachigen, englischsprachigen und französischsprachigen Diskurses in deutscher Übersetzung und zeigt erstmals die Vielfalt der verschiedenen Urbanitätsvorstellungen.

„Diese Anthologie verkörpert in ihrer Welt- und Sprachgewandtheit Urbanität in doppeltem Sinne. Mit der erfrischend polyglotten Textauswahl schillert das Büchlein facettenreicher als so mancher großer Klassiker der städtischen Anthologien – und öffnet neue Blicke auf einen vermeintlich alt bekannten Begriff."
Martin Müller, Universität Lausanne

„Ein zeitloser und bedeutender Beitrag zur sozialwissenschaftlichen Stadtforschung in der Geographie und eine wichtige Hilfestellung zum Verständnis des Urbanitätsdiskurses der Moderne. Den Herausgebern ist zu danken für diese sorgfältige Zusammenstellung sowohl klassischer Texte – z.T. neu übersetzt – als auch wichtigen Originalbeiträgen aus der aktuellen Stadt- und Urbanitätsforschung."
Marit Rosol, University of Calgary

DIE HERAUSGEBER
Peter Dirksmeier lehrt Kultur- und Sozialgeographie an der Leibniz Universität Hannover.

Mathis Stock lehrt Tourismusgeographie an der Universität Lausanne.

Franz Steiner Verlag

Hier bestellen:
service@steiner-verlag.de

Anton Escher / Sandra Petermann (Hg.)

Raum und Ort

BASISTEXTE GEOGRAPHIE – BAND 1
214 Seiten mit 16 s/w–Abbildungen und 2 Tabellen
978-3-515-09121-3 KARTONIERT

Seit Beginn der wissenschaftlichen Geographie gelten „Raum und Ort" als die zentralen Schlüsselbegriffe und Basiskategorien geographischer Forschung. Ende des 20. Jahrhunderts vollzieht sich in nahezu allen Geistes-, Sozial- und Gesellschaftswissenschaften eine Hinwendung (*spatial turn*) zur Thematisierung von „Raum und Ort". Inzwischen liegen nicht nur in der Geographie, sondern auch außerhalb der Fachdisziplin zahlreiche Publikationen mit einer Fülle von unterschiedlichen Definitionen und vielfältigen Konzepten dazu vor.

Der in der Reihe *Basistexte Geographie* erscheinende Band umfasst eine Auswahl der für die Entwicklung der Geographie wegweisenden und Impulse gebenden Texte, welche die Vielfalt der unterschiedlichen Raumverständnisse und Vorstellungen für Orte widerspiegelt. Ein besonderes Augenmerk wird hierbei auf aktuelle sozialgeographische Ansätze und die kulturalistische Geographie

gelegt, die davon ausgehen, dass Räume soziale Konstrukte sind.

AUS DEM INHALT

Anton Escher / Sandra Petermann: Einleitung | Gerhard Hard / Dietrich Bartels: Eine „Raum"-Klärung für aufgeweckte Studenten | Benno Werlen: Gibt es eine Geographie ohne Raum? Zum Verhältnis von traditioneller Geographie und zeitgenössischen Gesellschaften | Peter Weichhart: Die Räume zwischen den Welten und die Welt der Räume. Zur Konzeption eines Schlüsselbegriffs der Geographie | Andreas Pott: Systemtheoretische Raumkonzeption | Pierre Bourdieu: Ortseffekte | Michel Foucault: Andere Räume | Yi-Fu Tuan: Space and Place: Humanistic Perspective | David Harvey: Zwischen Raum und Zeit: Reflektionen zur Geographischen Imagination | Doreen Massey: A Global Sense Of Place

Franz Steiner
Verlag

Hier bestellen:
service@steiner-verlag.de